Probability and Statistics
for Engineers and Scientists

MACMILLAN PUBLISHING CO., INC.
New York

COLLIER MACMILLAN PUBLISHERS
London

Probability and Statistics for Engineers and Scientists

RONALD E. WALPOLE

Professor of Mathematics and Statistics
ROANOKE COLLEGE

RAYMOND H. MYERS

Professor of Statistics
VIRGINIA POLYTECHNIC INSTITUTE
and
STATE UNIVERSITY

Macmillan Publishing Co., Inc.
866 Third Avenue, New York, New York 10022

Collier-Macmillan Canada, Ltd.

Library of Congress catalog card number: 73-160075

AMS (MOS) subject classification number: 6280

Printing   9   Year   7 8 9

TO JANICE

PREFACE

This book has been written as an introductory probability and statistics text for students majoring in engineering, mathematics, statistics, or one of the natural sciences. For those students who have time for only a one-semester course, based upon a prerequisite of a year of differential and integral calculus, the authors recommend the study of Chapters 1 through 5 and selected topics from Chapters 6 and 7. Chapter 1 introduces the basic concepts of probability theory using set notation. Chapter 2 presents an introduction to discrete and continuous random variables and their probability distributions, joint probability distributions, and mathematical expectations. Chapters 3 and 4 are devoted to a discussion of the particular discrete and continuous probability distributions that the engineer or scientist is most likely to apply in solving the various problems in his field of specialization. Perhaps unusual for a text at this level is the extensive use of transformation theory in our derivations of the sampling distributions in Chapter 5. However, the treatment of estimation procedures and hypothesis testing in Chapters 6 and 7 can only be appreciated and properly understood if one has gained an insight into the mathematical derivation of the test statistics involved.

For those students who wish to continue their training in statistics for an additional semester, the remainder of the text provides an excellent introduction to the study of regression theory, linear models, analysis-of-variance procedures, and the planning and analysis of factorial experiments. As a rule, students who take more than one semester of statistics also enroll in additional courses in mathematics and, perhaps, computer science. We have therefore assumed a knowledge of matrix theory and the availability of a desk calculator or computer in our treatment of multiple and polynomial regression in Chapters 9 and 10. However, since the essential material on matrix theory is limited in scope and could easily be injected into the lecture sequence by the professor, we do not feel that a formal course in matrix theory or linear algebra is a prerequisite.

Throughout the text we have demonstrated each new idea by an example. Only by solving a large number of exercises can the student be expected to develop an understanding of the basic concepts of probability theory and statistics. Therefore, we have included at the end of each chapter numerous exercises, both theoretical and applied, all of which are keyed to answers at the back of the book.

The authors wish to acknowledge their appreciation to all those who assisted in the preparation of this textbook. We are particularly grateful to Miss Diane Milan, Mrs. Janet Parsons, and Miss Lynn Watson for typing the final manuscript; to Miss Kathleen Webster for giving so much of her time to the tedious chore of proofreading; to Mrs. Sharon Crews and Mrs. Susan Crandall for performing many of the computations throughout the text; to the reviewers for their helpful suggestions and criticisms; and to our wives for their patience and encouragement.

The authors are indebted to the literary executor of the late Sir Ronald A. Fisher, F.R.S., Cambridge, and to Oliver & Boyd Ltd., Edinburgh, for their permission to reprint a table from their book *Statistical Methods for Research Workers;* to Professor E. S. Pearson and the Biometrika trustees for permission to reprint in abridged form Tables 8 and 18 from *Biometrika Tables for Statisticians*, Vol. I; to Oliver & Boyd Ltd. for permission to reproduce tables from their book *Design and Analysis of Industrial Experiments* by O. L. Davies; to C. Eisenhart, M. W. Hastay, and W. A. Wallis for permission to reproduce a table from their book *Techniques of Statistical Analysis*. We wish also to express our appreciation for permission to reproduce tables from the *Annals of Mathematical Statistics*, from the *Bulletin of the Educational Research at Indiana University*, from a publication by the American Cyanamid Company, from *Biometrics*, and from *Biometrika*, Vol. 38.

R. E. W.

R. H. M.

CONTENTS

DETAILED CONTENTS

Probability and Statistics
for Engineers and Scientists

1 SETS AND PROBABILITY

A fundamental concept in all branches of mathematics that is necessary for the study of probability and statistics is that of a *set*. We may think of a *set* as a well-defined collection of objects. For example, the rivers in Virginia, the chemical elements of the periodic table, the members of the senate, and the monthly income from shoe sales all constitute sets. One can even think of a line segment as an infinite set of points. Each object in a set is called an *element* of the set or a *member* of the set.

Usually sets will be denoted by capital letters such as A, B, X, or Y whereas small letters such as a, b, x, or y will be used to indicate the elements of a set. To have a well-defined set we must have some means of indicating whether a given object is or is not a member of the set.

There are two different ways to describe a set. First, if the set has a finite number of elements we may *list* the members separated by commas and enclosed in brackets. Thus the set A consisting of the numbers of accidents at a certain industrial plant that resulted in loss of time during the past 4 years may be written

$$A = \{3,\ 1,\ 4,\ 2\},$$

or the set B of physical conditions when items are inspected as they come off an assembly line may be written as

$$B = \{D, N\},$$

where D and N correspond to *defective* and *nondefective*, respectively.

Second, a set may be described by a *statement* or *rule*. For example, we may let C be the set of cities in the world with a population over one million. If x is an arbitrary element of C, we write

$$C = \{x \mid x \text{ is a city with a population over one million}\},$$

which reads "C is the set of all x such that x is a city with a population over one million." The vertical bar is read "such that." Similarly, if P is the set of points (x,y) on the boundary of a circle of radius 2, we write

$$P = \{(x,y) \mid x^2 + y^2 = 4\}.$$

Whether we describe the set by the rule method or by listing the elements will depend on the specific problem at hand. It would be very difficult to list all elements in the set of people with blue eyes. On the other hand, there is no easy rule to specify the set

$$Y = \{\text{catalyst, precipitate, engineer, rivet}\}.$$

In set notation the symbol $\in$ means "is an element of" or "belongs to" and $\notin$ means "is not an element of" or "does not belong to." If x is an element of the set A while y is not, we write

$$x \in A \quad \text{and} \quad y \notin A.$$

Example 1.1 Let $A = \{2, 4, 6, 8\}$ and $B = \{x \mid x \text{ is an integer divisible by 3}\}$. Then $4 \in A$, $9 \in B$, $8 \notin B$, and $3 \notin A$.

DEFINITION 1.1 *Two sets are equal if they have exactly the same elements in them.*

If set A is equal or identical to set B, then every element that belongs to A also belongs to B and every element that belongs to B also belongs to A. We denote this equality by writing $A = B$. However, if either of the sets A or B contains at least one element that is not common to both, the sets are said to be unequal, and then we write $A \neq B$.

Example 1.2 Let $A = \{1, 3, 5\}$, $B = \{3, 1, 5\}$, and $C = \{1, 3, 5, 7\}$. Then $A = B$, $A \neq C$, and $B \neq C$.

Note that a set does not change when the order of the elements is rearranged.

DEFINITION 1.2 *The* null set *or* empty set *is a set that contains no elements. We denote this set by the symbol* $\varnothing$.

If we let A be the set of microscopic organisms detected by the naked eye in a biological experiment, then A must be the null set. Also if $B = \{x \mid x \text{ is a nonprime factor of } 7\}$, then B must be the null set since the only possible factors of 7 are the prime numbers 1 and 7.

Let us consider the set $A = \{x \mid 0 < x < 5\}$ and the set $B = \{y \mid 2 \leq y \leq 4\}$, where x and y are integers. We notice that every element in the set B is also an element of the set A. The set B is said to be a *subset* of A. Symbolically we write this as $B \subset A$, where $\subset$ means "is a subset of" or "is contained in."

DEFINITION 1.3 *If every element of a set A is also an element of a set B, then A is called a* subset *of B.*

According to this definition every set is a subset of itself. Any subset of a set that is not the set itself is called a *proper subset* of the set. Therefore B is a proper subset of A if $B \subset A$ and $B \neq A$.

Example 1.3 The set $B = \{2, 4\}$ is a proper subset of $A = \{1, 2, 3, 4, 5\}$. However, the set $C = \{3, 2, 5, 1, 4\}$ is a subset of A but not a proper subset since $A = C$.

In many discussions all sets are subsets of one particular set. This set is called the *universal set* and is usually denoted by U. In a mathematical discussion the real numbers could be used as the universal set. The set of I.Q.'s of all college students could be used as the universal set that has the I.Q.'s of students of a certain college as a subset.

Example 1.4 All the subsets of the universal set $U = \{1, 2, 3\}$ are $\{1, 2, 3\}$, $\{1, 2\}$, $\{1, 3\}$, $\{2, 3\}$, $\{1\}$, $\{2\}$, $\{3\}$, $\varnothing$.

Note that the number of subsets of a universal set containing three elements is $2^3 = 8$. In general a set with n elements has 2^n subsets.

The relationship between subsets and the corresponding universal set can be illustrated by means of *Venn diagrams*. In a Venn diagram we let the universal set be a rectangle and let subsets be circles drawn inside the rectangle. Thus in Fig. 1.1 we see that A, B, and C are all subsets of

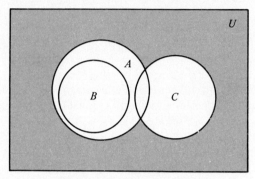

Figure 1.1 Subsets of U.

the universal set U. It is also clear that $B \subset A$; B and C have no elements in common; A and C have at least one element in common.

Sometimes it is convenient to shade various areas of the diagram as in Figure 1.2. In this case we take all the students of a certain college

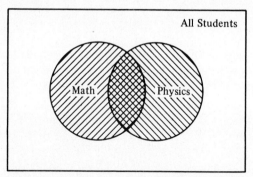

Figure 1.2 Subsets indicated by shading.

to be our universal set. The subset representing those students taking mathematics has been shaded by drawing straight lines in one direction and the subset representing those students studying physics has been shaded in by drawing lines in a different direction. The doubly shaded or crosshatched area represents the subset of students enrolled in both mathematics and physics and the unshaded part of the diagram corre-

sponds to those students who are studying subjects other than mathematics or physics.

1.2 Set Operations

We now consider certain operations on sets that will result in the formation of new sets. These new sets will be subsets of the same universal
set as the given sets.

DEFINITION 1.4 *The intersection of two sets A and B is the set of
elements that are common to A and B.*

Symbolically, we write $A \cap B$ for the intersection of A and B. The
elements in the set $A \cap B$ must be those and only those that belong to
both A and B. These elements may either be listed or defined by the
rule method, namely, $A \cap B = \{x \mid x \in A \text{ and } x \in B\}$. In the Venn
diagram in Figure 1.3 the shaded area corresponds to the intersection
$A \cap B$.

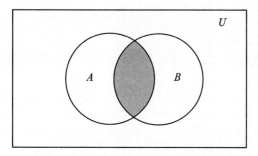

Figure 1.3 Intersection of A and B.

Example 1.5 Let $A = \{1, 2, 3, 4, 5\}$ and $B = \{2, 4, 6, 8\}$; then
$A \cap B = \{2, 4\}$.

Example 1.6 If R is the set of all taxpayers and S is the set of all people
over 65 years of age, then $R \cap S$ is the set of all taxpayers who are over
65 years of age.

Example 1.7 Let $P = \{a, e, i, o, u\}$ and $Q = \{r, s, t\}$; then $P \cap Q = \varnothing$.
That is, P and Q have no elements in common.

Two disjoint sets A and B are illustrated in the Venn diagram in
Figure 1.4. By shading the areas corresponding to the sets A and B we
find no doubly shaded area representing the set $A \cap B$. Hence $A \cap B$
is empty.

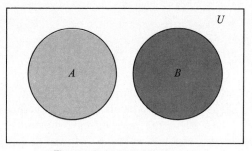

Figure 1.4 Disjoint sets.

Symbolically we write $A \cup B$ for the union of A and B. The elements
of $A \cup B$ may be listed or defined by the rule $A \cup B = \{x \mid x \in A$ or
$x \in B\}$. In the Venn diagram in Figure 1.5 the area representing the
elements of the set $A \cup B$ has been shaded.

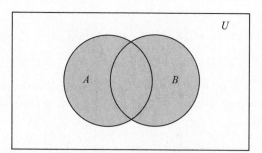

Figure 1.5 Union of A and B.

Example 1.8 Let $A = \{2, 3, 5, 8\}$ and $B = \{3, 6, 8\}$; then $A \cup B = \{2, 3, 5, 6, 8\}$.

Example 1.9 If $M = \{x \mid 3 < x < 9\}$ and $N = \{y \mid 5 < y < 12\}$, then $M \cup N = \{z \mid 3 < z < 12\}$.

Suppose we consider the employees of some manufacturing firm as the universal set. Let all the smokers form a subset. Then all the non-smokers form a set, also a subset of U, that is called the complement of the set of smokers.

DEFINITION 1.7 *If A is a subset of the universal set U, then the* complement *of A with respect to U is the set of all elements of U that are not in A. We denote the complement of A by A'.*

The elements of A' may be listed or defined by the rule $A' = \{x \mid x \in U \text{ and } x \notin A\}$. In the Venn diagram in Figure 1.6 the area representing the elements of the set A' has been shaded.

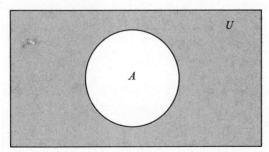

Figure 1.6 Complement of A.

Example 1.10 Let R be the set of red cards in an ordinary deck of playing cards and let U be the entire deck. Then R' is the set of cards in the deck that are not red, namely, the black cards.

Example 1.11 Consider the universal set $U = \{$book, catalyst, cigarette, precipitate, engineer, rivet$\}$. Let $A = \{$catalyst, rivet, book, cigarette$\}$. Then $A' = \{$precipitate, engineer$\}$.

Several results that follow from the above definitions that may easily be verified by means of Venn diagrams are

1. $A \cap \emptyset = \emptyset$.
2. $A \cup \emptyset = A$.
3. $A \cap A' = \emptyset$.
4. $A \cup A' = U$.
5. $U' = \emptyset$.
6. $\emptyset' = U$.
7. $(A')' = A$.

1.3 Sample Space

The scientist performs experiments in order to produce observations or measurements that will assist him in arriving at conclusions. We shall refer to the recorded information in its original collected form as *raw data*. In statistics we use the word *experiment* to describe any process that generates raw data. If a chemist runs an analysis several times under the same conditions, in most cases he will obtain different measurements, indicating an element of chance in the experimental procedure. It is these chance outcomes that occur in scientific investigation with which the statistician is basically concerned.

An example of a statistical experiment might be the tossing of a coin. This experiment consists only of the two outcomes heads or tails. Another experiment might be the launching of a missile and observing the velocity at specified times. The opinions of voters concerning a new sales tax can also be considered as observations of an experiment.

DEFINITION 1.8 *A set whose elements represent all possible outcomes of an experiment is called the* sample space *and is represented by the symbol S.*

The sample space is also sometimes called the universal set. In this book we shall use the notation S rather than U whenever we are dealing with a set whose members represent all possible outcomes of an experiment involving an element of chance.

DEFINITION 1.9 *An element of a sample space is called a* sample point.

Example 1.12 Consider the experiment of tossing a die. If we are interested in the number that shows on the top face, then the sample space would be

$$S_1 = \{1, 2, 3, 4, 5, 6\}.$$

If we are interested only in whether the number is even or odd, then the sample space is simply

$$S_2 = \{\text{even, odd}\}.$$

This example illustrates the fact that more than one sample space can be used to describe the outcomes of an experiment. In this case S_1 provides more information than S_2. If we know which element in S_1 occurs, we can tell which outcome in S_2 occurs; however, a knowledge of what happens in S_2 in no way helps us to know which element in S_1 occurs. In general it is desirable to use a sample space that gives the most information concerning the outcomes of the experiment.

Example 1.13 Suppose three items are selected at random from a manufacturing process. Each item is inspected and classified defective or nondefective. The sample space providing the most information would be

$$S_1 = \{\text{NNN, NDN, DNN, NND, DDN, DND, NDD, DDD}\}.$$

A second sample space, although it provides less information, might be

$$S_2 = \{0, 1, 2, 3\},$$

where the elements represent no defectives, one defective, two defectives, or three defectives in our random selection of three items.

In any given experiment we may be interested in the occurrence of certain *events* rather than in the outcome of a specific element in the sample space. For instance, we might be interested in the event A that the outcome when a die is tossed is divisible by 3. This will occur if the outcome is an element of the subset $A = \{3, 6\}$ of the sample space S_1 in Example 1.12. As a further illustration, we might be interested in the event B that the number of defectives is greater than 1 in Example 1.13. This will occur if the outcome is an element of the subset $B = \{\text{DDN, DND, NDD, DDD}\}$ of the sample space S_1.

To each event we assign a collection of sample points, which constitute a subset of the sample space. This subset represents all the elements for which the event is true.

DEFINITION 1.10 *An* event *is a subset of a sample space.*

Example 1.14 Given the subset $A = \{t \mid t < 5\}$ of the sample space $S = \{t \mid t \geq 0\}$, where t is the life in years of a certain electronic component, A is the event that the component fails before the end of the fifth year.

This example illustrates the fact that corresponding to any subset we can state an event whose elements are the given subset. In practice we usually state the event first and then determine its set.

DEFINITION 1.11 *If an event is a set containing only one element of the sample space, then it is called a* simple event. *A* compound event *is one that can be expressed as the union of simple events.*

Example 1.15 The event of drawing a heart from a deck of cards is the subset $A = \{\text{heart}\}$ of the sample space $S = \{\text{heart, spade, club, diamond}\}$. Therefore A is a simple event. Now the event B of drawing a red card is a compound event since $B = \{\text{heart} \cup \text{diamond}\} = \{\text{heart, diamond}\}$.

Note that the union of simple events produces a compound event that is still a subset of the sample space. We should also note that if the 52 cards of the deck were the elements of the sample space rather than the four suits, then the event A of Example 1.15 would be a compound event.

1.4 Counting Sample Points

One of the problems that the statistician must consider and attempt to evaluate is the element of chance associated with the occurrence of certain events when an experiment is performed. These problems belong in the field of probability, a subject to be introduced in Section 1.5. In many cases we will be able to solve a probability problem by counting the number of points in the sample space. A knowledge of the actual elements or a listing is not always required. The fundamental principle of counting is stated in the following theorem.

> THEOREM 1.1 *If an operation can be performed in n_1 ways, and if for each of these a second operation can be performed in n_2 ways, then the two operations can be performed together in $n_1 n_2$ ways.*

Example 1.16 How many sample points are in the sample space when a pair of dice are thrown once?

Solution. The first die can land in any 1 of 6 ways. For each of these 6 ways the second die can also land in 6 ways. Therefore, the pair of dice can land in $(6)(6) = 36$ ways. The student is asked to list these 36 elements in Exercise 12.

Theorem 1.1 may be extended to cover any number of events. The general case is stated in the following theorem.

> THEOREM 1.2 *If an operation can be performed in n_1 ways, and if for each of these a second operation can be performed in n_2 ways, and for each of the first two a third operation can be performed in n_3 ways, etc., then the sequence of k operations can be performed in $n_1 n_2 \cdots n_k$ ways.*

Example 1.17 How many lunches are possible consisting of a soup, a sandwich, a dessert, and a drink if one can select from four different soups, three kinds of sandwiches, five desserts, and four drinks?

Solution. The total number of lunches would be $(4)(3)(5)(4) = 240$.

Example 1.18 How many even three-digit numbers can be formed from the digits 1,2,5,6,9 if each digit can be used only once?

Solution. Since the number must be even we have only two choices for the units position. For each of these we have four choices for the hundreds position and then three choices for the tens position. Therefore we can form a total of $(2)(4)(3) = 24$ even three-digit numbers.

Frequently we are interested in a sample space that contains as elements all possible orders or arrangements of a group of objects. For example, we might want to know how many different arrangements are possible for sitting six people around a table, or we might ask how many different orders are possible for drawing 2 lottery tickets from a total of 20. The different arrangements are called *permutations*.

DEFINITION 1.12 *A* permutation *is an arrangement of all or part of a set of objects.*

Consider the three letters *a*, *b*, and *c*. The possible permutations are *abc*, *acb*, *bac*, *bca*, *cab*, and *cba*. Thus we see that there are six distinct arrangements. Using Theorem 1.2 we could have arrived at the answer six without actually listing the different orders. There are three positions to be filled from the letters *a*, *b*, and *c*. Therefore we have three choices for the first position, then two for the second, leaving only one choice for the last position, giving a total of $(3)(2)(1) = 6$ permutations. In general *n* distinct objects can be arranged in $n(n-1)(n-2)\cdots(3)(2)(1)$ ways. We represent this product by the symbol *n*!, which is read *"n* factorial." Three objects can be arranged in $3! = (3)(2)(1) = 6$ ways. By definition $1! = 1$ and $0! = 1$.

THEOREM 1.3 *The number of permutations of n distinct objects is n!.*

The number of permutations of the four letters *a*, *b*, *c*, and *d* will be $4! = 24$. Let us now consider the number of permutations that are possible by taking the four letters two at a time. These would be *ab*, *ac*, *ad*, *ba*, *ca*, *da*, *bc*, *cb*, *bd*, *db*, *cd*, and *dc*. Using Theorem 1.2 again, we have two positions to fill with four choices for the first and then three choices for the second for a total of $(4)(3) = 12$ permutations. In general *n* distinct objects taken *r* at a time can be arranged in $n(n-1)(n-2)\cdots(n-r+1)$ ways. We represent this product by the symbol $_nP_r = n!/(n-r)!$.

THEOREM 1.4 *The number of permutations of n distinct objects taken r at a time is*

$$_nP_r = \frac{n!}{(n-r)!}.$$

Example 1.19 Two lottery tickets are drawn from 20 for first and second prizes. Find the number of sample points in the space S.

Solution. The total number of sample points is

$$_{20}P_2 = \frac{20!}{18!} = (20)(19) = 380.$$

Example 1.20 How many ways can a local chapter of the American Chemical Society schedule three speakers for three different meetings if they are all available on any of five possible dates?

Solution. The total number of possible schedules is

$$_5P_3 = \frac{5!}{2!} = (5)(4)(3) = 60.$$

Permutations that occur by arranging objects in a circle are called *circular permutations.* Two circular permutations are not considered different unless corresponding objects in the two arrangements are preceded or followed by a different object as we proceed in a clockwise direction. For example, if four people are playing bridge, we do not have a new permutation if they all move one position in a clockwise direction. By considering one person in a fixed position and arranging the other three in 3! ways we find that there are six distinct arrangements for the bridge game.

THEOREM 1.5 *The number of permutations of n distinct objects arranged in a circle is* $(n-1)!$.

So far we have considered permutations of distinct objects. That is, all the objects were completely different or distinguishable. Obviously if the letters b and c are both equal to x, then the six permutations of the letters a, b, c, become axx, axx, xax, xax, xxa, and xxa, of which only three are distinct. Therefore with three letters, two being the same, we have $3!/2! = 3$ distinct permutations. With four different letters a, b, c, and d we had 24 distinct permutations. If we let $a = b = x$ and $c = d = y$, we can list only the following: $xxyy$, $xyxy$, $yxxy$, $yyxx$, $xyyx$, and $yxyx$. Thus we have $4!/2!2! = 6$ distinct permutations.

THEOREM 1.6 *The number of distinct permutations of n things of which* n_1 *are of one kind,* n_2 *of a second kind,* $\cdots$, n_k *of a kth kind is*

$$\frac{n!}{n_1!n_2!\cdots n_k!}.$$

Example 1.21 How many different ways can three red, four yellow, and two blue bulbs be arranged in a string of Christmas tree lights with nine sockets?

Solution. The total number of distinct arrangements is

$$\frac{9!}{3!4!2!} = 1260.$$

Often we are concerned with the number of ways of partitioning a set of n objects into r subsets called cells. A partition has been achieved if the intersection of every possible pair of the r subsets is the empty set $\emptyset$ and if the union of all subsets gives the original set. The order of the elements within a cell is of no importance. Consider the set $\{a,e,i,o,u\}$. The possible partitions into two cells in which the first cell contains four elements and the second cell one element are $\{(a, e, i, o),(u)\}$, $\{(a, i, o, u),(e)\}$, $\{(e, i, o, u),(a)\}$, $\{(a, e, o, u),(i)\}$, and $\{(a, e, i, u),(o)\}$. We see that there are five such ways to partition a set of five elements into two subsets or cells containing four elements in the first cell and one element in the second.

The number of partitions for this illustration is denoted by the symbol

$$\binom{5}{4,1} = \frac{5!}{4!1!} = 5,$$

where the top number represents the total number of elements and the bottom numbers represent the number of elements going into each cell. We state this more generally in the following theorem.

THEOREM 1.7 *The number of ways of partitioning a set of n objects into r cells with n_1 elements in the first cell, n_2 elements in the second, etc., is*

$$\binom{n}{n_1, n_2, \cdots, n_r} = \frac{n!}{n_1!n_2!\cdots n_r!},$$

where $n_1 + n_2 + \cdots + n_r = n$.

Example 1.22 How many ways can seven scientists be assigned to one triple and two double hotel rooms?

Solution. The total number of possible partitions would be

$$\binom{7}{3,\,2,\,2} = \frac{7!}{3!2!2!} = 210.$$

In many problems we are interested in the number of ways of *selecting r* objects from n without regard to order. These selections are called *combinations.* A combination is actually a partition with two cells, the one cell containing the r objects selected and the other cell containing the $(n-r)$ objects that are left.

The number of such combinations, denoted by $\binom{n}{r,\,n-r}$, is usually shortened to $\binom{n}{r}$ since the number of elements in the second cell must be $n-r$.

> **THEOREM 1.8** *The number of combinations of n distinct objects taken r at a time is*
>
> $$\binom{n}{r} = \frac{n!}{r!(n-r)!}.$$

Example 1.23 From four chemists and three physicists find the number of committees of three that can be formed consisting of two chemists and one physicist.

Solution. The number of ways of selecting 2 chemists from 4 is $\binom{4}{2} = \frac{4!}{2!2!} = 6$. The number of ways of selecting one physicist from 3 is $\binom{3}{1} = \frac{3!}{1!2!} = 3$. Using Theorem 1.1 we find the number of committees that can be formed with two chemists and one physicist to be $(6)(3) = 18$.

1.5 Probability

The statistician is basically concerned with drawing conclusions or inferences from experiments involving uncertainties. For these conclusions and inferences to be reasonably accurate, an understanding of probability theory is essential.

What do we mean when we make the statements "John will probably win the tennis match," "I have a fifty-fifty chance of getting an even number when a die is tossed," "I am not likely to win at bingo tonight," or "Most of our graduating class will likely be married within three years"? In each case we are expressing an outcome of which we are not certain, but owing to past information or from an understanding of the structure of the experiment we have some degree of confidence in the validity of the statement.

The mathematical theory of probability for finite sample spaces provides a set of numbers called *weights*, ranging from zero to 1, which provide a means of evaluating the likelihood of occurrence of events resulting from a statistical experiment. To every point in the sample space we assign a weight such that the sum of all the weights is 1. If we have reason to believe that a certain sample point is quite likely to occur when the experiment is conducted, the weight assigned should be close to 1. On the other hand, a weight closer to zero is assigned to a sample point that is unlikely to occur. In many experiments, such as tossing a coin or a die, all the sample points have the same chance of occurring and are assigned equal weights. For points outside the sample space, i.e., for simple events that cannot possibly occur, we assign a weight of zero.

To find the probability of any event A we sum all the weights assigned to the sample points in A. This sum is called the *measure* of A or the probability of A and is denoted by $Pr(A)$. Thus the measure of the set $\varnothing$ is zero and the measure of S is 1.

DEFINITION 1.13 *The probability of any event A is the sum of the weights of all sample points in A. Therefore*

$$0 \le Pr(A) \le 1, \quad Pr(\varnothing) = 0, \quad \text{and} \quad Pr(S) = 1.$$

Example 1.24 A coin is tossed twice. What is the probability that at least one head occurs?

Solution. The sample space for this experiment is

$$S = \{HH, HT, TH, TT\}.$$

If the coin is balanced, each of these outcomes would be equally likely to occur. Therefore we assign a weight of w to each sample point. Then $4w = 1$ or $w = 1/4$. If A represents the event of at least one head occurring, then $Pr(A) = 3/4$.

Example 1.25 A die is loaded in such a way that an even number is twice as likely to occur as an odd number. If E is the event that a number less than 4 occurs on a single toss of the die, find $Pr(E)$.

Solution. The sample space is $S = \{1, 2, 3, 4, 5, 6\}$. We assign a weight of w to each odd number and a weight of $2w$ to each even number. Since the sum of the weights must be 1, we have $9w = 1$ or $w = 1/9$. Hence weights of $1/9$ and $2/9$ are assigned to each odd number and even number, respectively. Therefore,

$$Pr(E) = \tfrac{1}{9} + \tfrac{2}{9} + \tfrac{1}{9} = \tfrac{4}{9}.$$

We can think of weights as being probabilities associated with simple events. If the experiment is of such a nature that we can assume equal weights for the sample points of S, then the probability of any event A is the ratio of the number of elements in A to the number of elements in S.

THEOREM 1.9 *If an experiment can result in any one of N different equally likely outcomes, and if exactly n of these outcomes correspond to event A, then the probability of event A is*

$$Pr(A) = \frac{n}{N}.$$

Example 1.26 If a card is drawn from an ordinary deck, find the probability that it is a heart.

Solution. The number of possible outcomes is 52 of which 13 are hearts. Therefore the probability of event A of getting a heart is $Pr(A) = 13/52 = 1/4$.

If the weights cannot be assumed equal, they must be assigned on the basis of prior knowledge or experimental evidence. For example, if a coin is not balanced, we would estimate the two weights by tossing the coin a large number of times and recording the outcomes. The true weights would be the fractions of heads and tails that occur in the long run. This method of arriving at weights is known as the *relative frequency* definition of probability.

To find a numerical value that represents adequately the probability of winning at tennis, we must depend on our past performance at the game as well as that of our opponent. Similarly, to find the probability that a horse will win a race we must arrive at a weight based on the previous records of all the horses entered in the race.

1.6 Some Probability Laws

Often it is easier to calculate the probability of some event from known probabilities of other events. This may well be true if the event in question can be represented as the union of two other events or as the complement of some event. Several important laws that frequently simplify the computation of probabilities are listed below.

Before proceeding to these laws we state the following definition.

DEFINITION 1.14 *Two events* A *and* B *are* mutually exclusive *if* $A \cap B = \varnothing$.

Using set language we could say that two events are mutually exclusive if they are disjoint or if they have no points in common. In a practical situation we would be more likely to say A and B are mutually exclusive if they cannot both occur at the same time.

Example 1.27 Suppose a die is tossed. Let A be the event that an even number turns up and let B be the event that an odd number shows. The intersection of the sets $A = \{2, 4, 6\}$ and $B = \{1, 3, 5\}$ is $A \cap B = \varnothing$ since they have no points in common. Therefore A and B are mutually exclusive events. Since both an even and an odd number could not occur at the same time on a single toss of a die we could have concluded that the events were mutually exclusive without finding their intersection.

THEOREM 1.10 *If A and B are any two events, then*

$$Pr(A \cup B) = Pr(A) + Pr(B) - Pr(A \cap B).$$

Proof. Consider the Venn diagram in Figure 1.7. The $Pr(A \cup B)$ is the sum of the weights of the sample points in $A \cup B$. Now $Pr(A) + Pr(B)$ is the sum of all the weights in A plus the sum of all the weights in B. Therefore we have added the weights in $A \cap B$ twice. Since these weights

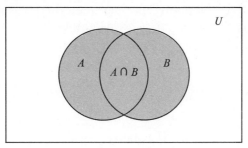

Figure 1.7.

add up to give $Pr(A \cap B)$ we must subtract this probability once to obtain the sum of the weights in $A \cup B$, which is $Pr(A \cup B)$.

COROLLARY 1. *If A and B are mutually exclusive, then*

$$Pr(A \cup B) = Pr(A) + Pr(B).$$

Corollary 1 is an immediate result of Theorem 1.10, since if A and B are mutually exclusive, $A \cap B = \emptyset$ and then $Pr(A \cap B) = Pr(\emptyset) = 0$. In general we write

COROLLARY 2. *If A_1, A_2, A_3, ..., A_n are mutually exclusive, then*

$$Pr(A_1 \cup A_2 \cup \cdots \cup A_n) = Pr(A_1) + Pr(A_2) + \cdots + Pr(A_n).$$

Note that if A_1, A_2, ..., A_n is a partition of a sample space S, then

$$
\begin{aligned}
Pr(A_1 \cup A_2 \cup \cdots \cup A_n) &= Pr(A_1) + Pr(A_2) + \cdots + Pr(A_n) \\
&= Pr(S) \\
&= 1.
\end{aligned}
$$

Example 1.28 The probability that a student passes mathematics is $2/3$, and the probability that he passes biology is $4/9$. If the probability of passing at least one course is $4/5$, what is the probability that he will pass both courses?

Solution. If M is the event "passing mathematics" and B the event "passing biology," then by transposing the terms in Theorem 1.10 we have

$$Pr(M \cap B) = Pr(M) + Pr(B) - Pr(M \cup B)$$
$$= \tfrac{2}{3} + \tfrac{4}{9} - \tfrac{4}{5}$$
$$= \tfrac{14}{45}.$$

Example 1.29 What is the probability of getting a total of 7 or 11 when a pair of dice are tossed?

Solution. Let A be the event that 7 occurs and B the event that 11 comes up. Now a total of 7 occurs for 6 of the 36 sample points and a total of 11 occurs for only 2 of the sample points. Since all sample points are equally likely, we have $Pr(A) = 1/6$ and $Pr(B) = 1/18$. The events A and B are mutually exclusive since a total of 7 and 11 cannot both occur on the same toss. Therefore

$$Pr(A \cup B) = Pr(A) + Pr(B)$$
$$= \tfrac{1}{6} + \tfrac{1}{18}$$
$$= \tfrac{2}{9}.$$

THEOREM 1.11 *If A and A' are complementary events, then*

$$Pr(A') = 1 - Pr(A).$$

Proof. Since $A \cup A' = S$ and the sets A and A' are disjoint, then

$$1 = Pr(S)$$
$$= Pr(A \cup A')$$
$$= Pr(A) + Pr(A').$$

Therefore

$$Pr(A') = 1 - Pr(A).$$

Example 1.30 A coin is tossed six times in succession. What is the probability that at least one head occurs?

Solution. Let E be the event that at least one head occurs. The sample space S consists of $2^6 = 64$ sample points since each toss can result in two outcomes. Now, $Pr(E) = 1 - Pr(E')$, where E' is the event that no head occurs. This can happen in only one way, namely, when all tosses result in a tail. Therefore $Pr(E') = 1/64$ and $Pr(E) = 1 - (1/64) = 63/64$.

1.7 Conditional Probability

The probability of an event B occurring when it is known that some event A has occurred is called a *conditional probability* and is denoted by $Pr(B \mid A)$. The symbol $Pr(B \mid A)$ is usually read "the probability that B occurs given that A occurs" or simply "the probability of B, given A."

DEFINITION 1.15 *The* conditional probability *of B, given A, denoted by $Pr(B \mid A)$, is defined by*

$$Pr(B \mid A) = \frac{Pr(A \cap B)}{Pr(A)} \qquad if\ Pr(A) > 0.$$

Consider the event B of getting a 4 when a die is tossed. The die is constructed so that the even numbers are twice as likely to occur as the odd numbers. Based on the sample space $S = \{1, 2, 3, 4, 5, 6\}$, with weights of $1/9$ and $2/9$ assigned, respectively, to the odd and even numbers, the probability of B occurring is $2/9$. Now suppose that it is known that the toss of the die resulted in a number greater than 3. We are now dealing with a reduced sample space $A = \{4, 5, 6\}$, which is a subset of S. To find the probability that B occurs, relative to the space A, we must first assign new weights to the elements of A proportional to their original weights such that their sum is 1. Assigning a weight of w to the odd number in A and a weight of $2w$ to each of the two even numbers, we have $5w = 1$ or $w = 1/5$. Relative to the space A, we find

$$Pr(B \mid A) = \frac{2}{5},$$

which can also be written

$$Pr(B \mid A) = \frac{2}{5} = \frac{2/9}{5/9} = \frac{Pr(A \cap B)}{Pr(A)},$$

where $Pr(A \cap B)$ and $Pr(A)$ are found from the original sample space S. This example illustrates that events may have different probabilities when considered relative to different sample spaces.

As a further illustration suppose our sample space S represents the adults in a small town who have completed the requirements for a college degree. We shall categorize them according to sex and employment status:

	Employed	*Unemployed*
Male	460	40
Female	140	260

One of these individuals is to be selected at random for a tour throughout the country to publicize the advantages of establishing new industries in the town. We shall be concerned with the following events:

$$M: \text{a man is chosen,}$$
$$E: \text{the chosen one is employed.}$$

Using the reduced sample space E, we find

$$Pr(M \mid E) = \frac{460}{600} = \frac{23}{30}.$$

Let $n(A)$ denote the number of elements in any set A. Using this notation we can write

$$Pr(M \mid E) = \frac{n(E \cap M)}{n(E)} = \frac{n(E \cap M)/n(S)}{n(E)/n(S)} = \frac{Pr(E \cap M)}{Pr(E)},$$

where $Pr(E \cap M)$ and $Pr(E)$ are found from the original sample space S. To verify this result, note that

$$Pr(E) = \frac{600}{900} = \frac{2}{3},$$

and

$$Pr(E \cap M) = \frac{460}{900} = \frac{23}{45}.$$

Hence

$$Pr(M \mid E) = \frac{23/45}{2/3} = \frac{23}{30},$$

as before.

Multiplying the formula in Definition 1.15 by $Pr(A)$, we obtain the following important *multiplication theorem*.

THEOREM 1.12 *If in an experiment the events A and B can both occur, then*

$$Pr(A \cap B) = Pr(A)Pr(B \mid A).$$

Thus the probability that both A and B occur is equal to the probability that A occurs multiplied by the probability that B occurs, given that A occurs.

To illustrate the use of Theorem 1.12, suppose we have a fuse box containing 20 fuses of which 5 are defective. If 2 fuses are selected at random and removed from the box in succession without replacing the first, what is the probability that both fuses are defective? To answer this question we shall let A be the event that the first fuse is defective and B the event that the second fuse is defective and then interpret $A \cap B$ as the event that A occurs and then B occurs after A has occurred. The probability of first removing a defective fuse is $1/4$ and then the probability of removing a second defective fuse from the remaining 4 is $4/19$. Hence $Pr(A \cap B) = (1/4)(4/19) = 1/19$.

Generalizing Theorem 1.12 we write

THEOREM 1.13 *If in an experiment the events A_1, A_2, A_3, etc., can occur, then*

$$Pr(A_1 \cap A_2 \cap A_3 \cap \cdots)$$
$$= Pr(A_1)Pr(A_2 \mid A_1)Pr(A_3 \mid A_1 \cap A_2) \cdots$$

If in the above illustration the first fuse is replaced and the fuses thoroughly rearranged before the second is removed, then the probability of a defective fuse on the second selection is still $1/4$. That is, $Pr(B \mid A) = Pr(B)$. When this is true the events A and B are said to be *independent*.

DEFINITION 1.16 *The events A and B are independent if, and only if,*

$$Pr(A \cap B) = Pr(A)Pr(B).$$

Example 1.31 A pair of dice are thrown twice. What is the probability of getting totals of 7 and 11?

Solution. Let A_1, A_2, B_1, and B_2 be the respective independent events that a 7 occurs on the first throw, a 7 occurs on the second throw, an 11 occurs on the first throw, and an 11 occurs on the second throw. We are interested in the probability of the union of the mutually exclusive events $A_1 \cap B_2$ and $B_1 \cap A_2$. Therefore,

$$
\begin{aligned}
Pr[(A_1 \cap B_2) \cup (B_1 \cap A_2)] &= Pr(A_1 \cap B_2) + Pr(B_1 \cap A_2) \\
&= Pr(A_1)Pr(B_2) + Pr(B_1)Pr(A_2) \\
&= (1/6)(1/18) + (1/18)(1/6) \\
&= 1/54.
\end{aligned}
$$

1.8 Bayes' Rule

Let us return to the illustration of Section 1.7, where an individual is being selected at random from the adults of a small town to tour the country and publicize the advantages of establishing new industries in the town. At that time we had no difficulty in establishing the fact that $Pr(E) = 2/3$, where E is the event that the one chosen is employed. Suppose we are given the additional information that 36 of those employed and 12 of those unemployed are members of the Rotary Club. What is the probability that the individual selected is employed if it is known that the person belongs to the Rotary Club?

Let A be the event that the person selected is a member of the Rotary Club. The conditional probability that we seek is then given by

$$
Pr(E \mid A) = \frac{Pr(E \cap A)}{Pr(A)}.
$$

Referring to Figure 1.8, we can write A as the union of the two mutually

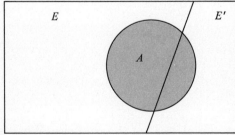

Figure 1.8 Venn diagram showing the events A, E, and E'.

exclusive events $E \cap A$ and $E' \cap A$. Hence

$$A = (E \cap A) \cup (E' \cap A),$$

and by Corollary 1 of Theorem 1.10

$$Pr(A) = Pr(E \cap A) + Pr(E' \cap A).$$

We can now write

$$Pr(E \mid A) = \frac{Pr(E \cap A)}{Pr(E \cap A) + Pr(E' \cap A)}.$$

The data of Section 1.7, together with the additional information about the set A, enable one to compute

$$Pr(E \cap A) = \frac{36}{900} = \frac{1}{25}$$

$$Pr(E' \cap A) = \frac{12}{900} = \frac{1}{75}$$

$$Pr(E \mid A) = \frac{1/25}{1/25 + 1/75} = \frac{3}{4}.$$

A generalization of the foregoing procedure leads to the following theorem, called *Bayes' rule*.

THEOREM 1.14 (BAYES' RULE) *Let* $\{B_1, B_2, \ldots, B_n\}$ *be a set of events forming a partition of the sample space* S, *where* $Pr(B_i) \neq 0$, *for* $i = 1, 2, \ldots, n$. *Let* A *be any event of* S *such that* $Pr(A) \neq 0$. *Then, for* $k = 1, 2, \ldots, n,$

$$Pr(B_k \mid A) = \frac{Pr(B_k \cap A)}{\sum\limits_{i=1}^{n} Pr(B_i \cap A)}.$$

Proof. Consider the Venn diagram in Figure 1.9. The event A is seen to be the union of the mutually exclusive events $B_1 \cap A$, $B_2 \cap A$, $\ldots$,

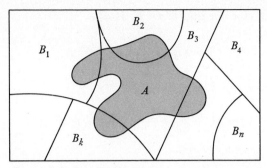

Figure 1.9 Partitioning of the sample space S.

$B_n \cap A$. That is,

$$A = (B_1 \cap A) \cup (B_2 \cap A) \cup \cdots \cup (B_n \cap A).$$

Using Corollary 2 of Theorem 1.10, we have

$$Pr(A) = Pr(B_1 \cap A) + Pr(B_2 \cap A) + \cdots + Pr(B_n \cap A)$$

$$= \sum_{i=1}^{n} Pr(B_i \cap A).$$

By the definition of conditional probability,

$$Pr(B_k \mid A) = \frac{Pr(B_k \cap A)}{Pr(A)}$$

$$= \frac{Pr(B_k \cap A)}{\sum_{i=1}^{n} Pr(B_i \cap A)},$$

which completes the proof.

Example 1.32 Three members of a private country club have been nomi-
nated for the office of president. The probability that Mr. Adams will be
elected is 0.3, the probability that Mr. Brown will be elected is 0.5, and
the probability that Mr. Cooper will be elected is 0.2. Should Mr. Adams
be elected, the probability for an increase in membership fees is 0.8.
Should Mr. Brown or Mr. Cooper be elected, the corresponding probabilities for an increase in fees are 0.1 and 0.4. If someone is considering
joining the club but delays his decision for several weeks only to find
out that the fees have been increased, what is the probability that Mr.
Cooper was elected president of the club?

Solution. We consider the following events:

A: The person elected increased fees

B_1: Mr. Adams is elected

B_2: Mr. Brown is elected

B_3: Mr. Cooper is elected.

Using Bayes' rule, we write

$$Pr(B_3 \mid A) = \frac{Pr(B_3 \cap A)}{Pr(B_1 \cap A) + Pr(B_2 \cap A) + Pr(B_3 \cap A)}.$$

Now

$$Pr(B_1 \cap A) = Pr(B_1)Pr(A \mid B_1) = (0.3)(0.8) = 0.24$$
$$Pr(B_2 \cap A) = Pr(B_2)Pr(A \mid B_2) = (0.5)(0.1) = 0.05$$
$$Pr(B_3 \cap A) = Pr(B_3)Pr(A \mid B_3) = (0.2)(0.4) = 0.08.$$

Hence

$$Pr(B_3 \mid A) = \frac{0.08}{0.24 + 0.05 + 0.08} = \frac{8}{37}.$$

In view of the fact that fees have increased, this result suggests that Mr. Cooper is probably not the president of the club.

EXERCISES

1. List the elements of each of the following sets:
   (a) The set of integers between 1 and 50 divisible by 7.
   (b) The set $A = \{x \mid x^2 + x - 6 = 0\}$.
   (c) The set of outcomes when a die and a coin are tossed simultaneously.
   (d) The set $B = \{x \mid x \text{ is a continent}\}$.
   (e) The set $C = \{x \mid 2x - 4 = 0 \text{ and } x > 5\}$.

2. Let $E = \{3, 5, 6, 8, 9\}$. Which of the following statements are true and which are false?
   (a) $5 \in E$.
   (b) $7 \in E$.
   (c) $9 \notin E$.
   (d) $\{5, 8, 3\} \subset E$.
   (e) $\{6, 8, 3, 4\} \not\subset E$.
   (f) $\emptyset \subset E$.

3. Which of the following sets are equal?
   (a) $A = \{1, 3\}$.
   (b) $B = \{x \mid x \text{ is a number on a die}\}$.
   (c) $C = \{x \mid x^2 - 4x + 3 = 0\}$.
   (d) $D = \{x \mid x \text{ is the number of heads when six coins are tossed}\}$.

4. How many subsets can be formed from the set $M = \{p, q, r, s\}$? List all the subsets that have exactly three elements.

5. List all the proper subsets of the universal set $U = \{$air, land, sea$\}$.

6. Construct a Venn diagram to illustrate the following subsets of all college students: all juniors, all mathematics majors, all women.

7. Construct a Venn diagram to illustrate the following subsets of

$\qquad U = \{$copper, sodium, nitrogen, potassium, uranium, oxygen, zinc$\}$

$\qquad A = \{$copper, sodium, zinc$\}$

$\qquad B = \{$sodium, nitrogen, potassium$\}$

$\qquad C = \{$oxygen$\}$.

8. If $U = \{0,1,2,3,4,5,6,7,8,9\}$ and $A = \{0,2,4,6,8\}$, $B = \{1,3,5,7,9\}$, $C = \{2,3,4,5\}$, and $D = \{1,6,7\}$, list the elements in the following sets:
   (a) $A \cup C$.
   (b) $A \cap B$.
   (c) C'
   (d) $(C' \cap D) \cup B$.
   (e) $(U \cap C)'$.
   (f) $A \cap C \cap D'$.

9. Referring to Exercise 7, list the elements in the following sets:
   (a) A'.
   (b) $A \cup C$.
   (c) $(A \cap B') \cup C'$.
   (d) $(B' \cap C')$.
   (e) $A \cap B \cap C$.
   (f) $(A' \cup B') \cap (A' \cap C)$.

10. If $P = \{x \mid 1 < x < 9\}$ and $Q = \{y \mid y < 5\}$, find $P \cup Q$ and $P \cap Q$.

11. Let A, B, and C be subsets of the universal set U. Using Venn diagrams shade the areas representing the following sets:
    (a) $A \cap B'$.
    (b) $(A \cup B)'$.
    (c) $(A \cap C) \cup B$.

12. An experiment involves tossing a pair of dice, one green and one red, and recording the numbers that come up.
    (a) List the elements of the sample space S.
    (b) List the elements of S corresponding to event A that the sum is less than 5.
    (c) List the elements of S corresponding to event B that a 6 occurs on either die.
    (d) List the elements of S corresponding to event C that a 2 comes up on the green die.

13. An experiment consists of flipping a coin and then flipping it a second time if a head occurs. If a tail occurs on the first flip, then a die is tossed once.
    (a) List the elements of the sample space S.
    (b) List the elements of S corresponding to event A that a number less than 4 occurred on the die.
    (c) List the elements of S corresponding to event B that two tails occurred.

14. An experiment consists of asking three women at random if they wash their dishes with brand X detergent.
    (a) List the elements of a sample space S using the letter Y for "yes" and N for "no."
    (b) List the elements of S corresponding to event E that at least two of the women use brand X.

(c) Define an event that has as its elements the points $\{YYY, NYY, YYN, NYN\}$.

15. (a) How many ways can five people be lined up to get on a bus?
    (b) If a certain two persons refuse to follow each other, how many ways are possible?

16. A college freshman must take a science course, a social studies course, and a mathematics course. If he may select any of three sciences, any of four social studies, and any of two mathematics courses, how many ways can he arrange his program?

17. In how many different ways can an eight-question true-false examination be answered?

18. How many distinct permutations can be made from the word *statistics?*

19. How many ways can the five starting positions on a basketball team be filled with nine men who can play any of the positions?

20. (a) How many three-digit numbers can be formed from the digits 0, 1, 2, 3, 4, 5 if each digit can be used only once?
    (b) How many of these are odd numbers?
    (c) How many are greater than 330?

21. A contractor wishes to build five houses, each different in design. In how many ways can he place these homes on a street if three lots are on one side of the street and two lots are on the opposite side?

22. In how many ways can four boys and three girls sit in a row if the boys and girls must alternate?

23. In how many ways can six trees be planted in a circle?

24. In how many ways can two oaks, three pines, and two maples be arranged in a straight line if one does not distinguish between trees of the same kind?

25. A college plays eight football games during a season. In how many ways can the team end the season with four wins, three losses, and one tie?

26. Ten people are going on a skiing trip in three cars that will hold 2, 4, and 5 passengers, respectively. How many ways is it possible to transport the 10 people to the ski lodge?

27. From a group of five men and three women how many committees of three people are possible
    (a) With no restrictions?
    (b) With two men and one woman?
    (c) With one man and two women if a certain woman must be on the committee?

28. How many bridge hands are possible containing five spades, three diamonds, three clubs, and two hearts?

29. From three red, four green, and five yellow apples, how many selections consisting of six apples are possible if two of each color are to be selected?

30. A shipment of 10 television sets contains 3 defective sets. In how many ways can a hotel purchase 4 of these sets and receive at least 2 of the defective sets?

31. Three men are seeking public office. Candidates A and B are given about the same chance of winning but candidate C is given twice the chance of either A or B. What is the probability that C wins? What is the probability that A does not win?

32. Find the probability of event A in Exercise 12.

33. A box contains 500 envelopes of which 50 contain $100 in cash, 100 contain $25, and 350 contain $10. An envelope may be purchased for $25. What is the sample space for the different amounts of money? Assign weights to the sample points and then find the probability that the first envelope purchased contains less than $100.

34. A pair of dice are tossed. What is the probability of getting a total of 5? At most a total of 4?

35. In a poker hand consisting of five cards, what is the probability of holding
    (a) Two aces and two kings?
    (b) Five spades?

36. If three books are picked at random from a shelf containing four novels, three books of poems, and a dictionary, what is the probability that
    (a) The dictionary is selected?
    (b) Two novels and one book of poems are selected?

37. Two cards are drawn in succession from a deck without replacement. What is the probability that both cards are greater than 2 and less than 9?

38. A basketball player sinks 50% of his shots. What is the probability that he makes exactly three of his next four shots?

39. If A and B are mutually exclusive events and $Pr(A) = 0.4$ and $Pr(B) = 0.5$, find
    (a) $Pr(A \cup B)$.   (b) $Pr(A')$.   (c) $Pr(A' \cap B)$.

40. From a box containing five black balls and three green balls, three balls are drawn in succession, each ball being replaced in the box before the next draw is made. What is the probability that all three balls are the same color? What is the probability that each color is represented?

41. A town has two fire engines operating independently. The probability that a specific fire engine is available when needed is 0.99.
    (a) What is the probability that neither is available when needed?
    (b) What is the probability that a fire engine is available when needed?

42. A real estate man has eight master keys to open several new homes. Only one master key will open any given house. If 40% of these homes are usually left unlocked, what is the probability that the real estate man can get into a specific home if he selects three master keys at random before leaving the office?

43. One bag contains four white balls and three black balls, and a second bag contains three white balls and five black balls. One ball is drawn from the first bag and placed unseen in the second bag. What is the probability that a ball now drawn from the second bag is black?

44. A pair of dice are thrown. If it is known that one die shows a 4, what is the probability that
    (a) The other die shows a 5?
    (b) The total of both dice is greater than 7?

45. A coin is biased so that a head is twice as likely to occur as a tail. If the coin is tossed three times, what is the probability of getting exactly two tails?

46. If the probability that Tom will be alive in 20 years is 0.6 and the probability that Jim will be alive in 20 years is 0.9, what is the probability that neither will be alive in 20 years?

47. In a high school graduating class of 100 students, 42 studied mathematics, 68 studied psychology, 54 studied history, 22 studied both mathematics and history, 25 studied both mathematics and psychology, 7 studied history and neither mathematics nor psychology, 10 studied all three subjects, and 8 did not take any of the three subjects. If a student is selected at random, find
 (a) The probability that he takes history and psychology but not mathematics.
 (b) The probability that if he is enrolled in history he takes all three subjects.
 (c) The probability that he takes mathematics only.

48. The probability that a married man watches a certain television show is 0.4 and the probability that a married woman watches the show is 0.5. The probability that a man watches the show, given that his wife does, is 0.7. Find
 (a) The probability that a married couple watch the show.
 (b) The probability that a wife watches the show given that her husband does.
 (c) The probability that at least one person of a married couple will watch the show.

49. Suppose that colored balls are distributed in three indistinguishable boxes as follows:

	Box 1	Box 2	Box 3
Red	2	4	3
White	3	1	4
Blue	5	3	3

A box is selected at random from which a ball is selected at random and it is observed to be red. What is the probability that box 3 was selected?

50. A commuter owns two cars, one a compact and one a standard model. About three fourths of the time he uses the compact to travel to work and about one fourth of the time the larger car is used. When he uses the compact car he usually gets home by 5:30 P.M. about 75% of the time; if he uses the standard-sized car he gets home by 5:30 P.M. about 60% of the time (but he enjoys the air conditioner in the larger car). If he gets home at 5:35 P.M., what is the probability that he used the compact car?

51. A truth serum given to a suspect is known to be 90% reliable when the person is guilty and 99% reliable when the person is innocent. In other words, 10% of the guilty are judged innocent by the serum and 1% of the innocent are judged guilty. If the suspect was selected from a group of suspects of which only 5% have ever committed a crime, and the serum indicates that he is guilty, what is the probability that he is innocent?

2 RANDOM VARIABLES

2.1 Concept of a Random Variable

The term *statistical experiment* has been used to describe any process by which several chance measurements are obtained. All possible outcomes of an experiment comprise a set that we have called the *sample space*. Often we are not interested in the details associated with each sample point but only in some numerical description of the outcome. For example, when one tosses a coin three times there are eight sample points in the sample space that give, in complete detail, the outcome of each toss. If one is concerned with the number of heads that fall, then a numerical value of 0, 1, 2, or 3 will be assigned to each sample point.

The numbers 0, 1, 2, and 3 are random *observations* determined by the outcome of the experiment. They may be looked upon as the values assumed by some *random variable X*, which in this case represents the number of heads when a coin is tossed three times.

DEFINITION 2.1 *A function whose value is a real number determined by each element in the sample space is called a* random variable.

We shall use a capital letter, say X, to denote a random variable and its corresponding small letter, x in this case, for one of its values. Each

32

possible value x of X then represents an event that is a subset of the sample space.

Example 2.1 Two balls are drawn in succession without replacement from an urn containing four red balls and three black balls. The possible outcomes and the values y of the random variable Y, where Y is the number of red balls, are

Simple event	y
RR	2
RB	1
BR	1
BB	0

Example 2.2 A stockroom clerk returns three safety helmets at random to three steel mill employees who had previously checked them. If Smith, Jones, and Brown, in that order, receive one of the three hats, list the sample points for the possible orders of returning the helmets and find the values m of the random variable M that represents the number of correct matches.

Solution. If S, J, and B stand for Smith's, Jones', and Brown's helmets, respectively, then the possible arrangements in which the helmets may be returned and the number of correct matches are

Simple event	m
SJB	3
SBJ	1
JSB	1
JBS	0
BSJ	0
BJS	1

If a sample space contains a finite number of points, as in the two examples above, or an unending sequence with as many elements as there are whole numbers, such as in the case of a die being thrown until a 5 occurs, it is called a *discrete sample space*. A random variable defined over a discrete sample space is called a *discrete random variable*.

Also, if the elements of a sample space are infinite in number, or as many as the number of points on a line segment, such as all possible heights, weights, temperatures, life periods, etc., we say we have a *continuous sample space*, and a variable defined over this space is called a *continuous random variable*.

In most practical problems, continuous random variables represent *measured* data and discrete random variables represent *count* data, such as the number of defectives in a sample of k items or the number of accidents per year.

2.2 Discrete Probability Distributions

A discrete random variable assumes each of its values with a certain probability. In the case of tossing a coin three times, the variable X, representing the number of heads, assumes the value 2 with probability 3/8, since three of the eight equally likely sample points result in two heads and one tail. Assuming equal weights for the simple events in Example 2.2, the probability that no employee gets back his right helmet, that is, the probability that M assumes the value zero, is 1/3. The possible values m of M and their probabilities are given by

m	0	1	3
$Pr(M = m)$	$\frac{1}{3}$	$\frac{1}{2}$	$\frac{1}{6}$

Note that the values of m exhaust all possible cases and hence the probabilities add to 1.

Frequently it is convenient to represent all the probabilities of a random variable X by a formula. Such a formula would necessarily be a function of the numerical values x that we shall denote by $f(x)$, $g(x)$, $r(x)$, and so forth. Hence we write $f(x) = Pr(X = x)$; that is, $f(3) = Pr(X = 3)$.

> DEFINITION 2.2 *The function $f(x)$ is a* probability function *or a* probability distribution *of the discrete random variable X if, for each possible outcome x,*
>
> 1. $f(x) \geq 0$
> 2. $\sum_x f(x) = 1$
> 3. $Pr(X = x) = f(x)$.

Example 2.3 Find the probability distribution of the sum of the numbers when a pair of dice are tossed.

Solution. Let X be a random variable whose values x are the possible totals. Then x can be any integer from 2 to 12. Two dice can fall in

(6)(6) = 36 ways, each with probability 1/36. The $Pr(X = 3) = 2/36$, since a total of 3 can occur in only two ways. Consideration of the other cases leads to the following probability distribution:

x	2	3	4	5	6	7	8	9	10	11	12
$f(x)$	$\frac{1}{36}$	$\frac{2}{36}$	$\frac{3}{36}$	$\frac{4}{36}$	$\frac{5}{36}$	$\frac{6}{36}$	$\frac{5}{36}$	$\frac{4}{36}$	$\frac{3}{36}$	$\frac{2}{36}$	$\frac{1}{36}$

Example 2.4 Find a formula for the probability distribution of the number of heads when a coin is tossed four times.

Solution. Since there are $2^4 = 16$ points in the sample space representing equally likely outcomes, the denominator for all probabilities, and therefore for our function, will be 16. To obtain the number of ways of getting, say three heads, we need to consider the number of ways of partitioning four outcomes into two cells with three heads assigned to one cell and a tail assigned to the other. This can be done in $\binom{4}{3} = 4$ ways. In general, x heads and $4 - x$ tails can occur in $\binom{4}{x}$ ways, where x can be 0, 1, 2, 3, or 4. Thus the probability distribution $f(x) = Pr(X = x)$ is

$$f(x) = \frac{\binom{4}{x}}{16}, \qquad x = 0, 1, 2, 3, 4.$$

DEFINITION 2.3 *The cumulative distribution $F(x)$ of a discrete random variable X with probability distribution $f(x)$ is given by*

$$F(x) = Pr(X \le x) = \sum_{t \le x} f(t).$$

For the random variable M, the number of correct matches, we have

$$F(2.4) = Pr(M \le 2.4) = f(0) + f(1) = (1/3) + (1/2) = 5/6.$$

The cumulative distribution of M is given by

$$F(m) = \begin{cases} 0 & \text{for } m < 0 \\ \frac{1}{3} & \text{for } 0 \le m < 1 \\ \frac{5}{6} & \text{for } 1 \le m < 3 \\ 1 & \text{for } m \ge 3. \end{cases}$$

One should pay particular notice to the fact that the cumulative distribution is defined not only for the values assumed by the given random variable but for all real numbers.

Example 2.5 Find the cumulative distribution of the random variable X in Example 2.4. Using $F(x)$, verify that $f(2) = 3/8$.

Solution. Direct calculations of the probability distribution of Example 2.4 give $f(0) = 1/16, f(1) = 1/4, f(2) = 3/8, f(3) = 1/4$, and $f(4) = 1/16$. Therefore

$$F(0) = f(0) = \tfrac{1}{16}$$
$$F(1) = f(0) + f(1) = \tfrac{5}{16}$$
$$F(2) = f(0) + f(1) + f(2) = \tfrac{11}{16}$$
$$F(3) = f(0) + f(1) + f(2) + f(3) = \tfrac{15}{16}$$
$$F(4) = f(0) + f(1) + f(2) + f(3) + f(4) = 1.$$

Hence

$$F(x) = \begin{cases} 0 & \text{for } x < 0 \\ \tfrac{1}{16} & \text{for } 0 \le x < 1 \\ \tfrac{5}{16} & \text{for } 1 \le x < 2 \\ \tfrac{11}{16} & \text{for } 2 \le x < 3 \\ \tfrac{15}{16} & \text{for } 3 \le x < 4 \\ 1 & \text{for } x \ge 4. \end{cases}$$

Now

$$f(2) = F(2) - F(1) = \tfrac{11}{16} - \tfrac{5}{16} = \tfrac{3}{8}.$$

It is often helpful to look at a probability distribution in graphic form. One might plot the points $(x, f(x))$ of Example 2.4 to obtain Figure 2.1. By joining the points to the x axis either with a dashed or solid line, we obtain what is commonly called a *bar chart*. Figure 2.1 makes it very

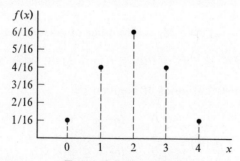

Figure 2.1 Bar chart.

easy to see what values of X are most likely to occur, and it also indicates a perfectly symmetric situation in this case.

Instead of plotting the points $(x,f(x))$, we more frequently construct rectangles, as in Figure 2.2. Here the rectangles are constructed so that their bases of equal width are centered at each value x and their heights are equal to the corresponding probabilities given by $f(x)$. The bases are constructed so as to leave no space between the rectangles. Figure 2.2 is called a *probability histogram*.

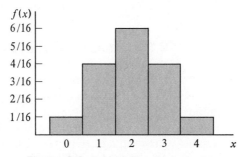

Figure 2.2 Probability histogram.

Since each base in Figure 2.2 has unit width, the $Pr(X = x)$ is equal to the area of the rectangle centered at x. Even if the bases were not of unit width we could adjust the heights of the rectangles to give areas that would still equal the probabilities of X assuming any of its values x. This concept of using areas to represent probabilities is necessary for our consideration of the probability distribution of a continuous random variable.

The graph of the cumulative distribution of Example 2.5, which appears as a step function in Figure 2.3, is obtained by plotting the points $(x,F(x))$.

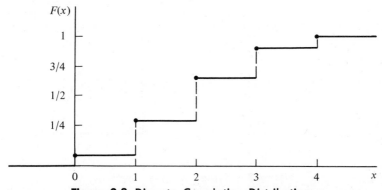

Figure 2.3 Discrete Cumulative Distribution.

Certain probability distributions are applicable to more than one physical situation. The probability distribution of Example 2.4, for example, also applies to the random variable Y, where Y is the number of red cards that occur when four cards are drawn at random from a deck in succession with each card replaced and the deck shuffled before the next drawing. Special discrete distributions that can be applied to many different experimental situations will be considered in Chapter 3.

2.3 Continuous Probability Distributions

A continuous random variable has a probability of zero of assuming exactly any of its values. Consequently, its probability distribution cannot be given in tabular form. At first this may seem startling, but it becomes more plausible when we consider a particular example. Let us discuss a random variable whose values are the heights of all people over 21 years of age. Between any two values, say 63.5 and 64.5 inches, or even 63.99 and 64.01 inches, there are an infinite number of heights, one of which is 64 inches. The probability of selecting a person at random exactly 64 inches tall and not one of the infinitely large set of heights so close to 64 inches that you cannot humanly measure the difference is

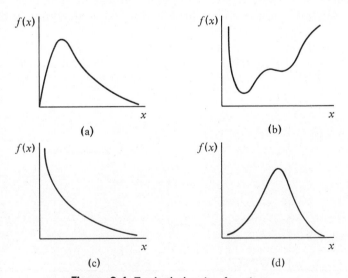

Figure 2.4 Typical density functions.

remote, and thus we assign a probability of zero to the event. It follows that

$$Pr(a < X \leq b) = Pr(a < X < b) + Pr(X = b)$$
$$= Pr(a < X < b).$$

That is, it does not matter whether we include an end point of the interval or not.

While the probability distribution of a continuous random variable cannot be presented in tabular form, it does have a formula. As before we shall designate the probability distribution of X by the functional notation $f(x)$. In dealing with continuous variables, $f(x)$ is usually called the *density function*. Since X is defined over a continuous sample space, the graph of $f(x)$ will be continuous and may, for example, take one of the forms shown in Figure 2.4.

A probability density function is constructed so that the area under its curve bounded by the x axis is equal to 1 when computed over the range of X for which $f(x)$ is defined. If $f(x)$ is represented in Figure 2.5,

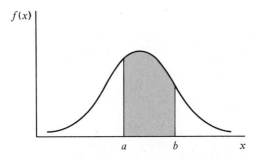

Figure 2.5 $Pr(a < X < b)$.

then the probability that X assumes a value between a and b is equal to the shaded area under the density function between the ordinates at $x = a$ and $x = b$, and from integral calculus is given by

$$Pr(a < X < b) = \int_a^b f(x)\,dx.$$

Areas have been computed and put in tabular form for those density functions that are used most frequently in conducting experiments. Because areas represent probabilities and probabilities are positive numerical values, the density function must be entirely above the x axis. If the range of X for which $f(x)$ is defined is a finite interval, it is always possible to extend the interval to include the entire set of real numbers by defining $f(x)$ to be zero at all points in the extended portions of the interval.

DEFINITION 2.4 *The function $f(x)$ is a probability density function for the continuous random variable X, defined over the set of real numbers R, if*

1. $f(x) \geq 0 \qquad$ *for all $x \in R$*

2. $\int_{-\infty}^{\infty} f(x)\, dx = 1$

3. $Pr(a < X < b) = \int_{a}^{b} f(x)\, dx.$

Example 2.6 Let the random variable X have the probability density function

$$f(x) = \frac{x^2}{3}, \qquad -1 < x < 2$$

$$= 0, \qquad \text{elsewhere.}$$

(1) Verify condition 2 of Definition 2.4. (2) Find $Pr(0 < X \leq 1)$.

Solution.

(1) $\quad \int_{-\infty}^{\infty} f(x)\, dx = \int_{-1}^{2} \frac{x^2}{3}\, dx = \frac{x^3}{9} \Big|_{-1}^{2} = \frac{8}{9} + \frac{1}{9} = 1.$

(2) $\quad Pr(0 < X \leq 1) = \int_{0}^{1} \frac{x^2}{3}\, dx = \frac{x^3}{9} \Big|_{0}^{1} = \frac{1}{9}.$

DEFINITION 2.5 *The cumulative distribution $F(x)$ of a continuous random variable X with density function $f(x)$ is given by*

$$F(x) = Pr(X \leq x) = \int_{-\infty}^{x} f(t)\, dt.$$

As an immediate consequence of Definition 2.5 one can write the two results

$$Pr(a < X < b) = F(b) - F(a)$$

and

$$f(x) = \frac{dF(x)}{dx}$$

if the derivative exists.

Example 2.7 For the density function of Example 2.6 find $F(x)$ and use it to evaluate $Pr(0 < X \le 1)$.

Solution.

$$F(x) = \int_{-\infty}^{x} f(t) \, dt = \int_{-1}^{x} \frac{t^2}{3} \, dt = \frac{t^3}{9} \Big|_{-1}^{x} = \frac{x^3 + 1}{9}.$$

Therefore

$$Pr(0 < X \le 1) = F(1) - F(0) = \tfrac{2}{9} - \tfrac{1}{9} = \tfrac{1}{9},$$

which agrees with the result obtained by using the density function in Example 2.6.

2.4 Empirical Distributions

Usually in an experiment involving a continuous random variable the density function $f(x)$ is unknown and its form is assumed. For the choice of $f(x)$ to be reasonably valid, good judgment based on all available information is needed in its selection. Statistical data, generated in large masses, can be very useful in studying the behavior of the distribution if presented in the form of a *relative frequency distribution*. Such an arrangement is obtained by grouping the data into classes and determining the proportion of measurements in each of the classes.

To illustrate the construction of a relative frequency distribution consider the data of Table 2.1, which represent the lives of 40 similar car batteries recorded to the nearest tenth of a year. The batteries were guaranteed to last 3 years.

Table 2.1 Car Battery Lives

2.2	4.1	3.5	4.5	3.2	3.7	3.0	2.6
3.4	1.6	3.1	3.3	3.8	3.1	4.7	3.7
2.5	4.3	3.4	3.6	2.9	3.3	3.9	3.1
3.3	3.1	3.7	4.4	3.2	4.1	1.9	3.4
4.7	3.8	3.2	2.6	3.9	3.0	4.2	3.5

We must first decide on the number of classes into which the data are to be grouped. This is done arbitrarily, although we are guided by the amount of data available. Usually we choose between 5 and 20 class intervals. The smaller the number of data available, the smaller is our choice for the number of classes. For the data of Table 2.1 let us choose 7 class intervals. The class width must be large enough so that 7 class intervals accommodate all the data. To determine the approximate class width, we divide the difference between the largest and smallest measurements by the number of intervals. Therefore in our example the class width can be no less than $(4.7 - 1.6)/7 = 0.443$. In practice it is desirable to choose equal class widths having the same number of significant places as the given data. Denoting this width by c, we choose $c = 0.5$. If we begin the lowest interval at 1.5, the second class would begin at 2.0 and so forth. The relative frequency distribution for the data of Table 2.1, showing the midpoints of each class interval, is given in Table 2.2.

Table 2.2 Relative Frequency Distribution
of Battery Lives

Class interval	Class midpoint	Frequency f	Relative frequency
1.5–1.9	1.7	2	0.050
2.0–2.4	2.2	1	0.025
2.5–2.9	2.7	4	0.100
3.0–3.4	3.2	15	0.375
3.5–3.9	3.7	10	0.250
4.0–4.4	4.2	5	0.125
4.5–4.9	4.7	3	0.075

The information provided by a relative frequency distribution in tabular form is easier to grasp if presented graphically. Using the midpoints of each interval and the corresponding relative frequencies we construct a *relative frequency histogram* (Figure 2.6) in exactly the same manner that we constructed the probability histogram of Section 2.2.

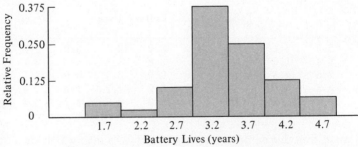

Figure 2.6 Relative frequency histogram.

In Section 2.2 we suggested that the heights of the rectangles be adjusted so that the areas would represent probabilities. Once this is done the vertical axis may be omitted. If we wish to estimate the probability distribution $f(x)$ of a continuous random variable X by a smooth curve as in Figure 2.7, it is important that the rectangles of the relative frequency histogram be adjusted so that the total area is equal to 1.

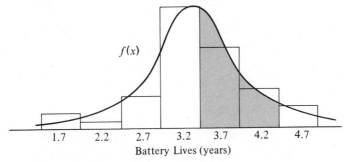

Figure 2.7 Estimating the probability density function.

The probability that a battery lasts between 3.45 and 4.45 years when selected at random from the infinite line of production of such batteries is given by the shaded area under the curve. Our estimated probability based on the recorded lives of the 40 batteries would be the sum of the areas contained in the rectangles between 3.45 and 4.45.

Although we have drawn an estimate of the shape of $f(x)$ in Figure 2.7, we still have no knowledge of its formula or equation and therefore cannot find the area that has been shaded. To help understand the method of estimating the formula for $f(x)$ let us recall some elementary analytic geometry. Parabolas, hyperbolas, circles, ellipses, and so forth all have well-known forms of equations, and in each case we would recognize their graphs. Thinking in reverse, if we had only their graphs, but recognized their form, then it is not difficult to estimate the unknown constants or parameters and arrive at the exact equation. For example, if the curve appeared to have the form of a parabola, then we know it has an equation of the form $f(x) = ax^2 + bx + c$, where a, b, and c are parameters that can be determined by various estimation procedures.

Many continuous distributions can be represented graphically by the characteristic bell-shaped curve of Figure 2.7. The equation of the probability density function $f(x)$ in this case is as well known as that of a parabola or circle and depends only on the determination of two parameters. Once these parameters are estimated from the data, we can write the estimated equation, and then, using appropriate tables, find any probabilities we choose.

A distribution is said to be symmetric if it can be folded along an axis so that the two sides coincide. A distribution that lacks symmetry with respect to a vertical axis is said to be *skewed*. The distribution illustrated in Figure 2.8(a) is said to be skewed to the right, since it has a long right tail and a much shorter left tail. In Figure 2.8(b) we see that the distribution is symmetric, while in Figure 2.8(c) it is skewed to the left.

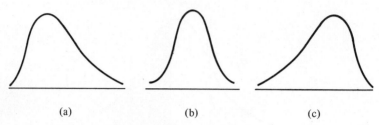

(a)   (b)   (c)

Figure 2.8 Skewness of data.

Histograms can have almost any shape or form. Some of the possible density functions that might arise were illustrated in Figure 2.4. In Chapter 4 we shall consider most of the important density functions that are used in engineering and scientific investigations.

The cumulative distribution of X, where X represents the life of the car battery, can be estimated geometrically using the data of Table 2.2. To construct such a graph we first arrange our data as in Table 2.3, a *relative cumulative frequency distribution*, and then plot the relative cumulative frequency less than any upper class boundary against the upper class boundary as in Figure 2.9. We estimate $F(x)$ by drawing a smooth curve through the points.

Percentile, decile, and quartile points may be read quickly from the cumulative distribution. In Figure 2.9 the dashed lines indicate that the twenty-fifth percentile or first quartile and the seventh decile are approx-

Table 2.3 Relative Cumulative Frequency
Distribution of Battery Lives

Class boundaries	*Relative cumulative frequency*
Less than 1.45	0.000
Less than 1.95	0.050
Less than 2.45	0.075
Less than 2.95	0.175
Less than 3.45	0.550
Less than 3.95	0.800
Less than 4.45	0.925
Less than 4.95	1.000

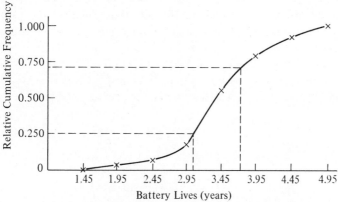

Figure 2.9 Continuous cumulative distribution.

imately 3.05 and 3.70 years, respectively. This means that 25% or one fourth of all the batteries of this type are expected to last less than 3.05 years, while 70% of such batteries can be expected to last less than 3.70 years.

2.5 Joint Probability Distributions

Our study of random variables and their probability distributions in the preceding sections was restricted to one-dimensional sample spaces in that we recorded outcomes of an experiment assumed by a single random variable. There will be many situations, however, where we may find it desirable to record the simultaneous outcomes of several random variables. For example, we might measure the amount of precipitate P and volume V of gas released from a controlled chemical experiment giving rise to a two-dimensional sample space consisting of the outcomes (p,v), or one might be interested in the hardness H and tensile strength T of cold-drawn copper resulting in the outcomes (h,t). In a study to determine the likelihood of success in college, based on high school data, one might use a three-dimensional sample space and record for each individual his aptitude test score, high school rank in class, and grade-point average at the end of the freshman year in college.

If X and Y are two random variables, the probability distribution for their simultaneous occurrence can be represented in functional notation by $f(x,y)$. It is customary to refer to $f(x,y)$ as the *joint probability distribution* of X and Y. Hence in the discrete case where a listing is possible, $f(x,y) = Pr(X = x, Y = y)$. That is, $f(x,y)$ gives the probability that the outcomes x and y occur at the same time. For example, if a television set is to be serviced and X represents the age of the set and Y represents the number of defective tubes in the set, then $f(5,3)$ is the probability that the television set is five years old and needs three new tubes.

DEFINITION 2.6 *The function $f(x,y)$ is a* joint probability function *of the* discrete *random variables X and Y if*

$$1. \quad f(x,y) \geq 0 \qquad for\ all\ (x,y)$$

$$2. \quad \sum_x \sum_y f(x,y) = 1$$

$$3. \quad Pr[(X,Y) \in A] = \sum_A \sum f(x,y)$$

for any region A in the xy plane.

Example 2.8 Two refills for a ballpoint pen are selected at random from a box that contains three blue refills, two red refills, and three green refills. If X is the number of blue refills and Y is the number of red refills selected, find (1) the joint probability function $f(x,y)$ and (2) $Pr[(X,Y) \in A]$, where A is the region $\{(x,y) \mid x + y \leq 1\}$.

Solution. (1) The possible pairs of values (x,y) are (0,0), (0,1), (1,0), (1,1), (0,2), and (2,0). Now, $f(0,1)$, for example, represents the probability that a red and a green refill are selected. The total number of equally likely ways of selecting any two refills from the eight is $\binom{8}{2} = 28$. The number of ways of selecting one red from two red refills and one green from three green refills is $\binom{2}{1}\binom{3}{1} = 6$. Hence $f(0,1) = 6/28 = 3/14$. Similar calculations yield the probabilities for the other cases, which are presented in Table 2.4. In Chapter 3 it will become clear that the joint probability distribution of Table 2.4 can be represented by the formula

$$f(x,y) = \frac{\binom{3}{x}\binom{2}{y}\binom{3}{2-x-y}}{\binom{8}{2}}, \qquad \begin{matrix} x = 0,\ 1,\ 2 \\ y = 0,\ 1,\ 2 \\ 0 \leq x + y \leq 2. \end{matrix}$$

$$(2) \qquad Pr[(X,Y) \in A] = Pr(X + Y \leq 1)$$
$$= f(0,0) + f(0,1) + f(1,0)$$
$$= \tfrac{3}{28} + \tfrac{3}{14} + \tfrac{9}{28}$$
$$= \tfrac{9}{14}.$$

Table 2.4 Joint Probability Distribution
for Example 2.8

y \ x	0	1	2
0	$\frac{3}{28}$	$\frac{9}{28}$	$\frac{3}{28}$
1	$\frac{3}{14}$	$\frac{3}{14}$	
2	$\frac{1}{28}$		

DEFINITION 2.7 *The function $f(x,y)$ is a* joint density function *of the* continuous *random variables X and Y if*

1. $f(x,y) \geq 0$ *for all* (x,y)

2. $\int_{-\infty}^{\infty} \int_{-\infty}^{\infty} f(x,y)\, dx\, dy = 1$

3. $Pr[(X,Y) \in A] = \iint_A f(x,y)\, dx\, dy$

for any region A in the xy plane.

Example 2.9 Consider the joint density function

$$f(x,y) = \frac{x(1 + 3y^2)}{4}, \qquad 0 < x < 2,\, 0 < y < 1$$
$$= 0, \qquad\qquad\qquad \text{elsewhere.}$$

(1) Verify condition 2 of Definition 2.7. (2) Find $Pr[(X,Y) \in A]$, where A is the region $\{(x,y) \mid 0 < x < 1,\, 1/4 < y < 1/2\}$.

Solution.

(1) $\displaystyle \int_{-\infty}^{\infty} \int_{-\infty}^{\infty} f(x,y)\, dx\, dy = \int_0^1 \int_0^2 \frac{x(1 + 3y^2)}{4}\, dx\, dy$

$$= \int_0^1 \frac{x^2}{8} + \frac{3x^2 y^2}{8}\, \Big|_{x=0}^{x=2}\, dy$$

$$= \int_0^1 \left(\frac{1}{2} + \frac{3y^2}{2} \right) dy = \frac{y}{2} + \frac{y^3}{2}\, \Big|_0^1$$

$$= \frac{1}{2} + \frac{1}{2} = 1.$$

(2) $Pr[(X,Y) \in A] = Pr(0 < X < 1, 1/4 < Y < 1/2)$

$$= \int_{1/4}^{1/2} \int_{0}^{1} \frac{x(1 + 3y^2)}{4} \, dx \, dy$$

$$= \int_{1/4}^{1/2} \frac{x^2}{8} + \frac{3x^2y^2}{8} \Big|_{x=0}^{x=1} \, dy$$

$$= \int_{1/4}^{1/2} \left(\frac{1}{8} + \frac{3y^2}{8} \right) dy = \frac{y}{8} + \frac{y^3}{8} \Big|_{1/4}^{1/2}$$

$$= \left(\frac{1}{16} + \frac{1}{64} \right) - \left(\frac{1}{32} + \frac{1}{512} \right) = \frac{23}{512}.$$

Given the probability distribution $f(x,y)$ of the random variables X and Y, the probability distributions of X alone and Y alone are given by

$$g(x) = \sum_{y} f(x,y)$$

$$h(y) = \sum_{x} f(x,y)$$

for the discrete case and by

$$g(x) = \int_{-\infty}^{\infty} f(x,y) \, dy$$

$$h(y) = \int_{-\infty}^{\infty} f(x,y) \, dx$$

for the continuous case. We define $g(x)$ and $h(y)$ to be the *marginal distributions* of X and Y, respectively. The fact that these marginal distributions are indeed the probability distributions of the individual variables can easily be verified by showing that the conditions of Definition 2.2 or Definition 2.4 are satisfied. For example, in the continuous case,

$$\int_{-\infty}^{\infty} g(x) \, dx = \int_{-\infty}^{\infty} \int_{-\infty}^{\infty} f(x,y) \, dy \, dx = 1$$

and

$$Pr(a < X < b) = Pr(a < X < b, -\infty < Y < \infty)$$

$$= \int_{a}^{b} \int_{-\infty}^{\infty} f(x,y) \, dy \, dx$$

$$= \int_{a}^{b} g(x) \, dx.$$

In Section 2.1 we stated that the value x of the random variable X represents an event that is a subset of the sample space. Using the definition of conditional probability as given in Chapter 1, namely,

$$Pr(B \mid A) = \frac{Pr(A \cap B)}{Pr(A)}, \qquad Pr(A) > 0,$$

where A and B are now the events defined by $X = x$ and $Y = y$, respectively,

$$Pr(Y = y \mid X = x) = \frac{Pr(X = x, Y = y)}{Pr(X = x)}$$

$$= \frac{f(x,y)}{g(x)}, \qquad g(x) > 0,$$

when X and Y are discrete random variables.

It is not difficult to show that the function $f(x,y)/g(x)$, which is strictly a function of y with x fixed, satisfies all the conditions of a probability distribution. Writing this probability distribution as $f(y \mid x)$, we have

$$f(y \mid x) = \frac{f(x,y)}{g(x)}, \qquad g(x) > 0,$$

which is called the *conditional distribution* of the *discrete* random variable Y, given that $X = x$. Similarly we define $f(x \mid y)$ to be the conditional distribution of the *discrete* random variable X, given that $Y = y$, and write

$$f(x \mid y) = \frac{f(x,y)}{h(y)}, \qquad h(y) > 0.$$

The *conditional probability density function* of the *continuous* random variable X, given that $Y = y$, is by definition

$$f(x \mid y) = \frac{f(x,y)}{h(y)}, \qquad h(y) > 0,$$

while the conditional probability density function of the *continuous* random variable Y, given that $X = x$, is defined to be

$$f(y \mid x) = \frac{f(x,y)}{g(x)}, \qquad g(x) > 0.$$

If one wished to find the probability that the continuous random variable X falls between a and b when it is known that $Y = y$, we evaluate

$$Pr(a < X < b \mid Y = y) = \int_a^b f(x \mid y) \, dx.$$

Example 2.10 Referring to Example 2.8, find $f(x \mid 1)$ and

$$Pr(X = 0 \mid Y = 1).$$

Solution. First we find

$$h(1) = \sum_{x=0}^{2} f(x,1) = \frac{3}{14} + \frac{3}{14} + 0 = \frac{3}{7}.$$

Now

$$f(x \mid 1) = \frac{f(x,1)}{h(1)} = \frac{7}{3} f(x,1), \qquad x = 0, 1, 2.$$

Therefore

$$f(0 \mid 1) = \frac{7}{3} f(0,1) = \left(\frac{7}{3}\right)\left(\frac{3}{14}\right) = \frac{1}{2}$$

$$f(1 \mid 1) = \frac{7}{3} f(1,1) = \left(\frac{7}{3}\right)\left(\frac{3}{14}\right) = \frac{1}{2}$$

$$f(2 \mid 1) = \frac{7}{3} f(2,1) = \left(\frac{7}{3}\right)(0) = 0$$

and the conditional distribution of X, given that $Y = 1$, is

x	0	1	2
$f(x \mid 1)$	$\frac{1}{2}$	$\frac{1}{2}$	0

Finally,

$$Pr(X = 0 \mid Y = 1) = f(0 \mid 1) = \tfrac{1}{2}.$$

Example 2.11 Find $g(x)$, $h(y)$, $f(x \mid y)$, and

$$Pr(1/4 < X < 1/2 \mid Y = 1/3)$$

for the density function of Example 2.9.

Solution. By definition,

$$g(x) = \int_{-\infty}^{\infty} f(x,y) \, dy = \int_{0}^{1} \frac{x(1 + 3y^2)}{4} \, dy$$

$$= \frac{xy}{4} + \frac{xy^3}{4} \Big|_{y=0}^{y=1} = \frac{x}{2}, \qquad 0 < x < 2,$$

and

$$h(y) = \int_{-\infty}^{\infty} f(x,y) \, dx = \int_{0}^{2} \frac{x(1 + 3y^2)}{4} \, dx$$

$$= \frac{x^2}{8} + \frac{3x^2y^2}{8} \Big|_{x=0}^{x=2} = \frac{1 + 3y^2}{2}, \qquad 0 < y < 1.$$

Therefore

$$f(x \mid y) = \frac{f(x,y)}{h(y)} = \frac{x(1 + 3y^2)/4}{(1 + 3y^2)/2} = \frac{x}{2}, \qquad 0 < x < 2,$$

and

$$Pr\left(\frac{1}{4} < X < \frac{1}{2} \mid Y = \frac{1}{3}\right) = \int_{1/4}^{1/2} \frac{x}{2} \, dx = \frac{3}{64}.$$

If $f(x \mid y)$ does not depend on y, as was the case in Example 2.11, then $f(x \mid y) = g(x)$ and $f(x,y) = g(x)h(y)$. The proof follows by substituting

$$f(x,y) = f(x \mid y)h(y)$$

into the marginal distribution of X. That is,

$$g(x) = \int_{-\infty}^{\infty} f(x,y) \, dy = \int_{-\infty}^{\infty} f(x \mid y)h(y) \, dy.$$

If $f(x \mid y)$ does not depend on y, we may write

$$g(x) = f(x \mid y) \int_{-\infty}^{\infty} h(y) \, dy.$$

Now

$$\int_{-\infty}^{\infty} h(y) \, dy = 1$$

since $h(y)$ is the probability density function of Y. Therefore

$$g(x) = f(x \mid y)$$

and then

$$f(x,y) = g(x)h(y),$$

which leads to the following definition.

DEFINITION 2.8 *Let X and Y be two random variables, discrete or continuous, with joint probability distribution $f(x,y)$ and marginal distributions $g(x)$ and $h(y)$, respectively. The random variables X and Y are said to be* statistically independent *if and only if*

$$f(x,y) = g(x)h(y)$$

for all (x,y).

The continuous random variables of Example 2.11 are statistically independent since the product of the two marginal distributions gives the joint density function. This is not the case, however, for the discrete random variables of Example 2.8. Consider the three probabilities $f(0,1)$, $g(0)$, and $h(1)$. From Table 2.4 we find

$$f(0,1) = \frac{3}{14}$$

$$g(0) = \sum_{y=0}^{2} f(0,y) = \frac{3}{28} + \frac{3}{14} + \frac{1}{28} = \frac{5}{14}$$

$$h(1) = \sum_{x=0}^{2} f(x,1) = \frac{3}{14} + \frac{3}{14} + 0 = \frac{3}{7}.$$

Clearly,

$$f(0,1) \neq g(0)h(1)$$

and therefore X and Y are not statistically independent.

All the preceding definitions concerning two random variables can be generalized to the case of n random variables. Let $f(x_1,x_2,\ldots,x_n)$ be the joint probability function of the random variables X_1, X_2, ..., X_n. The marginal distribution of X_1, for example, is given by

$$g(x_1) = \sum_{x_2} \cdots \sum_{x_n} f(x_1,x_2,\ldots,x_n)$$

for the discrete case and by

$$g(x_1) = \int_{-\infty}^{\infty} \cdots \int_{-\infty}^{\infty} f(x_1,x_2,\ldots,x_n) \, dx_2 \, dx_3 \cdots dx_n$$

for the continuous case. We can now obtain *joint marginal distributions* such as $\phi(x_1,x_2)$ where

$$\phi(x_1,x_2) = \sum_{x_3} \cdots \sum_{x_n} f(x_1,x_2,\ldots,x_n) \qquad \text{(discrete case)}$$

$$= \int_{-\infty}^{\infty} \cdots \int_{-\infty}^{\infty} f(x_1,x_2,\ldots,x_n) \, dx_3 \, dx_4 \cdots dx_n \quad \text{(continuous case)}$$

One could consider numerous conditional distributions. For example, the *joint conditional distribution* of X_1, X_2, and X_3 given that $X_4 = x_4$, $X_5 = x_5$, ..., $X_n = x_n$ is written

$$f(x_1,x_2,x_3 \mid x_4,x_5,\ldots,x_n) = \frac{f(x_1,x_2,\ldots,x_n)}{g(x_4,x_5,\ldots,x_n)},$$

where $g(x_4, x_5, \ldots, x_n)$ is the joint marginal distribution of the random variables X_4, $X_5, \ldots, X_n$.

A generalization of Definition 2.8 leads to the following definition for the mutually statistical independence of the variables X_1, X_2, $\ldots$, X_n.

DEFINITION 2.9 *Let X_1, X_2, $\ldots$, X_n be n random variables, discrete or continuous, with joint probability distribution $f(x_1, x_2, \ldots, x_n)$ and marginal distributions $f_1(x_1)$, $f_2(x_2)$, $\ldots$, $f_n(x_n)$, respectively. The random variables X_1, X_2, $\ldots$, X_n are said to be mutually statistically independent if and only if*

$$f(x_1, x_2, \ldots, x_n) = f_1(x_1) f_2(x_2) \cdots f_n(x_n).$$

Example 2.12 Let X_1, X_2, and X_3 be three mutually statistically independent random variables and let each have probability density function

$$f(x) = e^{-x}, \qquad x > 0$$
$$= 0, \qquad \text{elsewhere.}$$

Find $Pr(X_1 < 2, 1 < X_2 < 3, X_3 > 2)$.

Solution. The joint probability density function of X_1, X_2, and X_3 is

$$f(x_1, x_2, x_3) = f(x_1) f(x_2) f(x_3)$$
$$= e^{-x_1} e^{-x_2} e^{-x_3}$$
$$= e^{-x_1 - x_2 - x_3}, \qquad x_1 > 0, \, x_2 > 0, \, x_3 > 0.$$

Hence

$$Pr(X_1 < 2, 1 < X_2 < 3, X_3 > 2) = \int_2^\infty \int_1^3 \int_0^2 e^{-x_1 - x_2 - x_3} \, dx_1 \, dx_2 \, dx_3$$
$$= (1 - e^{-2})(e^{-1} - e^{-3}) e^{-2}$$
$$= 0.0376.$$

2.6 Mathematical Expectation

If two coins are tossed 16 times and X is the number of heads that occur per toss, then the values of X can be 0, 1, and 2. Suppose the experiment yields no heads, one head, and two heads a total of 4, 7, and 5 times, respectively. The average number of heads per toss of the two coins is then

$$\frac{(0)(4) + (1)(7) + (2)(5)}{16} = (0)\left(\frac{4}{16}\right) + (1)\left(\frac{7}{16}\right) + (2)\left(\frac{5}{16}\right)$$

$$= 1.06.$$

This is an average value and is not necessarily a possible outcome for the experiment. For instance, a salesman's average monthly income is not likely to be equal to any of his monthly pay checks.

The numbers 4/16, 7/16, and 5/16 are the fractions of the total tosses resulting in zero, one, and two heads, respectively. These fractions are also the relative frequencies for the different outcomes.

Let us now consider the problem of calculating the average number of heads per toss that we might expect in the long run. We denote this expected value or mathematical expectation by $E(X)$. From the relative frequency definition of probability we can, in the long run, expect no heads about one fourth of the time, one head about one half of the time, and two heads about one fourth of the time. Therefore

$$E(X) = (0)(\tfrac{1}{4}) + (1)(\tfrac{1}{2}) + 2(\tfrac{1}{4}) = 1.$$

This means that a person who throws two coins over and over again will, on the average, get one head per toss.

The above illustration suggests that the mean or expected value of any random variable may be obtained by multiplying each value of the random variable by its corresponding probability and summing the results. This is true, of course, only if the variable is discrete. In the case of continuous random variables the definition of mathematical expectation is essentially the same with summations being replaced by integrals.

DEFINITION 2.10 *Let X be a random variable with probability distribution $f(x)$. The* expected value *of X or the* mathematical expectation *of X is*

$$E(X) = \sum_{x} xf(x) \qquad \text{if } X \text{ is discrete}$$

$$= \int_{-\infty}^{\infty} xf(x)\, dx \qquad \text{if } X \text{ is continuous.}$$

Example 2.13 Find the expected number of chemists on a committee of 3 selected at random from four chemists and three biologists.

Solution. Let X represent the number of chemists on the committee. The probability distribution of X is given by

$$f(x) = \frac{\binom{4}{x}\binom{3}{3-x}}{\binom{7}{3}}, \qquad x = 0, 1, 2, 3.$$

A few simple calculations yield $f(0) = 1/35$, $f(1) = 12/35$, $f(2) = 18/35$, and $f(3) = 4/35$. Therefore

$$E(X) = (0)(\tfrac{1}{35}) + (1)(\tfrac{12}{35}) + (2)(\tfrac{18}{35}) + (3)(\tfrac{4}{35})$$
$$= \tfrac{12}{7} = 1.7.$$

Thus if a committee of 3 is selected at random over and over again from four chemists and three biologists, it would contain on the average 1.7 chemists.

Example 2.14 In a gambling game a man is paid $5 if he gets all heads or all tails when three coins are tossed and he pays out $3 if either one or two heads show. What is his expected gain?

Solution. The random variable of interest is Y, the amount he can win. The possible values of Y are 5 and -3 with probabilities 1/4 and 3/4, respectively. Therefore,

$$E(Y) = (5)(\tfrac{1}{4}) + (-3)(\tfrac{3}{4}) = -1.$$

In this game the gambler will, on the average, lose $1 per toss.

A game is considered "fair" if the gambler will, on the average, come out even. Therefore an expected gain of zero defines a fair game.

Example 2.15 Let X be the random variable that denotes the life in hours of a certain type of tube. The probability density function is given by

$$f(x) = \frac{20{,}000}{x^3}, \qquad x > 100$$
$$= 0, \qquad \text{elsewhere.}$$

Find the expected life of this type of tube.

Solution. Using Definition 2.10, we have

$$E(X) = \int_{100}^{\infty} x \frac{20,000}{x^3}\, dx$$

$$= \int_{100}^{\infty} \frac{20,000}{x^2}\, dx$$

$$= 200.$$

Therefore, we can expect this type of tube to last, on the average, 200 hours.

Now let us consider a function $g(X)$ of the random variable X. That is, each value of $g(X)$ is determined by knowing the values of X. For instance, $g(X)$ might be X^2 or $3X - 1$, so that whenever X assumes the value 2, $g(X)$ assumes the value $g(2)$. In particular, if X is a discrete random variable with probability distribution $f(x)$, $x = -1, 0, 1, 2$, and $g(X) = X^2$, then

$$Pr[g(X) = 0] = Pr(X = 0) = f(0)$$
$$Pr[g(X) = 1] = Pr(X = -1) + Pr(X = 1) = f(-1) + f(1)$$
$$Pr[g(X) = 4] = Pr(X = 2) = f(2).$$

By Definition 2.10,

$$E[g(X)] = \sum_{g(x)} g(x) Pr[g(X) = g(x)]$$

$$= 0 Pr[g(X) = 0] + 1 Pr[g(X) = 1] + 4 Pr[g(X) = 4]$$

$$= 0 f(0) + 1[f(-1) + f(1)] + 4 f(2)$$

$$= \sum_x g(x) f(x).$$

This result is generalized in Theorem 2.1 for both discrete and continuous random variables.

THEOREM 2.1 *Let X be a random variable with probability distribution $f(x)$. The expected value of the function $g(X)$ is*

$$E[g(X)] = \sum_x g(x) f(x) \qquad \text{if } X \text{ is discrete}$$

$$= \int_{-\infty}^{\infty} g(x) f(x)\, dx \qquad \text{if } X \text{ is continuous.}$$

Example 2.16 Let X be a random variable with probability distribution as follows:

x	0	1	2	3
$f(x)$	$\frac{1}{3}$	$\frac{1}{2}$	0	$\frac{1}{6}$

Find the expected value of $Y = (X - 1)^2$.

Solution. By Theorem 2.1, we write

$$E[(X - 1)^2] = \sum_{x=0}^{3} (x - 1)^2 f(x)$$
$$= (-1)^2 f(0) + (0)^2 f(1) + (1)^2 f(2) + (2)^2 f(3)$$
$$= (1)(\tfrac{1}{3}) + (0)(\tfrac{1}{2}) + (1)(0) + (4)(\tfrac{1}{6})$$
$$= 1.$$

Example 2.17 Let X be a random variable with density function

$$f(x) = \frac{x^2}{3}, \qquad -1 < x < 2$$
$$= 0, \qquad \text{elsewhere.}$$

Find the expected value of $g(X) = 2X - 1$.

Solution. By Theorem 2.1, we have

$$E(2X - 1) = \int_{-1}^{2} \frac{(2x - 1)x^2}{3} \, dx$$
$$= \frac{1}{3} \int_{-1}^{2} (2x^3 - x^2) \, dx$$
$$= \frac{3}{2}.$$

We shall now extend our concept of mathematical expectation to the case of two random variables X and Y with joint probability distribution $f(x,y)$.

DEFINITION 2.11 *Let X and Y be random variables with joint probability distribution $f(x,y)$. The expected value of the function $g(X,Y)$ is*

$$E[g(X,Y)] = \sum_x \sum_y g(x,y)f(x,y) \qquad \textit{if X and Y are discrete}$$
$$= \int_{-\infty}^{\infty} \int_{-\infty}^{\infty} g(x,y)f(x,y) \, dx \, dy \qquad \textit{if X and Y are continuous.}$$

Generalization of Definition 2.11 for the calculation of mathematical expectations of functions of several random variables is straightforward.

Example 2.18 Let X and Y be random variables with joint probability distribution given by Table 2.4. Find the expected value of $g(X,Y) = XY$.

Solution. By Definition 2.11, we write

$$E(XY) = \sum_{x=0}^{2} \sum_{y=0}^{2} xyf(x,y)$$

$$= (0)(0)f(0,0) + (0)(1)f(0,1) + (0)(2)f(0,2)$$
$$+ (1)(0)f(1,0) + (1)(1)f(1,1)$$
$$+ (2)(0)f(2,0)$$

$$= f(1,1) = \tfrac{3}{14}.$$

Example 2.19 Find $E(Y/X)$ for the density function

$$f(x,y) = \frac{x(1 + 3y^2)}{4}, \qquad 0 < x < 2, \, 0 < y < 1$$

$$= 0, \qquad\qquad\quad \text{elsewhere.}$$

Solution. We have

$$E(Y/X) = \int_0^1 \int_0^2 \frac{y(1 + 3y^2)}{4} \, dx \, dy$$

$$= \int_0^1 \frac{(y + 3y^3)}{2} \, dy$$

$$= \tfrac{5}{8}.$$

Note that if $g(x,y) = X$ in Definition 2.11, we have

$$E(X) = \sum_x \sum_y xf(x,y) = \sum_x xg(x) \qquad\qquad \text{(discrete case)}$$

$$= \int_{-\infty}^{\infty} \int_{-\infty}^{\infty} xf(x,y) \, dx \, dy = \int_{-\infty}^{\infty} xg(x) \, dx \qquad \text{(continuous case),}$$

where $g(x)$ is the marginal distribution of X. Therefore, in calculating $E(X)$ over a two-dimensional space, one may use either the joint probability distribution of X and Y or the marginal distribution of X.

Similarly we define

$$E(Y) = \sum_x \sum_y yf(x,y) = \sum_y yh(y) \qquad \text{(discrete case)}$$

$$= \int_{-\infty}^{\infty} \int_{\infty}^{\infty} yf(x,y)\, dx\, dy = \int_{-\infty}^{\infty} yh(y)\, dy \qquad \text{(continuous case)},$$

where $h(y)$ is the marginal distribution of the random variable Y.

2.7 Laws of Expectation

We shall now develop some useful laws that will simplify the calculations of mathematical expectations. These laws or theorems will permit us to calculate expectations in terms of other expectations that are either known or easily computed. All the results are valid for both discrete and continuous random variables. Proofs will be given only for the continuous case.

THEOREM 2.2 *If a and b are constant, then*

$$E(aX + b) = aE(X) + b.$$

Proof. By the definition of an expected value,

$$E(aX + b) = \int_{-\infty}^{\infty} (ax + b)f(x)\, dx$$

$$= a \int_{-\infty}^{\infty} xf(x)\, dx + b \int_{-\infty}^{\infty} f(x)\, dx.$$

The first integral on the right is $E(X)$ and the second integral equals 1. Therefore, we have

$$E(aX + b) = aE(X) + b.$$

COROLLARY 1 *Setting $a = 0$, we see that $E(b) = b$.*

COROLLARY 2 *Setting $b = 0$, we see that $E(aX) = aE(X)$.*

THEOREM 2.3 *The expected value of the sum or difference of two or more functions of a random variable X is the sum or difference of the expected values of the functions. That is,*

$$E[g(X) \pm h(X)] = E[g(X)] \pm E[h(X)].$$

Proof. By definition,

$$E[g(X) \pm h(X)] = \int_{-\infty}^{\infty} [g(x) \pm h(x)]f(x)\, dx$$

$$= \int_{-\infty}^{\infty} g(x)f(x)\, dx \pm \int_{-\infty}^{\infty} h(x)f(x)\, dx$$

$$= E[g(X)] \pm E[h(X)].$$

Example 2.20 In Example 2.16 we could write

$$E[(X - 1)^2] = E(X^2 - 2X + 1) = E(X^2) - 2E(X) + E(1).$$

From Corollary 1, $E(1) = 1$, and by direct computation

$$E(X) = (0)(\tfrac{1}{3}) + (1)(\tfrac{1}{2}) + (2)(0) + (3)(\tfrac{1}{6}) = 1$$
$$E(X^2) = (0)(\tfrac{1}{3}) + (1)(\tfrac{1}{2}) + (4)(0) + (9)(\tfrac{1}{6}) = 2.$$

Hence,

$$E[(X - 1)^2] = 2 - (2)(1) + 1 = 1,$$

as before.

Example 2.21 In Example 2.17 we may prefer to write

$$E(2X - 1) = 2E(X) - 1.$$

Now

$$E(X) = \int_{-1}^{2} x\left(\frac{x^2}{3}\right) dx = \int_{-1}^{2} \frac{x^3}{3}\, dx = \frac{5}{4}.$$

Therefore

$$E(2X - 1) = (2)\left(\frac{5}{4}\right) - 1 = \frac{3}{2},$$

as before.

Suppose we have two random variables X and Y with joint probability distribution $f(x,y)$. Two additional laws that will be very useful in succeeding chapters involve the expected values of the sum, difference, and product of these two random variables. First, however, let us prove a theorem on the expected value of the sum or difference of functions of the given variables. This, of course, is merely an extension of Theorem 2.3.

THEOREM 2.4 *The expected value of the sum or difference of two or more functions of the random variables X and Y is the sum or difference of the expected values of the functions. That is,*

$$E[g(X,Y) \pm h(X,Y)] = E[g(X,Y)] \pm E[h(X,Y)].$$

Proof. By Definition 2.11,

$E[g(X,Y) \pm h(X,Y)]$

$$= \int_{-\infty}^{\infty} \int_{-\infty}^{\infty} [g(x,y) \pm h(x,y)]f(x,y)\ dx\ dy$$

$$= \int_{-\infty}^{\infty} \int_{-\infty}^{\infty} g(x,y)f(x,y)\ dx\ dy \pm \int_{-\infty}^{\infty} \int_{-\infty}^{\infty} h(x,y)f(x,y)\ dx\ dy$$

$$= E[g(X,Y)] \pm E[h(X,Y)].$$

COROLLARY *Setting $g(X,Y) = X$ and $h(X,Y) = Y$, we see that*

$$E(X \pm Y) = E(X) \pm E(Y).$$

If X represents the daily production of some item from machine A and Y the daily production of the same kind of item from machine B, then $X + Y$ represents the total number of items produced daily from both machines. The Corollary of Theorem 2.4 states that the average daily production for both machines is equal to the sum of the average daily production of each machine.

THEOREM 2.5 *Let X and Y be two independent random variables. Then*

$$E(XY) = E(X)E(Y).$$

Proof. By Definition 2.11,

$$E(XY) = \int_{-\infty}^{\infty} \int_{-\infty}^{\infty} xyf(x,y)\ dx\ dy.$$

Since X and Y are independent, we may write

$$f(x,y) = g(x)h(y),$$

where $g(x)$ and $h(y)$ are the marginal distributions of X and Y, respectively. Hence

$$E(XY) = \int_{-\infty}^{\infty} \int_{-\infty}^{\infty} xyg(x)h(y)\ dx\ dy$$

$$= \int_{-\infty}^{\infty} xg(x)\ dx \int_{-\infty}^{\infty} yh(y)\ dy$$

$$= E(X)E(Y).$$

Example 2.22 Let X and Y be independent random variables with joint probability distribution

$$f(x,y) = \frac{x(1 + 3y^2)}{4}, \qquad 0 < x < 2, 0 < y < 1$$

$$= 0, \qquad\qquad\qquad \text{elsewhere.}$$

Verify Theorem 2.5.

Solution. Now

$$E(XY) = \int_0^1 \int_0^2 \frac{x^2 y(1 + 3y^2)}{4}\ dx\ dy$$

$$= \int_0^1 \frac{x^3 y(1 + 3y^2)}{12} \Big|_{x=0}^{x=2} dy$$

$$= \int_0^1 \frac{2y(1 + 3y^2)}{3}\ dy$$

$$= \frac{5}{6}$$

$$E(X) = \int_0^1 \int_0^2 \frac{x^2(1 + 3y^2)}{4}\ dx\ dy$$

$$= \int_0^1 \frac{x^3(1 + 3y^2)}{12} \Big|_{x=0}^{x=2} dy$$

$$= \int_0^1 \frac{2(1 + 3y^2)}{3}\ dy$$

$$= \frac{4}{3}$$

$$E(Y) = \int_0^1 \int_0^2 \frac{xy(1 + 3y^2)}{4} \, dx \, dy$$

$$= \int_0^1 \frac{x^2 y(1 + 3y^2)}{8} \Big|_{x=0}^{x=2} dy$$

$$= \int_0^1 \frac{y(1 + 3y^2)}{2} \, dy$$

$$= \frac{5}{8}.$$

Hence

$$E(X)E(Y) = \left(\frac{4}{3}\right)\left(\frac{5}{8}\right) = \frac{5}{6} = E(XY).$$

2.8 Special Mathematical Expectations

If $g(X) = X^k$, Theorem 2.1 yields an expected value called the *kth moment about the origin* of the random variable X, which we denote by μ'_k. Therefore

$$\mu'_k = E(X^k) = \sum_x x^k f(x) \qquad \text{if } X \text{ is discrete}$$

$$= \int_{-\infty}^{\infty} x^k f(x) \, dx \qquad \text{if } X \text{ is continuous.}$$

Note that when $k = 0$ we have $\mu'_0 = E(X^0) = E(1) = 1$, since

$$E(1) = \sum_x f(x) = 1 \qquad \text{if } X \text{ is discrete}$$

$$= \int_{-\infty}^{\infty} f(x) \, dx = 1 \qquad \text{if } X \text{ is continuous.}$$

When $k = 1$ we have $\mu'_1 = E(X)$, which is just the expected value of the random variable X itself. Because the first moment about the origin of a random variable X is somewhat special in that it represents the *mean* of the random variable, we shall write it as μ_X or simply μ. Thus

$$\mu = \mu'_1 = E(X).$$

If $g(X) = (X - \mu)^k$, Theorem 2.1 yields an expected value called the *kth moment about the mean* of the random variable X, which we denote by μ_k. Therefore

$$\mu_k = E[(X - \mu)^k] = \sum_x (x - \mu)^k f(x) \qquad \text{if } X \text{ is discrete}$$

$$= \int_{-\infty}^{\infty} (x - \mu)^k f(x) \, dx \qquad \text{if } X \text{ is continuous.}$$

The second moment about the mean, μ_2, is of special importance because it tells us something about the variability of the measurements about the mean. We shall henceforth call μ_2 the *variance* of the random variable X and denote it by σ_X^2, or simply σ^2. Thus

$$\sigma^2 = \mu_2 = E[(X - \mu)^2].$$

The positive square root of the variance is a measure called the *standard deviation*.

An alternate and preferred formula for σ^2 is given in the following theorem.

THEOREM 2.6 *The variance of a random variable X is given by*

$$\sigma^2 = E(X^2) - \mu^2.$$

Proof.

$$\sigma^2 = E[(X - \mu)^2]$$
$$= E(X^2 - 2\mu X + \mu^2)$$
$$= E(X^2) - 2\mu E(X) + E(\mu^2)$$
$$= E(X^2) - \mu^2,$$

since $\mu = E(X)$ by definition and $E(\mu^2) = \mu^2$ by Theorem 2.2, Corollary 1.

Example 2.23 Calculate the variance of X, where X is the number of chemists on a committee of 3 selected at random from four chemists and three biologists.

Solution. In Example 2.13 we showed that $\mu = 12/7$. Now

$$E(X^2) = (0)(\tfrac{1}{35}) + (1)(\tfrac{12}{35}) + (4)(\tfrac{18}{35}) + (9)(\tfrac{4}{35})$$
$$= \tfrac{24}{7}.$$

Therefore

$$\sigma^2 = \tfrac{24}{7} - (\tfrac{12}{7})^2 = \tfrac{24}{49}.$$

Example 2.24 Find the mean and variance of the random variable X, where X has the density function

$$f(x) = 2(x - 1), \qquad 1 < x < 2$$
$$= 0, \qquad\qquad \text{elsewhere.}$$

Solution.

$$\mu = E(X) = 2 \int_1^2 x(x - 1) \, dx = \frac{5}{3},$$

and

$$E(X^2) = 2 \int_1^2 x^2(x - 1) \, dx = \frac{17}{6}.$$

Therefore,

$$\sigma^2 = \frac{17}{6} - \left(\frac{5}{3}\right)^2 = \frac{1}{18}.$$

If $g(X,Y) = (X - \mu_X)(Y - \mu_Y)$, where $\mu_X = E(X)$ and $\mu_Y = E(Y)$, Definition 2.11 yields an expected value called the *covariance* of X and Y, which we denote by σ_{XY} or $\text{cov}(X,Y)$. Therefore

$$\sigma_{XY} = E[(X - \mu_X)(Y - \mu_Y)]$$
$$= \sum_x \sum_y (x - \mu_X)(y - \mu_Y)f(x,y) \qquad \text{if } X \text{ and } Y \text{ are discrete}$$
$$= \int_{-\infty}^{\infty} \int_{-\infty}^{\infty} (x - \mu_X)(y - \mu_Y)f(x,y) \, dx \, dy$$
$$\qquad\qquad\qquad\qquad\qquad \text{if } X \text{ and } Y \text{ are continuous.}$$

The covariance will be positive when high values of X are associated with high values of Y and low values of X are associated with low values of Y. If low values of X are associated with high values of Y, and vice versa, then the covariance will be negative. When X and Y are statistically independent it can be shown that the covariance is zero (see Theorem 2.11, Corollary 1). The converse, however, is not generally true. Two variables may have zero covariance and still not be statistically independent.

The alternate and preferred formula for σ_{XY} is given in the following theorem.

THEOREM 2.7 *The* covariance *of two random variables* X *and* Y *with means* μ_X *and* μ_Y, *respectively, is given by*

$$\sigma_{XY} = E(XY) - \mu_X\mu_Y.$$

Proof.

$$\sigma_{XY} = E[(X - \mu_X)(Y - \mu_Y)]$$
$$= E(XY - \mu_X Y - \mu_Y X + \mu_X\mu_Y)$$
$$= E(XY) - \mu_X E(Y) - \mu_Y E(X) + E(\mu_X\mu_Y)$$
$$= E(XY) - \mu_X\mu_Y,$$

since $\mu_X = E(X)$ and $\mu_Y = E(Y)$ by definition and $E(\mu_X\mu_Y) = \mu_X\mu_Y$ by Theorem 2.2, Corollary 1.

Example 2.25 Referring to the joint probability distribution of Example 2.8 and to the computations of Example 2.18 we see that $E(XY) = \frac{3}{14}$. Now,

$$\mu_X = E(X) = \sum_{x=0}^{2} \sum_{y=0}^{2} xf(x,y) = \sum_{x=0}^{2} xg(x)$$
$$= (0)(\tfrac{10}{28}) + (1)(\tfrac{15}{28}) + (2)(\tfrac{3}{28})$$
$$= \tfrac{3}{4},$$

and

$$\mu_Y = E(Y) = \sum_{x=0}^{2} \sum_{y=0}^{2} yf(x,y) = \sum_{y=0}^{2} yh(y)$$
$$= (0)(\tfrac{15}{28}) + (1)(\tfrac{3}{7}) + (2)(\tfrac{1}{28})$$
$$= \tfrac{1}{2}.$$

Therefore

$$\sigma_{XY} = E(XY) - \mu_X\mu_Y$$
$$= \tfrac{3}{14} - (\tfrac{3}{4})(\tfrac{1}{2})$$
$$= -\tfrac{9}{56}.$$

Example 2.26 Let the random variables X and Y have the joint probability density function

$$f(x,y) = 2, \qquad 0 < x < y, 0 < y < 1$$
$$= 0, \qquad \text{elsewhere.}$$

Find σ_{XY}.

Solution. We compute

$$\mu_X = E(X) = \int_0^1 \int_0^y 2x \, dx \, dy = \frac{1}{3}$$

$$\mu_Y = E(Y) = \int_0^1 \int_0^y 2y \, dx \, dy = \frac{2}{3}$$

and

$$E(XY) = \int_0^1 \int_0^y 2xy \, dx \, dy = \frac{1}{4}.$$

Then

$$\sigma_{XY} = E(XY) - \mu_X \mu_Y$$

$$= \frac{1}{4} - \left(\frac{1}{3}\right)\left(\frac{2}{3}\right)$$

$$= \frac{1}{36}$$

2.9 Properties of the Variance

We shall now prove four theorems that are useful in calculating variances or standard deviations. If we let $g(X)$ be a function of the random variable X, then the mean and variance of $g(X)$ will be denoted by $\mu_{g(X)}$ and $\sigma^2_{g(X)}$, respectively.

THEOREM 2.8 *Let X be a random variable with probability distribution $f(x)$. The variance of the function $g(X)$ is*

$$\sigma^2_{g(X)} = E[\{g(X) - \mu_{g(X)}\}^2].$$

Proof. Since $g(X)$ is a random variable, the result follows from the definition of the variance.

THEOREM 2.9 *If X is a random variable and b is a constant, then*

$$\sigma^2_{X+b} = \sigma^2_X = \sigma^2.$$

Proof.

$$\sigma_{X+b}^2 = E[\{(X + b) - \mu_{X+b}\}^2].$$

Now

$$\mu_{X+b} = E(X + b) = E(X) + b = \mu + b$$

by Theorem 2.2. Therefore

$$\sigma_{X+b}^2 = E[(X + b - \mu - b)^2]$$
$$= E[(X - \mu)^2]$$
$$= \sigma^2.$$

This theorem states that the variance is unchanged if a constant is added to or subtracted from a random variable. The addition or subtraction of a constant simply shifts the values of X to the right or to the left but does not change their variability.

THEOREM 2.10 *If X is a random variable and a is any constant, then*

$$\sigma_{aX}^2 = a^2\sigma_X^2 = a^2\sigma^2.$$

Proof.

$$\sigma_{aX}^2 = E[\{aX - \mu_{aX}\}^2].$$

Now

$$\mu_{aX} = E(aX) = aE(X) = a\mu$$

by Theorem 2.2, Corollary 2. Therefore

$$\sigma_{aX}^2 = E[(aX - a\mu)^2]$$
$$= a^2E[(X - \mu)^2]$$
$$= a^2\sigma^2.$$

Therefore, if a random variable is multiplied or divided by a constant the variance is multiplied or divided by the square of the constant.

THEOREM 2.11 *If X and Y are random variables with joint probability distribution $f(x,y)$, then*

$$\sigma_{aX+bY}^2 = a^2\sigma_X^2 + b^2\sigma_Y^2 + 2ab\sigma_{XY}.$$

Proof.

$$\sigma^2_{aX+bY} = E[(aX + bY) - \mu_{aX+bY}]^2.$$

Now

$$\mu_{aX+bY} = E(aX + bY) = aE(X) + bE(Y) = a\mu_X + b\mu_Y$$

by using Theorem 2.3 followed by Theorem 2.2, Corollary 2. Therefore

$$
\begin{aligned}
\sigma^2_{aX+bY} &= E\{[(aX + bY) - (a\mu_X + b\mu_Y)]^2\} \\
&= E\{[a(X - \mu_X) + b(Y - \mu_Y)]^2\} \\
&= a^2 E[(X - \mu_X)^2] + b^2 E[(Y - \mu_Y)^2] + 2abE[(X - \mu_X)(Y - \mu_Y)] \\
&= a^2\sigma^2_X + b^2\sigma^2_Y + 2ab\sigma_{XY}.
\end{aligned}
$$

COROLLARY 1 *If X and Y are independent random variables, then*

$$\sigma^2_{aX+bY} = a^2\sigma^2_X + b^2\sigma^2_Y.$$

The result given in Corollary 1 is obtained from Theorem 2.11 by proving the covariance of the independent variables X and Y to be zero. Hence, from Theorem 2.7

$$
\begin{aligned}
\sigma_{XY} &= E(XY) - \mu_X\mu_Y \\
&= 0,
\end{aligned}
$$

since $E(XY) = E(X)E(Y)$ for independent variables.

COROLLARY 2 *If X and Y are independent random variables, then*

$$\sigma^2_{aX-bY} = a^2\sigma^2_X + b^2\sigma^2_Y.$$

Corollary 2 follows by writing $aX - bY$ as $aX + (-bY)$. Then $\sigma^2_{aX-bY} = \sigma^2_{aX} + \sigma^2_{(-bY)}$. From Theorem 2.10 we know that $\sigma^2_{(-bY)} = (-b)^2\sigma^2_Y = b^2\sigma^2_Y$. Therefore

$$\sigma^2_{aX-bY} = a^2\sigma^2_X + b^2\sigma^2_Y.$$

Example 2.27 If X and Y are independent random variables with variances $\sigma_X^2 = 1$ and $\sigma_Y^2 = 2$, find the variance of the random variable $Z = 3X - 2Y + 5$.

Solution.

$$
\begin{aligned}
\sigma_Z^2 &= \sigma_{3X-2Y+5}^2 \\
&= \sigma_{3X-2Y}^2 & \text{by Theorem 2.9} \\
&= 9\sigma_X^2 + 4\sigma_Y^2 & \text{by Theorem 2.11, Corollary 2} \\
&= (9)(1) + (4)(2) \\
&= 17.
\end{aligned}
$$

2.10 Chebyshev's Theorem

In Section 2.8 we stated that the variance of a random variable tells us something about the variability of the observations about the mean. If a random variable has a small variance or standard deviation, we would expect most of the values to be grouped around the mean. Therefore, the probability that a random variable assumes a value within a certain interval about the mean is greater than for a similar random variable with a larger standard deviation. If we think of probability in terms of area, we would expect a continuous distribution with a small standard deviation to have most of its area close to μ, as in Figure 2.10(a). However, a large value of σ indicates a greater variability and therefore we would expect the area to be more spread out, as in Figure 2.10(b).

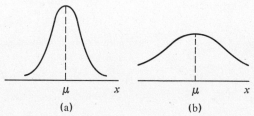

μ x μ x

(a)   (b)

Figure 2.10 Variability of continuous observations about the mean.

We can argue the same way for a discrete distribution. The area in the probability histogram in Figure 2.11(b) is spread out much more than that of Figure 2.11(a), indicating a more variable distribution of measurements or outcomes.

The Russian mathematician Chebyshev discovered that the fraction

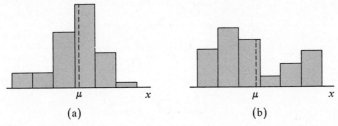

Figure 2.11 Variability of discrete observations about the mean.

of the area between any two values symmetric about the mean is related to the standard deviation. Since the area under a probability distribution curve or in a probability histogram adds to 1, the area between any two numbers is the probability of the random variable assuming a value between these numbers.

The following theorem, due to Chebyshev, gives a conservative estimate of the probability of a random variable falling within k standard deviations of its mean. We shall give the proof only for the continuous case, leaving the discrete case as an exercise.

CHEBYSHEV'S THEOREM. *The probability that any random variable X falls within k standard deviations of the mean is at least $(1 - 1/k^2)$. That is,*

$$Pr(\mu - k\sigma < X < \mu + k\sigma) \geq 1 - \frac{1}{k^2}.$$

Proof. By our previous definition of the variance of X we can write

$$\sigma^2 = E[(X - \mu)^2]$$

$$= \int_{-\infty}^{\infty} (x - \mu)^2 f(x)\, dx$$

$$= \int_{-\infty}^{\mu - k\sigma} (x - \mu)^2 f(x)\, dx + \int_{\mu - k\sigma}^{\mu + k\sigma} (x - \mu)^2 f(x)\, dx$$

$$+ \int_{\mu + k\sigma}^{\infty} (x - \mu)^2 f(x)\, dx$$

$$\geq \int_{-\infty}^{\mu - k\sigma} (x - \mu)^2 f(x)\, dx + \int_{\mu + k\sigma}^{\infty} (x - \mu)^2 f(x)\, dx,$$

since the second of the three integrals is positive. Now, since $|x - \mu| \geq k\sigma$ wherever $x \geq \mu + k\sigma$ or $x \leq \mu - k\sigma$, we have $(x - \mu)^2 \geq k^2\sigma^2$ in both remaining integrals. It follows that

$$\sigma^2 \geq \int_{-\infty}^{\mu - k\sigma} k^2\sigma^2 f(x)\ dx + \int_{\mu + k\sigma}^{\infty} k^2\sigma^2 f(x)\ dx$$

and that

$$\int_{-\infty}^{\mu - k\sigma} f(x)\ dx + \int_{\mu + k\sigma}^{\infty} f(x)\ dx \leq \frac{1}{k^2}.$$

Hence

$$Pr(\mu - k\sigma < X < \mu + k\sigma) = \int_{\mu - k\sigma}^{\mu + k\sigma} f(x)\ dx \geq 1 - \frac{1}{k^2}$$

and the theorem is established.

For $k = 2$ the theorem states that the random variable X has a probability of at least $1 - (1/2)^2 = 3/4$ of falling within two standard deviations of the mean. That is, three fourths or more of the observations of any distribution lie in the interval $\mu \pm 2\sigma$. Similarly, the theorem says that at least eight ninths of the observations of any distribution fall in the interval $\mu \pm 3\sigma$.

Example 2.28 A random variable X has a mean $\mu = 8$, a variance $\sigma^2 = 9$, and an unknown probability distribution. Find (1) $Pr(-4 < X < 20)$ and (2) $Pr(|X - 8| \geq 6)$.

Solution.

(1) $Pr(-4 < X < 20) = Pr[8 - (4)(3) < X < 8 + (4)(3)]$
$$\geq \tfrac{15}{16}.$$

(2) $Pr(|X - 8| \geq 6) = 1 - Pr(|X - 8| < 6)$
$$= 1 - Pr(-6 < X - 8 < 6)$$
$$= 1 - Pr[8 - (2)(3) < X < 8 + (2)(3)]$$
$$\leq \tfrac{1}{4}.$$

Chebyshev's theorem holds for any distribution of observations and, for this reason, the results are usually weak. The value given by the theorem is a lower bound only. That is, we know the probability of a random variable falling within two standard deviations of the mean can be *no less* than 3/4, but we never know how much more it might actually be. Only when the probability distribution is known can we determine exact probabilities.

EXERCISES

1. Classify the following random variables as discrete or continuous.
   X: The number of automobile accidents per year in Virginia.
   Y: The length of time to play 18 holes of golf.
   M: The amount of milk produced yearly by a particular cow.
   N: The number of eggs laid each month by one hen.
   P: The number of building permits issued each month in a certain city.
   Q: The weight of grain in pounds produced per acre.

2. From a box containing four black balls and two green balls, three balls are drawn in succession, each ball being replaced in the box before the next draw is made. Find the probability distribution for the number of green balls.

3. Find the probability distribution for the number of jazz records when four records are selected at random from a collection consisting of five jazz records, two classical records, and three polka records. Express your results by means of a formula.

4. Find a formula for the probability distribution of the random variable X representing the outcome when a single die is rolled once.

5. A shipment of six television sets contains two defective sets. A hotel makes a random purchase of three of the sets. If X is the number of defective sets purchased by the hotel, find the probability distribution of X. Express the results graphically as a probability histogram.

6. A coin is biased so that a head is twice as likely to occur as a tail. If the coin is tossed three times, find the probability distribution for the number of heads. Construct a probability histogram for the distribution.

7. Find the cumulative distribution of the random variable X in Exercise 5. Using $F(x)$, find
   (a) $Pr(X = 1)$.   (b) $Pr(0 < X \leq 2)$.

8. Construct a graph of the cumulative distribution of Exercise 7.

9. Find the cumulative distribution of the random variable X representing the number of heads in Exercise 6. Using $F(x)$, find
   (a) $Pr(1 \leq X < 3)$.   (b) $Pr(X > 2)$.

10. Construct a graph of the cumulative distribution of Exercise 9.

11. A continuous random variable X that can assume values between $x = 1$ and $x = 3$ has a density function given by $f(x) = 1/2$.
    (a) Show that the area under the curve is equal to 1.
    (b) Find $Pr(2 < X < 2.5)$.
    (c) Find $Pr(X \leq 1.6)$.

12. A continuous random variable X that can assume values between $x = 2$ and $x = 5$ has a density function given by $f(x) = 2(1 + x)/27$. Find
    (a) $Pr(X < 4)$.   (b) $Pr(3 \leq X < 4)$.

13. For the density function of Exercise 11, find $F(x)$ and use it to evaluate $Pr(2 < X < 2.5)$.

14. For the density function of Exercise 12, find $F(x)$ and use it to evaluate $Pr(3 \leq X < 4)$.

15. The following scores represent the final examination grade for an elementary statistics course:

$$
\begin{array}{cccccccccc}
23 & 60 & 79 & 32 & 57 & 74 & 52 & 70 & 82 & 36 \\
80 & 77 & 81 & 95 & 41 & 65 & 92 & 85 & 55 & 76 \\
52 & 10 & 64 & 75 & 78 & 25 & 80 & 98 & 81 & 67 \\
41 & 71 & 83 & 54 & 64 & 72 & 88 & 62 & 74 & 43 \\
60 & 78 & 89 & 76 & 84 & 48 & 84 & 90 & 15 & 79 \\
34 & 67 & 17 & 82 & 69 & 74 & 63 & 80 & 85 & 61
\end{array}
$$

Using 10 intervals with the lowest starting at 9,
(a) Set up a relative frequency distribution.
(b) Construct a relative frequency histogram.
(c) Construct a smoothed relative cumulative frequency distribution.
(d) Estimate the first quartile and the seventh decile.

16. The following data represent the length of life in years, measured to the nearest tenth, of a random sample of 30 similar fuel pumps:

$$
\begin{array}{cccccc}
2.0 & 3.0 & 0.3 & 3.3 & 1.3 & 0.4 \\
0.2 & 6.0 & 5.5 & 6.5 & 0.2 & 2.3 \\
1.5 & 4.0 & 5.9 & 1.8 & 4.7 & 0.7 \\
4.5 & 0.3 & 1.5 & 0.5 & 2.5 & 5.0 \\
1.0 & 6.0 & 5.6 & 6.0 & 1.2 & 0.2
\end{array}
$$

Using 6 intervals with the lowest starting at 0.1,
(a) Set up a relative frequency distribution.
(b) Construct a relative frequency histogram.
(c) Construct a smoothed relative cumulative frequency distribution.
(d) Estimate the value below which two thirds of the values fall.

17. From a sack of fruit containing three oranges, two apples, and three bananas a random sample of four pieces of fruit is selected. If X is the number of oranges and Y is the number of apples in the sample, find
(a) The joint probability distribution of X and Y.
(b) $Pr[(X,Y) \in A]$, where A is the region $\{ (x,y) \mid x + y \leq 2 \}$.

18. Two random variables have the joint density given by

$$f(x,y) = 4xy, \qquad 0 < x < 1, 0 < y < 1$$
$$= 0, \qquad \text{elsewhere.}$$

(a) Find the probability that $0 \leq X \leq 3/4$ and $1/8 \leq Y \leq 1/2$.
(b) Find the probability that $Y > X$.

19. Two random variables have the joint density given by

$$f(x,y) = k(x^2 + y^2), \qquad 0 < x < 2, 1 < y < 4$$
$$= 0, \qquad \text{elsewhere.}$$

(a) Find k.

(b) Find the probability that $1 < X < 2$ and $2 < Y \le 3$.

(c) Find the probability that $1 \le X \le 2$.

(d) Find the probability that $X + Y > 4$.

20. Referring to Exercise 17, find

    (a) $f(y \mid 2)$,   (b) $Pr(Y = 0 \mid X = 2)$.

21. Suppose X and Y have the following joint probability distribution:

y \ x	1	2	3
1	0	$\frac{1}{6}$	$\frac{1}{12}$
2	$\frac{1}{5}$	$\frac{1}{9}$	0
3	$\frac{2}{15}$	$\frac{1}{4}$	$\frac{1}{18}$

Evaluate the marginal and conditional probability distributions.

22. Suppose X and Y have the following joint probability function:

y \ x	2	4
1	0.10	0.15
3	0.20	0.30
5	0.10	0.15

Find the marginal probability distributions and determine whether X and Y are independent.

23. Determine whether the two random variables of Exercise 18 are dependent or independent.

24. Determine whether the two random variables of Exercise 19 are dependent or independent.

25. The joint probability density function of the random variables X, Y, and Z is given by

$$f(x,y,z) = \frac{4xyz^2}{9}, \qquad 0 < x < 1, 0 < y < 1, 0 < z < 3$$

$$= 0, \qquad \text{elsewhere.}$$

Find

(a) The joint marginal density function of Y and Z.

(b) The marginal density of Y.

(c) $Pr(1/4 < X < 1/2, Y > 1/3, 1 < Z < 2)$.

26. The joint probability density function of the random variables X and Y is given by

$$f(x,y) = 2, \qquad 0 < x < y < 1$$
$$= 0, \qquad \text{elsewhere.}$$

    (a) Determine if X and Y are independent.
    (b) Find $Pr(1/4 < X < 1/2 \mid Y = 3/4)$.

27. Find the expected value of the random variable X in Exercise 5.

28. By investing in a particular stock a man can make a profit in 1 year of $3000 with probability 0.3 or take a loss of $1000 with probability 0.7. What is his mathematical expectation?

29. In a gambling game a man is paid $2 if he draws a jack or queen and $5 if he draws a king or ace from an ordinary deck of 52 playing cards. If he draws any other card he loses. How much should he pay to play if the game is fair?

30. The probability distribution of the discrete random variable X is

$$f(x) = \binom{3}{x} \left(\frac{1}{4}\right)^x \left(\frac{3}{4}\right)^{3-x}, \qquad x = 0, 1, 2, 3.$$

Find $E(X)$.

31. Find the expected value of the random variable X having the density function

$$f(x) = 2(1 - x), \qquad 0 < x < 1$$
$$= 0, \qquad\qquad \text{elsewhere.}$$

32. The density function of coded measurements of pitch diameter of threads of a fitting is given by

$$f(x) = \frac{4}{\pi(1 + x^2)}, \qquad 0 < x < 1$$
$$= 0, \qquad\qquad \text{elsewhere.}$$

Find $E(X)$.

33. Let X represent the outcome when a balanced die is tossed. Find $E(Y)$, where $Y = 2X^2 - 5$.

34. A race car driver wishes to insure his car for the racing season for $10,000. The insurance company estimates a total loss may occur with probability 0.002, a 50% loss with probability 0.01, and a 25% loss with probability 0.1. Ignoring all other partial losses, what premium should the insurance company charge each season to make a profit of $100?

35. Find the expected value of $g(X) = X^2$, where X has the density function of Exercise 31.

36. Let X and Y have the joint probability distribution of Exercise 22. Find the expected value of $g(X,Y) = XY^2$.

37. Suppose X and Y are independent random variables with probability densities $g(x) = 8/x^3$, $x > 2$, and $h(y) = 2y$, $0 < y < 1$. Find the expected value of $Z = XY$.

38. Let X be a random variable with the following probability distribution:

x	-3	6	9
$Pr(X = x)$	$\frac{1}{6}$	$\frac{1}{2}$	$\frac{1}{3}$

Find $E(X)$ and $E(X^2)$ and then, using the laws of expectation, evaluate
(a) $E[(2X + 1)^2]$.   (b) $E[\{X - E(X)\}^2]$.

39. Let X represent the number that occurs when a red die is tossed and Y the number that occurs when a green die is tossed. Find
(a) $E(X + Y)$.   (b) $E(X - Y)$.   (c) $E(XY)$.

40. Referring to the random variables whose joint distribution is given in Exercise 22, find
(a) $E(2X - 3Y)$.   (b) $E(XY)$.

41. Find the variance of the random variable X of Exercise 5.

42. Let X be a random variable with the following probability distribution:

x	-2	3	5
$Pr(X = x)$	0.3	0.2	0.5

Find the standard deviation of X.

43. From a group of five men and three women a committee of 3 is selected at random. If X represents the number of women on the committee, find the mean and variance of X.

44. Find the variance of the random variable X of Exercise 31.

45. Compute the $Pr(\mu - 2\sigma < X < \mu + 2\sigma)$, where X has the density function

$$f(x) = 6x(1 - x), \quad 0 < x < 1$$
$$= 0, \qquad\qquad \text{elsewhere.}$$

46. Find the covariance of the random variables X and Y of Exercise 17.

47. Find the covariance of the random variables X and Y of Exercise 21.

48. Find the covariance of the random variables X and Y of Exercise 26.

49. Find the covariance of the random variables X and Y having the joint probability density function

$$f(x,y) = x + y, \quad 0 < x < 1, 0 < y < 1$$
$$= 0, \qquad\quad \text{elsewhere.}$$

50. Show that cov $(aX,bY) = ab$ cov (X,Y).

51. Let X represent the number that occurs when a green die is tossed and Y the number that occurs when a red die is tossed. Find the variance of the random variable
   (a) $2X - Y$.   (b) $X + 3Y - 5$.

52. If X and Y are independent random variables with variances $\sigma_X^2 = 5$ and $\sigma_Y^2 = 3$, find the variance of the random variable $Z = -2X + 4Y - 3$.

53. Repeat Exercise 52 if X and Y are not independent and $\sigma_{XY} = 1$.

54. A random variable X has a mean $\mu = 12$, a variance $\sigma^2 = 9$, and an unknown probability distribution. Using Chebyshev's theorem find
   (a) $Pr(6 < X < 18)$.   (b) $Pr(3 < X < 21)$.

55. A random variable X has a mean $\mu = 10$ and a variance $\sigma^2 = 4$. Using Chebyshev's theorem, find
   (a) $Pr(|X - 10| \geq 3)$.
   (b) $Pr(|X - 10| < 3)$.
   (c) $Pr(5 < X < 15)$.
   (d) The value of c such that $Pr(|X - 10| \geq c) \leq 0.04$.

56. Prove Chebyshev's theorem when X is a discrete random variable.

3 SOME DISCRETE PROBABILITY DISTRIBUTIONS

In Chapter 2 a discrete probability distribution was represented graphically, in tabular form, and, if convenient, by means of a formula. No matter what method of presentation is used, the behavior of a random variable is described. Many random variables associated with statistical experiments have similar properties and can be described by essentially the same probability distribution. For example, all random variables representing the number of successes in n independent trials of an experiment, where the probability of a success is constant for all n trials, have the same general type of behavior and therefore can be represented by a single formula. Thus if in firing a rifle at a target a direct hit is considered a success and the probability of a hit remains constant for successive firings, the formula for the distribution of hits in five firings of the rifle has the same structure as the formula for the distribution of 4's in seven tosses of a die, where the occurrence of a 4 is considered a success.

Frequent reference will be made throughout the text to the moments of a probability distribution. By this we shall mean the moments of any random variable having that particular probability distribution. Therefore, the mean or variance of a given probability distribution is defined to be the mean or variance of any random variable having that distribution.

Care should be exercised in choosing the probability distribution that correctly describes the observations being generated by the experiment. In this chapter we shall investigate several important discrete probability distributions that describe most random variables encountered in practice.

3.2 Uniform Distribution

The simplest of all discrete probability distributions is one where the random variable assumes all its values with equal probability. Such a probability distribution is called the *uniform distribution*.

> UNIFORM DISTRIBUTION *If the random variable X assumes the values $x_1, x_2, \ldots, x_k$, with equal probability, then the discrete uniform distribution is given by*
>
> $$f(x;k) = \frac{1}{k}, \qquad x = x_1, x_2, \ldots, x_k.$$

We have used the notation $f(x;k)$ instead of $f(x)$ to indicate that the uniform distribution depends on the *parameter k*.

Example 3.1 When a die is tossed, each element of the sample space $S = \{1, 2, 3, 4, 5, 6\}$ occurs with probability 1/6. Therefore we have a uniform distribution, with $f(x;6) = 1/6$, $x = 1, 2, 3, 4, 5, 6$.

Example 3.2 Suppose an employee is selected at random from a staff of 10 to supervise a certain project. Each employee has the same probability 1/10 of being selected. Assuming the employees have been numbered in some way from 1 to 10, the distribution is uniform with $f(x;10) = 1/10$, $x = 1, 2, \ldots, 10$.

The graphic representation of the uniform distribution by means of a histogram always turns out to be a set of rectangles with equal heights. The histogram for Example 3.1 is shown in Figure 3.1.

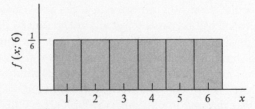

Figure 3.1 Histogram for the tossing of a die.

THEOREM 3.1 *The mean and variance of the discrete uniform distribution $f(x;k)$ are*

$$\mu = \frac{\sum_{i=1}^{k} x_i}{k} \quad and \quad \sigma^2 = \frac{\sum_{i=1}^{k} (x_i - \mu)^2}{k}.$$

Proof. By definition

$$\mu = E(X) = \sum_{i=1}^{k} x_i f(x_i, k) = \sum_{i=1}^{k} \frac{x_i}{k}.$$

Similarly,

$$E(X^2) = \sum_{i=1}^{k} \frac{x_i^2}{k}.$$

Hence

$$\sigma^2 = E(X^2) - \mu^2 = \sum_{i=1}^{k} \frac{x_i^2}{k} - \mu^2$$

$$= \frac{\sum_{i=1}^{k} x_i^2 - k\mu^2}{k} = \frac{\sum_{i=1}^{k} (x_i - \mu)^2}{k},$$

since

$$\sum_{i=1}^{k} (x_i - \mu)^2 = \sum_{i=1}^{k} (x_i^2 - 2\mu x_i + \mu^2) = \sum_{i=1}^{k} x_i^2 - 2k\mu^2 + k\mu^2$$

$$= \sum_{i=1}^{k} x_i^2 - k\mu^2.$$

Example 3.3 Referring to Example 3.1, we find that

$$\mu = \frac{1 + 2 + 3 + 4 + 5 + 6}{6} = 3.5$$

and

$$\sigma^2 = \frac{(1 - 3.5)^2 + (2 - 3.5)^2 + \cdots + (6 - 3.5)^2}{6} = \frac{35}{12}.$$

3.3 Binomial and Multinomial Distributions

An experiment often consists of repeated trials, each with two possible outcomes, which may be labeled *success* or *failure*. This is true in testing items as they come off an assembly line where each test or trial may indicate a defective or a nondefective item. We may choose to define either outcome as a success. It is also true if five cards are drawn in succession from an ordinary deck and each trial is labeled a success or failure depending on whether the card is red or black. If each card is replaced and the deck shuffled before the next drawing, then the two experiments described have similar properties in that the repeated trials are independent and the probability of a success remains constant, 1/2, from trial to trial. Experiments of this type are known as *binomial experiments*. Observe in the card-drawing example that the probabilities of a success for repeated trials change if the cards are not replaced. That is, the probability of selecting a red card on the first draw is 1/2, but on the second draw it is a conditional probability having a value of 26/51 or 25/51, depending on the color that occurred on the first draw. This then would no longer be considered a binomial experiment.

A binomial experiment is one that possesses the following properties:

1. The experiment consists of n repeated trials.
2. Each trial results in an outcome that may be classified as a success or a failure.
3. The probability of success, denoted by p, remains constant from trial to trial.
4. The repeated trials are independent.

Consider the binomial experiment where three items are selected at random from a manufacturing process, inspected, and classified defective or nondefective. A defective item is designated a success. The number of successes is a random variable X assuming integral values from zero through 3. The eight possible outcomes and the corresponding values of X are

Outcome	x
NNN	0
NDN	1
NND	1
DNN	1
NDD	2
DND	2
DDN	2
DDD	3

Since the items are selected independently from a process that we shall assume produces 25% defectives, the

$$Pr(\text{NDN}) = Pr(\text{N})Pr(\text{D})Pr(\text{N}) = (3/4)(1/4)(3/4) = 9/64.$$

Similar calculations yield the probabilities for the other possible out-comes. The probability distribution of X is therefore given by

x	0	1	2	3
$f(x)$	$\frac{27}{64}$	$\frac{27}{64}$	$\frac{9}{64}$	$\frac{1}{64}$

DEFINITION 3.1 *The number X of successes in n trials of a binomial experiment is called a* binomial random variable.

The probability distribution of the binomial variable X is called the *binomial distribution* and will be denoted by $b(x;n,p)$ since its values depend on the number of trials and the probability of a success on a given trial. Thus for the probability distribution of X, the number of defectives,

$$Pr(X = 2) = f(2) = b(2;3,\tfrac{1}{4}) = \tfrac{9}{64}.$$

Let us now generalize the above illustration to yield a formula for $b(x;n,p)$. That is, we wish to find a formula that gives the probability of x successes in n trials for a binomial experiment. First, consider the probability of x successes and $n - x$ failures in a specified order. Since the trials are independent, we can multiply all the probabilities corre-sponding to the different outcomes. Each success occurs with probability p and each failure with probability $q = 1 - p$. Therefore, the probability for the specified order is $p^x q^{n-x}$. We must now determine the total num-ber of sample points in the experiment that have x successes and $n - x$ failures. This number is equal to the number of partitions of n outcomes into two groups with x in one group and $n - x$ in the other and is given by $\binom{n}{x}$. Because these partitions are mutually exclusive, we add the probabilities of all the different partitions to obtain the general formula, or simply multiply $p^x q^{n-x}$ by $\binom{n}{x}$.

BINOMIAL DISTRIBUTION *If a binomial trial can result in a success with probability p and a failure with probability $q = 1 - p$, then the probability distribution of the binomial random variable X, the number of successes in n independent trials, is*

$$b(x;n,p) = \binom{n}{x} p^x q^{n-x}, \qquad x = 0, 1, 2, \ldots, n.$$

Note that when $n = 3$ and $p = 1/4$, the probability distribution of X, the number of defectives, may be written as

$$b(x;3,\tfrac{1}{4}) = \binom{3}{x}\left(\frac{1}{4}\right)^x\left(\frac{3}{4}\right)^{3-x}, \qquad x = 0, 1, 2, 3,$$

rather than in the tabular form above.

Example 3.4 The probability that a certain kind of component will survive a given shock test is 3/4. Find the probability that exactly two of the next four components tested survive.

Solution. Assuming the tests are independent and $p = 3/4$ for each of the four tests,

$$b(2;4,\tfrac{3}{4}) = \binom{4}{2}\left(\frac{3}{4}\right)^2\left(\frac{1}{4}\right)^2$$

$$= \frac{4!}{2!2!}\cdot\frac{3^2}{4^4}$$

$$= \frac{27}{128}.$$

The binomial distribution derives its name from the fact that the $n + 1$ terms in the binomial expansion of $(q + p)^n$ correspond to the values of $b(x;n,p)$ for $x = 0, 1, 2, \ldots, n$. That is,

$$(q + p)^n = \binom{n}{0}q^n + \binom{n}{1}pq^{n-1} + \binom{n}{2}p^2q^{n-2} + \cdots + \binom{n}{n}p^n$$
$$= b(0;n,p) + b(1;n,p) + b(2;n,p) + \cdots + b(n;n,p).$$

Since $p + q = 1$, we see that $\sum_{x=0}^{n} b(x;n,p) = 1$, a condition that must hold for any probability distribution.

Frequently we are interested in problems where it is necessary to find $Pr(X < r)$ or $Pr(a \leq X \leq b)$. Fortunately binomial sums $B(r;n,p) = \sum_{x=0}^{r} b(x;n,p)$ are available and are given in Table II (see Statistical Tables) for samples of size $n = 5, 10, 15,$ and 20, and selected values of p from 0.1 to 0.90. We illustrate the use of Table II with the following example.

Example 3.5 The probability that a patient recovers from a rare blood disease is 0.4. If 15 people are known to have contracted this disease, what is the probability that (1) at least 10 survive, (2) from 3 to 8 survive, (3) exactly 5 survive?

Solution.

(1) Let X be the number of people that survive. Then

$$Pr(X \geq 10) = 1 - Pr(X < 10)$$

$$= 1 - \sum_{x=0}^{9} b(x;15,0.4)$$

$$= 1 - 0.9662$$

$$= 0.0338.$$

(2) $\quad Pr(3 \leq X \leq 8) = \sum_{x=3}^{8} b(x;15,0.4)$

$$= \sum_{x=0}^{8} b(x;15,0.4) - \sum_{x=0}^{2} b(x;15,0.4)$$

$$= 0.9050 - 0.0271$$

$$= 0.8779.$$

(3) $\qquad Pr(x = 5) = b(5;15,0.4)$

$$= \sum_{x=0}^{5} b(x;15,0.4) - \sum_{x=0}^{4} b(x;15,0.4)$$

$$= 0.4032 - 0.2173$$

$$= 0.1859.$$

THEOREM 3.2 *The mean and variance of the binomial distribution* $b(x;n,p)$ *are*

$$\mu = np \quad and \quad \sigma^2 = npq.$$

Proof. Let the outcome on the jth trial be represented by the random variable I_j, which assumes the values zero and 1 with probabilities q and p, respectively. This is called an indicator variable since $I_j = 0$ indicates a failure and $I_j = 1$ indicates a success.

Therefore in a binomial experiment the number of successes can be written as the sum of the n independent indicator variables. Hence,

$$X = I_1 + I_2 + \cdots + I_n.$$

The mean of any I_j is $E(I_j) = 0 \cdot q + 1 \cdot p = p$. Therefore, using the corollary of Theorem 2.4, the mean of the binomial distribution is

$$\mu = E(X) = E(I_1) + E(I_2) + \cdots + E(I_n)$$
$$= \underbrace{p + p + \cdots + p}_{n \text{ terms}}$$
$$= np.$$

The variance of any I_j is given by

$$\sigma_{I_j}^2 = E[(I_j - p)^2] = E(I_j^2) - p^2 = (0)^2 q + (1)^2 p - p^2 = p(1 - p) = pq.$$

Therefore, by Theorem 2.11, Corollary 1, the variance of the binomial distribution is

$$\sigma_X^2 = \sigma_{I_1}^2 + \sigma_{I_2}^2 + \cdots + \sigma_{I_n}^2$$
$$= \underbrace{pq + pq + \cdots + pq}_{n \text{ terms}}$$
$$= npq.$$

Example 3.6 Using Chebyshev's theorem, find and interpret the interval $\mu \pm 2\sigma$ for Example 3.5.

Solution. Since Example 3.5 was a binomial experiment with $n = 15$ and $p = 0.4$, by Theorem 3.2 we have

$$\mu = (15)(0.4) = 6 \quad \text{and} \quad \sigma^2 = (15)(0.4)(0.6) = 3.6.$$

Taking the square root of 3.6 we find $\sigma = 1.897$. Hence the required interval is $6 \pm (2)(1.897)$, or from 2.206 to 9.794. Chebyshev's theorem states that the recovery rate of 15 patients subjected to the given disease has a probability of at least 3/4 of falling between 2.206 and 9.794.

The binomial experiment becomes a *multinomial experiment* if we let each trial have more than two possible outcomes. Hence the classification of a manufactured product as being light, heavy, or acceptable and the recording of accidents at a certain intersection according to the day of the week constitute multinomial experiments. The drawing of a card from a deck *with replacement* is also a multinomial experiment if the four suits are the outcomes of interest.

In general, if a given trial can result in any one of k possible outcomes $E_1, E_2, \ldots, E_k$ with probabilities $p_1, p_2, \ldots, p_k$, then the *multinomial distribution* will give the probability that E_1 occurs x_1 times, E_2 occurs x_2 times, $\ldots$, E_k occurs x_k times in n independent trials, where $x_1 + x_2$

$+ \cdots + x_k = n$. We shall denote this joint probability distribution by $f(x_1,x_2,\ldots,x_k;p_1,p_2,\ldots,p_k,n)$. Clearly $p_1 + p_2 + \cdots + p_k = 1$, since the result of each trial must be one of the k possible outcomes.

To derive the general formula we proceed as in the binomial case. Since the trials are independent, any specified order yielding x_1 outcomes for E_1, x_2 for E_2, ..., x_k for E_k will occur with probability $p_1^{x_1}p_2^{x_2} \cdots p_k^{x_k}$. The total number of orders yielding similar outcomes for the n trials is equal to the number of partitions of n items into k groups with x_1 in the first group, x_2 in the second group, ..., x_k in the kth group. This can be done in

$$\binom{n}{x_1,x_2,\ldots,x_k} = \frac{n!}{x_1!x_2! \cdots x_k!}$$

ways. Since all the partitions are mutually exclusive and occur with equal probability, we obtain the multinomial distribution by multiplying the probability for a specified order by the total number of partitions.

MULTINOMIAL DISTRIBUTION *If a given trial can result in the k outcomes E_1, E_2, ..., E_k with probabilities p_1, p_2, ..., p_k, then the probability distribution of the random variables X_1, X_2, ..., X_k, representing the number of occurrences for E_1, E_2, ..., E_k in n independent trials is*

$$f(x_1,x_2,\ldots,x_k;p_1,p_2,\ldots,p_k,n) = \binom{n}{x_1,x_2,\ldots,x_k} p_1^{x_1}p_2^{x_2} \cdots p_k^{x_k}$$

with

$$\sum_{i=1}^{k} x_i = n \quad and \quad \sum_{i=1}^{k} p_i = 1.$$

The multinomial distribution derives its name from the fact that the terms of the multinomial expansion of $(p_1 + p_2 + \cdots + p_k)^n$ correspond to all the possible values of $f(x_1,x_2,\ldots,x_k;p_1,p_2,\ldots,p_k,n)$.

Example 3.7 If a pair of dice is tossed six times, what is the probability of obtaining a total of 7 or 11 twice, a matching pair once, and any other combination three times?

Solution. We list the following possible events,

E_1: a total of 7 or 11 occurs

E_2: a matching pair occurs

E_3: neither a pair nor a total of 7 or 11 occurs.

The corresponding probabilities for a given trial are $p_1 = 2/9$, $p_2 = 1/6$, and $p_3 = 11/18$. These values remain constant for all six trials. Using the multinomial distribution with $x_1 = 2$, $x_2 = 1$, and $x_3 = 3$, the required probability is

$$f\left(2,1,3;\tfrac{2}{9},\tfrac{1}{6},\tfrac{11}{18},6\right) = \binom{6}{2,1,3}\left(\frac{2}{9}\right)^2\left(\frac{1}{6}\right)^1\left(\frac{11}{18}\right)^3$$

$$= \frac{6!}{2!1!3!}\cdot\frac{2^2}{9^2}\cdot\frac{1}{6}\cdot\frac{11^3}{18^3}$$

$$= 0.1127.$$

3.4 Hypergeometric Distribution

In Section 3.3 we saw that the binomial distribution did not apply if we wished to find the probability of observing 3 red cards in five draws from an ordinary deck of 52 playing cards unless each card is replaced and the deck reshuffled before the next drawing is made. To solve the problem of sampling without replacement let us restate the problem. If 5 cards are drawn at random we are interested in the probability of selecting 3 red cards from the 26 available and 2 black cards from the 26 black cards available in the deck. There are $\binom{26}{3}$ ways of selecting 3 red cards and for each of these ways we can choose 2 black cards in $\binom{26}{2}$ ways. Therefore, the total number of ways to select 3 red and 2 black cards in five draws is the product $\binom{26}{3}\binom{26}{2}$. The total number of ways to select any 5 cards from the 52 that are available is $\binom{52}{5}$. Hence the probability of selecting 5 cards without replacement of which 3 are red and 2 are black is given by

$$\frac{\binom{26}{3}\binom{26}{2}}{\binom{52}{5}} = \frac{(26!/3!23!)(26!/2!24!)}{(52!/5!47!)} = 0.3251.$$

In general we are interested in the probability of selecting x successes from the k items labeled success and $n - x$ failures from the $N - k$ items labeled failures when a random sample of size n is selected from N items. This is known as a *hypergeometric experiment*.

A hypergeometric experiment is one that possesses the following two properties:

1. A random sample of size n is selected from N items.
2. k of the N items may be classified as successes and $N - k$ are classified as failures.

DEFINITION 3.2 *The number X of successes in a hypergeometric experiment is called a* hypergeometric random variable.

The probability distribution of the hypergeometric variable X is called the *hypergeometric distribution* and will be denoted by $h(x;N,n,k)$ since its values depend on the number of successes k in the set N from which we select n items.

Example 3.8 A committee of size 5 is to be selected at random from three chemists and five physicists. Find the probability distribution for the number of chemists on the committee.

Solution. Let the random variable X be the number of chemists on the committee. The two properties of a hypergeometric experiment are satisfied. Hence

$$Pr(X = 0) = h(0;8,5,3) = \frac{\binom{3}{0}\binom{5}{5}}{\binom{8}{5}} = \frac{1}{56}$$

$$Pr(X = 1) = h(1;8,5,3) = \frac{\binom{3}{1}\binom{5}{4}}{\binom{8}{5}} = \frac{15}{56}$$

$$Pr(X = 2) = h(2;8,5,3) = \frac{\binom{3}{2}\binom{5}{3}}{\binom{8}{5}} = \frac{30}{56}$$

$$Pr(X = 3) = h(3;8,5,3) = \frac{\binom{3}{3}\binom{5}{2}}{\binom{8}{5}} = \frac{10}{56}.$$

In tabular form the hypergeometric distribution of X is as follows:

x	0	1	2	3
$h(x;8,5,3)$	$\frac{1}{56}$	$\frac{15}{56}$	$\frac{30}{56}$	$\frac{10}{56}$

It is not difficult to see that the probability distribution can be given by the formula

$$h(x;8,5,3) = \frac{\binom{3}{x}\binom{5}{5-x}}{\binom{8}{5}}, \qquad x = 0, 1, 2, 3.$$

Let us now generalize Example 3.8 to find a formula for $h(x;N,n,k)$. The total number of samples of size n chosen from N items is $\binom{N}{n}$. These samples are assumed to be equally likely. There are $\binom{k}{x}$ ways of selecting x successes from the k that are available and for each of these ways we can choose the $n - x$ failures in $\binom{N-k}{n-x}$ ways. Thus the total number of favorable samples among the $\binom{N}{n}$ possible samples is given by $\binom{k}{x}\binom{N-k}{n-x}$. Hence we have the following definition.

HYPERGEOMETRIC DISTRIBUTION *The probability distribution of the hypergeometric random variable X, the number of successes in a random sample of size n selected from N items of which k are labeled success and $N - k$ labeled failure, is*

$$h(x;N,n,k) = \frac{\binom{k}{x}\binom{N-k}{n-x}}{\binom{N}{n}}, \qquad x = 0, 1, 2, \ldots, n.$$

Example 3.9 Lots of 40 components each are called acceptable if they contain no more than 3 defectives. The procedure for sampling the lot is to select 5 components at random and to reject the lot if a defective is found. What is the probability that exactly 1 defective will be found in the sample if there are 3 defectives in the entire lot?

Solution. Using the hypergeometric distribution with $n = 5$, $N = 40$, $k = 3$, and $x = 1$, we find the probability of obtaining 1 defective to be

$$h(1;40,5,3) = \frac{\binom{3}{1}\binom{37}{4}}{\binom{40}{5}} = 0.3011.$$

THEOREM 3.3 *The mean and variance of the hypergeometric distribution $h(x;N,n,k)$ are*

$$\mu = \frac{nk}{N}$$

$$\sigma^2 = \frac{N-n}{N-1} \cdot n \cdot \frac{k}{N}\left(1 - \frac{k}{N}\right).$$

Proof. To find the mean of the hypergeometric distribution we write

$$E(X) = \sum_{x=0}^{n} x \frac{\binom{k}{x}\binom{N-k}{n-x}}{\binom{N}{n}}$$

$$= k \sum_{x=1}^{n} \frac{(k-1)!}{(x-1)!(k-x)!} \cdot \frac{\binom{N-k}{n-x}}{\binom{N}{n}}$$

$$= k \sum_{x=1}^{n} \frac{\binom{k-1}{x-1}\binom{N-k}{n-x}}{\binom{N}{n}}.$$

Letting $y = x - 1$, this becomes

$$E(X) = k \sum_{y=0}^{n-1} \frac{\binom{k-1}{y}\binom{N-k}{n-1-y}}{\binom{N}{n}}.$$

Writing

$$\binom{N-k}{n-1-y} = \binom{(N-1)-(k-1)}{n-1-y}$$

and

$$\binom{N}{n} = \frac{N!}{n!(N-n)!} = \frac{N}{n}\binom{N-1}{n-1},$$

then

$$E(X) = \frac{nk}{N} \sum_{y=0}^{n-1} \frac{\binom{k-1}{y}\binom{(N-1)-(k-1)}{n-1-y}}{\binom{N-1}{n-1}}$$

$$= \frac{nk}{N},$$

since the summation represents the total of all probabilities in a hyper-geometric experiment when $n-1$ items are selected at random from $N-1$ of which $k-1$ are labeled success.

To obtain the variance of the hypergeometric distribution we proceed along the same steps as above to obtain

$$E[X(X-1)] = \frac{k(k-1)n(n-1)}{N(N-1)}.$$

Now, by Theorem 2.6,

$$\begin{aligned}
\sigma^2 &= E(X^2) - \mu^2 \\
&= E[X(X-1)] + \mu - \mu^2 \\
&= \frac{k(k-1)n(n-1)}{N(N-1)} + \frac{nk}{N} - \frac{n^2k^2}{N^2} \\
&= \frac{nk(N-k)(N-n)}{N^2(N-1)} \\
&= \frac{N-n}{N-1} \cdot n \cdot \frac{k}{N}\left(1 - \frac{k}{N}\right).
\end{aligned}$$

Example 3.10 Using Chebyshev's theorem find and interpret the interval $\mu \pm 2\sigma$ for Example 3.9.

Solution. Since Example 3.9 was a hypergeometric experiment with $N = 40$, $n = 5$, and $k = 3$, then by Theorem 3.3 we have

$$\mu = \frac{(5)(3)}{40} = \frac{3}{8} = 0.375$$

and

$$\sigma^2 = \left(\frac{40 - 5}{39}\right) (5) \left(\frac{3}{40}\right) \left(1 - \frac{3}{40}\right)$$
$$= 0.3113.$$

Taking the square root of 0.3113 we find $\sigma = 0.558$. Hence the required interval is $0.375 \pm 2(0.558)$, or from -0.741 to 1.491. Chebyshev's theorem states that the number of defectives obtained when 5 components are selected at random from a lot of 40 components of which 3 are defective has a probability of at least 3/4 of falling between -0.741 and 1.491. That is, at least three fourths of the time, the 5 components include less than 2 defectives.

If n is small relative to N, the probability for each drawing will change only slightly. Hence we essentially have a binomial experiment and can approximate the hypergeometric distribution by using the binomial distribution with $p = k/N$. The mean and variance can also be approximated by the formulas

$$\mu = np = \frac{nk}{N}$$

$$\sigma^2 = npq = n \cdot \frac{k}{N} \left(1 - \frac{k}{N}\right).$$

Comparing these formulas with those of Theorem 3.3 we see that the mean is the same while the variance differs by a correction factor of $(N - n)/(N - 1)$. This is negligible when n is small relative to N.

Example 3.11 A manufacturer of automobile tires reports that among a shipment of 5000 sent to a local distributor 1000 are slightly blemished. If one purchases 10 of these tires at random from the distributor, what is the probability that exactly 3 will be blemished?

Solution. Since $N = 5000$ is large relative to the sample size $n = 10$, we shall approximate the desired probability by using the binomial distribution. The probability of obtaining a blemished tire is 0.2. Therefore, the probability of obtaining exactly 3 blemished tires is

$$h(3;5000,10,1000) \simeq b(3;10,0.2)$$

$$= \sum_{x=0}^{3} b(x;10,0.2) - \sum_{x=0}^{2} b(x;10,0.2)$$

$$= 0.8791 - 0.6778$$

$$= 0.2013.$$

The hypergeometric distribution can be extended to treat the case where the N items can be partitioned into k cells A_1, A_2, ..., A_k with a_1 elements in the first cell, a_2 elements in the second cell, ..., a_k elements in the kth cell. We are now interested in the probability that a random sample of size n yields x_1 elements from A_1, x_2 elements from A_2, ..., and x_k elements from A_k. Let us represent this probability by

$$f(x_1,x_2,\ldots,x_k;a_1,a_2,\ldots,a_k,N,n).$$

To obtain a general formula we note that the total number of samples that can be chosen of size n from N items is still $\binom{N}{n}$. There are $\binom{a_1}{x_1}$ ways of selecting x_1 items from the items in A_1 and for each of these we can choose x_2 items from the items in A_2 in $\binom{a_2}{x_2}$ ways. Therefore we can select x_1 items from A_1 and x_2 items from A_2 in $\binom{a_1}{x_1}\binom{a_2}{x_2}$ ways. Continuing in this way we can select all n items consisting of x_1 from A_1, x_2 from A_2, ..., and x_k from A_k in $\binom{a_1}{x_1}\binom{a_2}{x_2}\cdots\binom{a_k}{x_k}$ ways. The required probability distribution is now defined as follows.

EXTENSION OF THE HYPERGEOMETRIC DISTRIBUTION *If N items can be partitioned into the k cells A_1, A_2, ..., A_k with a_1, a_2, ..., a_k elements, respectively, then the probability distribution of the random variables X_1, X_2, ..., X_k, representing the number of elements selected from A_1, A_2, ..., A_k in a random sample of size n is*

$$f(x_1,x_2,\ldots,x_k;a_1,a_2,\ldots,a_k,N,n) = \frac{\binom{a_1}{x_1}\binom{a_2}{x_2}\cdots\binom{a_k}{x_k}}{\binom{N}{n}}$$

with

$$\sum_{i=1}^{k} x_i = n \quad and \quad \sum_{i=1}^{k} a_i = N.$$

Example 3.12 A group of 10 individuals are being used in a biological case study. The group contains 3 people with blood type O, 4 with blood type A, and 3 with blood type B. What is the probability that a random sample of 5 will contain 1 person with blood type O, 2 people with blood type A, and 2 people with blood type B?

Solution. Using the extension of the hypergeometric distribution with $x_1 = 1$, $x_2 = 2$, $x_3 = 2$, $a_1 = 3$, $a_2 = 4$, $a_3 = 3$, $N = 10$, and $n = 5$, the desired probability is

$$f(1,2,2;3,4,3,10,5) = \frac{\binom{3}{1}\binom{4}{2}\binom{3}{2}}{\binom{10}{5}} = \frac{3}{14}.$$

3.5 Poisson Distribution

Experiments yielding numerical values of a random variable X, the number of successes occurring during a given time interval or in a specified region, are often called *Poisson experiments*. The given time interval may be of any length, such as a minute, a day, a week, a month, or even a year. Hence a Poisson experiment might generate observations for the random variable X representing the number of telephone calls per hour received by an office, the number of days school is closed due to snow during the winter, or the number of postponed games due to rain during a baseball season. The specified region could be a line segment, an area, a volume, or perhaps a piece of material. In this case X might represent the number of field mice per acre, the number of bacteria in a given culture, or the number of typing errors per page.

A Poisson experiment is one that possesses the following properties:

1. The number of successes occurring in one time interval or specified region are independent of those occurring in any other disjoint time interval or region of space.
2. The probability of a single success occurring during a very short time interval or in a small region is proportional to the length of the time interval or the size of the region and does not depend on the number of successes occurring outside this time interval or region.
3. The probability of more than one success occurring in such a short time interval or falling in such a small region is negligible.

DEFINITION 3.3 *The number X of successes in a Poisson experiment is called a* Poisson *random variable.*

The probability distribution of the Poisson variable X is called the *Poisson distribution* and will be denoted by $p(x;\mu)$ since its values depend only on μ, the average number of successes occurring in the given time interval or specified region. The derivation of the formula for $p(x;\mu)$ based on the properties that have been listed for a Poisson experiment is beyond the scope of this text. We list the result in the following definition.

POISSON DISTRIBUTION *The probability distribution of the Poisson random variable X, representing the number of successes occurring in a given time interval or specified region, is*

$$p(x;\mu) = \frac{e^{-\mu}\mu^x}{x!}, \qquad x = 0, 1, 2, \ldots,$$

where μ is the average number of successes occurring in the given time interval or specified region and $e = 2.71828\ldots$

Table III (see Statistical Tables) contains Poisson probability sums $P(r;\mu) = \sum_{x=0}^{r} p(x;\mu)$ for a few selected values of μ ranging from 0.1 to 18. We illustrate the use of this table with the following two examples.

Example 3.13 The average number of radioactive particles passing through a counter during a millisecond in a laboratory experiment is 4. What is the probability that 6 particles enter the counter in a given millisecond?

Solution. Using the Poisson distribution with $x = 6$ and $\mu = 4$, we find from Table III that

$$p(6;4) = \frac{e^{-4}4^6}{6!} = \sum_{x=0}^{6} p(x;4) - \sum_{x=0}^{5} p(x;4) = 0.8893 - 0.7851$$

$$= 0.1042.$$

Example 3.14 The average number of oil tankers arriving each day at a certain port city is known to be 10. The facilities at the port can handle at most 15 tankers per day. What is the probability that on a given day tankers will have to be sent away?

Solution. Let X be the number of tankers arriving each day. Then using Table III we have

$$Pr(X > 15) = 1 - Pr(X \leq 15)$$

$$= 1 - \sum_{x=0}^{15} p(x;10)$$

$$= 1 - 0.9513$$

$$= 0.0487.$$

THEOREM 3.4 *The mean and variance of the Poisson distribution $p(x;\mu)$ both have the value μ.*

Proof. To verify that the mean is indeed μ, we write

$$E(X) = \sum_{x=0}^{\infty} x \cdot \frac{e^{-\mu}\mu^x}{x!} = \sum_{x=1}^{\infty} x \cdot \frac{e^{-\mu}\mu^x}{x!}$$

$$= \mu \sum_{x=1}^{\infty} \frac{e^{-\mu}\mu^{x-1}}{(x-1)!}.$$

Now, let $y = x - 1$ to give

$$E(X) = \mu \sum_{y=0}^{\infty} \frac{e^{-\mu}\mu^y}{y!} = \mu,$$

since

$$\sum_{y=0}^{\infty} \frac{e^{-\mu}\mu^y}{y!} = \sum_{y=0}^{\infty} p(y;\mu) = 1.$$

The variance of the Poisson distribution is obtained by first finding

$$E[X(X-1)] = \sum_{x=0}^{\infty} x(x-1) \cdot \frac{e^{-\mu}\mu^x}{x!} = \sum_{x=2}^{\infty} x(x-1) \frac{e^{-\mu}\mu^x}{x!}$$

$$= \mu^2 \sum_{x=2}^{\infty} \frac{e^{-\mu}\mu^{x-2}}{(x-2)!}.$$

Setting $y = x - 2$ we have

$$E[X(X-1)] = \mu^2 \sum_{y=0}^{\infty} \frac{e^{-\mu}\mu^y}{y!} = \mu^2.$$

Hence

$$\sigma^2 = E[X(X-1)] + \mu - \mu^2$$
$$= \mu^2 + \mu - \mu^2$$
$$= \mu.$$

In Example 3.13 where $\mu = 4$, we also have $\sigma^2 = 4$ and hence $\sigma = 2$. Using Chebyshev's theorem we can state that our random variable has a probability of at least 3/4 of falling in the interval $\mu \pm 2\sigma = 4 \pm 2(2)$, or from zero to 8. Therefore we conclude that at least three fourths of the time the number of radioactive particles entering the counter will be anywhere from zero to 8 during a given millisecond.

We shall now derive the Poisson distribution as a limiting form of the binomial distribution when $n \to \infty$, $p \to 0$, and np remains constant. Hence if n is large and p is close to zero, the Poisson distribution can be used, with $\mu = np$, to approximate binomial probabilities. If p is close to 1, we can still use the Poisson distribution to approximate binomial probabilities by interchanging what we have defined to be a success and a failure, thereby changing p to a value close to zero.

THEOREM 3.5 *Let X be a binomial random variable with probability distribution $b(x;n,p)$. When $n \to \infty$, $p \to 0$, and $\mu = np$ remains constant,*

$$b(x;n,p) \to p(x;\mu).$$

Proof. The binomial distribution can be written as

$$b(x;n,p) = \binom{n}{x} p^x q^{n-x}$$

$$= \frac{n!}{x!(n-x)!} p^x (1-p)^{n-x}$$

$$= \frac{n(n-1)\cdots(n-x+1)}{x!} p^x (1-p)^{n-x}.$$

Substituting $p = \mu/n$, we have

$$b(x;n,p) = \frac{n(n-1)\cdots(n-x+1)}{x!} \left(\frac{\mu}{n}\right)^x \left(1 - \frac{\mu}{n}\right)^{n-x}$$

$$= 1\left(1 - \frac{1}{n}\right)\cdots\left(1 - \frac{x-1}{n}\right)\frac{\mu^x}{x!}\left(1 - \frac{\mu}{n}\right)^n\left(1 - \frac{\mu}{n}\right)^{-x}.$$

As $n \to \infty$ while x and μ remain constant,

$$\lim_{n \to \infty} 1 \left(1 - \frac{1}{n}\right) \cdots \left(1 - \frac{x-1}{n}\right) = 1$$

$$\lim_{n \to \infty} \left(1 - \frac{\mu}{n}\right)^{-x} = 1,$$

and from the definition of the number e

$$\lim_{n \to \infty} \left(1 - \frac{\mu}{n}\right)^{n} = \lim_{n \to \infty} \left\{ \left[1 + \frac{1}{(-n)/\mu}\right]^{-n/\mu} \right\}^{-\mu} = e^{-\mu}.$$

Hence, under the given limiting conditions

$$b(x;n,p) \to \frac{e^{-\mu}\mu^{x}}{x!}, \qquad x = 0, 1, 2, \ldots.$$

Example 3.15 In a certain manufacturing process in which glass items are being produced, defects or bubbles occur, occasionally rendering the piece undesirable for marketing. It is known that on the average 1 in every 1000 of these items produced has one or more bubbles. What is the probability that a random sample of 8000 will yield fewer than 7 items possessing bubbles?

Solution. This is essentially a binomial experiment with $n = 8000$ and $p = 0.001$. Since p is very close to zero and n is quite large, we shall approximate with the Poisson distribution using $\mu = (8000)(0.001) = 8$. Hence if X represents the number of bubbles, we have

$$Pr(X < 7) = \sum_{x=0}^{6} b(x;8000,0.001)$$

$$\simeq \sum_{x=0}^{6} p(x;8)$$

$$= 0.3134.$$

3.6 Negative Binomial and Geometric Distributions

Let us consider an experiment in which the properties are the same as those listed for a binomial experiment, with the exception that the trials will be repeated until a *fixed* number of successes occur. Therefore, instead of finding the probability of x successes in n trials, where n is fixed, we are now interested in the probability that the kth success occurs on the xth trial. Experiments of this kind are called *negative binomial experiments*.

As an illustration, consider the use of a drug that is known to be effective in 60% of the cases in which is is used. The use of the drug will be considered a success if it is effective in bringing some degree of relief to the patient. We are interested in finding the probability that the fifth patient to experience relief is the seventh patient to receive the drug during a given week. Designating a success by S and a failure by F, a possible order of achieving the desired result is SFSSSFS, which occurs with probability $(0.6)(0.4)(0.6)(0.6)(0.6)(0.4)(0.6) = (0.6)^5(0.4)^2$. We could list all possible orders by rearranging the F's and S's except for the last outcome, which must be the fifth success. The total number of possible orders is equal to the number of partitions of the first six trials into two groups with two failures assigned to the one group and the four successes assigned to the other group. This can be done in $\binom{6}{4} = 15$ mutually exclusive ways. Hence, if X represents the outcome on which the fifth success occurs, then

$$Pr(X = 7) = \binom{6}{4} (0.6)^5(0.4)^2 = 0.1866.$$

DEFINITION 3.4 *The number X of trials to produce k successes in a negative binomial experiment is called a* negative binomial variable.

The probability distribution of the negative binomial variable X is called the *negative binomial distribution* and will be denoted by $b^*(x;k,p)$ since its values depend on the number of successes desired and the probability of a success on a given trial. To obtain the general formula for $b^*(x;k,p)$, consider the probability of a success on the xth trial preceded by $k - 1$ successes and $x - k$ failures in some specified order. Since the trials are independent, we can multiply all the probabilities corresponding to each desired outcome. Each success occurs with probability p and each failure with probability $q = 1 - p$. Therefore, the probability for the specified order, ending in a success, is $p^{k-1}q^{x-k}p = p^kq^{x-k}$. The total number of sample points in the experiment ending in a success, after the occurrence of $k - 1$ successes and $x - k$ failures in any order, is equal to the number of partitions of $x - 1$ trials into two groups with $k - 1$ successes corresponding to one group and $x - k$ failures corresponding to the other group. This number is given by the term $\binom{x - 1}{k - 1}$, each mutually exclusive and occurring with equal probability p^kq^{x-k}. We obtain the general formula by multiplying p^kq^{x-k} by $\binom{x - 1}{k - 1}$.

NEGATIVE BINOMIAL DISTRIBUTION *If repeated independent trials can result in a success with probability p and a failure with probability $q = 1 - p$, then the probability distribution of the random variable X, the number of the trial on which the kth success occurs, is given by*

$$b^*(x;k,p) = \binom{x-1}{k-1} p^k q^{x-k}, \qquad x = k,\ k+1,\ k+2,\ \dots.$$

Example 3.16 Find the probability that a person tossing three coins will get either all heads or all tails for the second time on the fifth toss.

Solution. Using the negative binomial distribution with $x = 5$, $k = 2$, and $p = 1/4$, we have

$$b^*(5;2,\tfrac{1}{4}) = \binom{4}{1}\left(\frac{1}{4}\right)^2 \left(\frac{3}{4}\right)^3$$

$$= \frac{4!}{1!\,3!} \cdot \frac{3^3}{4^5}$$

$$= \frac{27}{256}.$$

The negative binomial distribution derives its name from the fact that each term in the expansion of $p^k(1 - q)^{-k}$ corresponds to the values of $b^*(x;k,p)$ for $x = k,\ k+1,\ k+2,\ \dots.$

If we consider the special case of the negative binomial distribution where $k = 1$, we have a probability distribution for the number of trials required for a single success. An example would be the tossing of a coin until a head occurs. We might be interested in the probability that the first head occurs on the fourth toss. The negative binomial distribution reduces to the form $b^*(x;1,p) = pq^{x-1}$, $x = 1,\ 2,\ 3,\ \dots.$ Since the successive terms constitute a geometric progression, it is customary to refer to this special case as the *geometric distribution* and denote it by $g(x;p)$.

GEOMETRIC DISTRIBUTION *If repeated independent trials can result in a success with probability p and a failure with probability $q = 1 - p$, then the probability distribution of the random variable X, the number of the trial on which the first success occurs, is given by*

$$g(x;p) = pq^{x-1}, \qquad x = 1,\ 2,\ 3,\ \dots.$$

Example 3.17 In a certain manufacturing process it is known that on the average 1 in every 100 items is defective. What is the probability that 5 items are inspected before a defective item is found?

Solution. Using the geometric distribution with $x = 5$ and $p = 0.01$, we have

$$g(5;0.01) = (0.01)(0.99)^4$$
$$= 0.0096.$$

EXERCISES

1. Find a formula for the distribution of the random variable X representing the number on a tag drawn at random from a box containing 10 tags numbered 1 to 10. What is the probability that the number drawn is less than 4?

2. A roulette wheel is divided into 25 sectors of equal area numbered from 1 to 25. Find a formula for the probability distribution of X, the number that occurs when the wheel is spun.

3. Find the mean and variance of the random variable X of Exercise 1.

4. A baseball player's batting average is 0.250. What is the probability that he gets exactly one hit in his next four times at bat?

5. If we define the random variable X to be equal to the number of heads that occur when a balanced coin is flipped once, find the probability distribution of X. What two well-known distributions describe the values of X?

6. In testing a certain kind of truck tire over a rugged terrain, it is found that 25% of the trucks fail to complete the test run without a blowout. What is the probability that from 5 to 10 of the next 15 trucks tested have flat tires?

7. The probability that a patient recovers from a delicate heart operation is 0.9. What is the probability that exactly five of the next seven patients having this operation survive?

8. A traffic control engineer reports that 75% of the vehicles passing through a check point are from within the state. What is the probability that at least three of the next five vehicles are from out of the state?

9. A survey of the residents in a United States city showed that 20% preferred a white telephone over any other color available. What is the probability that more than half of the next 20 telephones installed in this city will be white?

10. It is known that 75% of mice inoculated with a serum are protected from a certain disease. If three mice are inoculated, what is the probability that at most two of the mice contract the disease?

11. If X represents the number of out-of-state vehicles in Exercise 8 when five vehicles are checked, find the probability distribution of X. Using Chebyshev's theorem, find and interpret $\mu \pm 2\sigma$.

12. Suppose that airplane engines operate independently in flight and fail with probability $q = 1/5$. Assuming a plane makes a safe flight if at least half of its engines run, determine whether a four-engine plane or a two-engine plane has the highest probability for a successful flight.

13. Repeat Exercise 12 for $q = 1/2$ and $q = 1/3$.

14. In Exercise 6, how many of the 15 trucks tested would you expect to have flat tires? Using Chebyshev's theorem find and interpret the interval $\mu \pm 2\sigma$.

15. A card is drawn from a well-shuffled deck of 52 playing cards, the result recorded, and the card replaced. If the experiment is repeated five times, what is the probability of obtaining two spades and one heart?

16. The surface of a circular dart board has a small center circle called the bull's-eye and 20 pie-shaped regions numbered from 1 to 20. Each of the pie-shaped regions is further divided into three parts such that a person throwing a dart that lands on a specified number scores the value of the number, double the number, or triple the number, depending on which of the three parts the dart falls. If a person hits the bull's-eye with probability 0.01, hits a double with probability 0.10, hits a triple with probability 0.05, and misses the dart board with probability 0.02, what is the probability that seven throws will result in no bull's-eyes, no triples, a double twice, and a complete miss once?

17. According to the theory of genetics a certain cross of guinea pigs will result in red, black, and white offspring in the ratio of $8:4:4$. Find the probability that among eight offspring five will be red, two black, and one white.

18. To avoid detection at customs, a traveler has placed six narcotic tablets in a bottle containing nine vitamin pills that are similar in appearance. If the customs official selects three of the tablets at random for analysis, what is the probability that the traveler will be arrested for illegal possession of narcotics?

19. A company is interested in evaluating its current inspection procedure on shipments of 50 identical items. The procedure is to take a sample of 5 and pass the shipment if no more than 2 are found to be defective. What proportion of 20 % defective shipments will be accepted?

20. A manufacturing company uses an acceptance scheme on production items before they are shipped. The plan is a two-stage one. Boxes of 25 are readied for shipment and a sample of 3 are tested for defectives. If any defectives are found, the entire box is sent back for 100 % screening. If no defectives are found, the box is shipped.
   (a) What is the probability that a box containing 3 defectives will be shipped?
   (b) What is the probability that a box containing only one defective will be sent back for screening?

21. Suppose the manufacturing company in Exercise 20 decided to change its acceptance scheme. Under the new scheme an inspector takes 1 at random, inspects it, and then replaces it in the box; a second inspector does likewise. Finally a third inspector goes through the same procedure. The box is not shipped if any of the three find a defective. Answer questions (a) and (b) in Exercise 20 under this new plan.

22. A homeowner plants six bulbs selected at random from a box containing five tulip bulbs and four daffodil bulbs. What is the probability that he planted two daffodil bulbs and four tulip bulbs?

23. Find the probability of obtaining 2 ones, 1 two, 1 three, 2 fours, 3 fives, and 1 six in 10 rolls of a balanced die?

24. A random committee of 3 is selected from four doctors and two nurses. Write a formula for the probability distribution of the random variable X representing the numbers of doctor on the committee. Find the $Pr(2 \leq X \leq 3)$.

25. From a lot of 10 missiles 4 are selected at random and fired. If the lot contains 3 defective missiles that will not fire, what is the probability that
    (a) All 4 will fire?
    (b) At most two will not fire?

26. In Exercise 25 how many defective missiles might we expect to be included among the 4 that are selected? Use Chebyshev's theorem to describe the variability of the number of defective missiles included when 4 are selected from several lots each of size 10 containing 3 defective missiles.

27. If a person is dealt 13 cards from an ordinary deck of 52 playing cards several times, how many hearts per hand can he expect? Between what two values would you expect the number of hearts to fall at least 75% of the time?

28. It is estimated that 4,000 of the 10,000 voting residents of a town are against a new sales tax. If 15 eligible voters are selected at random and asked their opinion, what is the probability that at most 7 favor the new tax?

29. An annexation suit is being considered against a county subdivision of 1200 residences by a neighboring city. If the occupants of half the residences object to being annexed, what is the probability that in a random sample of 10 at least 3 favor the annexation suit?

30. Find the probability of being dealt a bridge hand of 13 cards containing 5 spades, 2 hearts, 3 diamonds, and 3 clubs.

31. A foreign student club lists as its members two Canadians, three Japanese, five Italians, and two Germans. Find the probability that all nationalities are represented if a committee of 4 is selected at random.

32. An urn contains three green balls, two blue balls, and four red balls. In a random sample of five balls find the probability that both blue balls and at least one red ball are selected.

33. On the average a certain intersection results in three traffic accidents per week. What is the probability that exactly five accidents will occur at this intersection in any given week?

34. On the average a secretary makes two errors per page. What is the probability that she makes
    (a) Four or more errors on the next page she types?
    (b) No errors?

35. A certain area of the eastern United States is, on the average, hit by six hurricanes per year. Find the probability that in a given year
    (a) Fewer than four hurricanes will hit this area.
    (b) Anywhere from six to eight hurricanes will hit the area.

36. In an inventory study it was determined that on the average demands for a particular item at a warehouse were made five times per day. What is the probability that on a given day this item is requested
    (a) More than five times?
    (b) Not at all?

37. The probability that a person dies from a certain respiratory infection is 0.002. Find the probability that fewer than 5 of the next 2000 so infected will die.

38. Suppose that on the average 1 person in 1,000 makes a numerical error in preparing their income tax returns. If 10,000 forms are selected at random and examined, find the probability that 6, 7, or 8 of the forms will be in error.

39. Using Chebyshev's theorem find and interpret the interval $\mu \pm 2\sigma$ for Exercise 37.

40. The probability that a person will install a black telephone in a residence is estimated to be 0.3. Find the probability that the tenth phone installed in a new subdivision is the fifth black phone.

41. A scientist inoculates several mice, one at a time, with a disease germ until he finds two that have contracted the disease. If the probability of contracting the disease is 1/6, what is the probability that eight mice are required?

42. Three people toss a coin and the odd man pays for the coffee. If the coins all turn up the same they are tossed again. Find the probability that fewer than four tosses are needed.

43. Find the probability that a person flipping a coin gets the third head on the seventh flip.

44. The probability that a student pilot passes the written test for his private pilot's license is 0.7. Find the probability that a person passes the test
    (a) On the third try.
    (b) Before the fourth try.

4 SOME CONTINUOUS PROBABILITY DISTRIBUTIONS

4.1 Normal Distribution

The most important continuous probability distribution in the entire field of statistics is the *normal distribution*. Its graph, called the *normal curve*, is the bell-shaped curve of Figure 4.1 that describes the distribution of so many sets of data that occur in nature, industry, and research. In 1733 DeMoivre developed the mathematical equation of the normal curve. This provided a basis for much of the theory of inductive statistics. The normal distribution is often referred to as the *Gaussian distribution* in honor of Gauss (1777–1855), who also derived its equation from a study of errors in repeated measurements of the same quantity.

A random variable X having the bell-shaped distribution of Figure 4.1

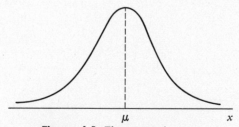

Figure 4.1 The normal curve.

is called a *normal random variable*. The mathematical equation for the probability distribution of the continuous normal variable depends on the two parameters μ and σ, its mean and standard deviation. Hence we shall denote the density function of X by $n(x;\mu,\sigma)$.

NORMAL DISTRIBUTION *The density function of the normal random variable X, with mean μ and variance σ^2, is*

$$n(x;\mu,\sigma) = \frac{1}{\sqrt{2\pi}\sigma} e^{-(1/2)[(x-\mu)/\sigma]^2}, \qquad -\infty < x < \infty,$$

where $\pi = 3.14159\cdots$ and $e = 2.71828\cdots$.

Once μ and σ are specified, the normal curve is completely determined. For example, if $\mu = 50$ and $\sigma = 5$, then the ordinates $n(x;50,5)$ can easily be computed for various values of x and the curve drawn. In Figure 4.2 we have sketched two normal curves having the same standard deviation but different means. The two curves are identical in form but are centered at different positions along the horizontal axis.

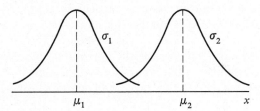

Figure 4.2 Normal curves with $\mu_1 < \mu_2$ and $\sigma_1 = \sigma_2$.

In Figure 4.3 we have sketched two normal curves with the same mean but different standard deviations. This time we see that the two curves are centered at exactly the same position on the horizontal axis, but the

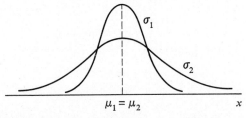

Figure 4.3 Normal curves with $\mu_1 = \mu_2$ and $\sigma_1 < \sigma_2$.

curve with the larger standard deviation is lower and spreads out further. Remember that the area under a probability curve must be equal to 1 and therefore the more variable the set of observations the lower and wider the corresponding curve will be.

Figure 4.4 shows the results of sketching two normal curves having different means and different standard deviations. Clearly they are centered at different positions on the horizontal axis and their shapes reflect the two different values of σ.

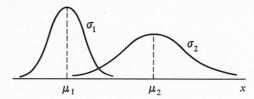

Figure 4.4 Normal curves with $\mu_1 < \mu_2$ and $\sigma_1 < \sigma_2$.

From an inspection of the graphs and by examination of the first and second derivatives of $n(x;\mu,\sigma)$, we list the following properties of the normal curve:

1. The *mode*, which is the point on the horizontal axis where the curve is a maximum, occurs at $x = \mu$.
2. The curve is symmetric about a vertical axis through the mean μ.
3. The curve has its points of inflection at $x = \mu \pm \sigma$, is concave downward if $\mu - \sigma < X < \mu + \sigma$, and is concave upward otherwise.
4. The normal curve approaches the horizontal axis asymptotically as we proceed in either direction away from the mean.
5. The total area under the curve and above the horizontal axis is equal to 1.

We shall now show that the parameters μ and σ^2 are indeed the mean and the variance of the normal distribution. To evaluate the mean, we write

$$E(X) = \frac{1}{\sqrt{2\pi}\sigma} \int_{-\infty}^{\infty} x e^{-(1/2)\,[(x-\mu)/\sigma]^2}\, dx.$$

Setting $z = (x - \mu)/\sigma$ and $dx = \sigma\, dz$, we obtain

$$E(X) = \frac{1}{\sqrt{2\pi}} \int_{-\infty}^{\infty} (\mu + \sigma z) e^{-z^2/2}\, dz$$

$$= \mu \frac{1}{\sqrt{2\pi}} \int_{-\infty}^{\infty} e^{-z^2/2}\, dz + \frac{\sigma}{\sqrt{2\pi}} \int_{-\infty}^{\infty} z e^{-z^2/2}\, dz.$$

The first integral is μ times the area under a normal curve with mean zero and variance 1, and hence equal to μ. By straightforward integration or from the fact that the integrand is an odd function, the second integral is equal to zero. Hence,

$$E(X) = \mu.$$

The variance of the normal distribution is given by

$$E[(X - \mu)^2] = \frac{1}{\sqrt{2\pi}\sigma} \int_{-\infty}^{\infty} (x - \mu)^2 e^{-(1/2)\,[(x-\mu)/\sigma]^2} \, dx.$$

Again setting $z = (x - \mu)/\sigma$ and $dx = \sigma\,dz$, we obtain

$$E[(X - \mu)^2] = \frac{\sigma^2}{\sqrt{2\pi}} \int_{-\infty}^{\infty} z^2 e^{-z^2/2} \, dz.$$

Integrating by parts with $u = z$ and $dv = ze^{-z^2/2}$ so that $du = dz$ and $v = -e^{-z^2/2}$, we find

$$E[(X - \mu)^2] = \frac{\sigma^2}{\sqrt{2\pi}} \left(-ze^{-z^2/2} \Big|_{-\infty}^{\infty} + \int_{-\infty}^{\infty} e^{-z^2/2} \, dz \right)$$

$$= \sigma^2(0 + 1)$$

$$= \sigma^2.$$

4.2 Areas Under the Normal Curve

The curve of any continuous probability distribution or density function is constructed so that the area under the curve bounded by the two ordinates $x = x_1$ and $x = x_2$ equals the probability that the random variable X assumes a value between $x = x_1$ and $x = x_2$. Thus for the normal curve in Figure 4.5,

$$Pr(x_1 < X < x_2) = \int_{x_1}^{x_2} n(x;\mu,\sigma) \, dx$$

$$= \frac{1}{\sqrt{2\pi}\sigma} \int_{x_1}^{x_2} e^{-(1/2)\,[(x-\mu)/\sigma]^2} \, dx$$

is represented by the area of the shaded region.

In Figures 4.2, 4.3, and 4.4 we saw how the normal curve is dependent on the mean and the standard deviation of the distribution under investigation. The area under the curve between any two ordinates must then

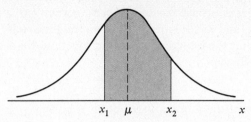

Figure 4.5 $Pr(x_1 < X < x_2)$ = area of the shaded region.

also depend on the values μ and σ. This is evident in Figure 4.6, where we have shaded regions corresponding to $Pr(x_1 < X < x_2)$ for two curves with different mean and variances. The $Pr(x_1 < X < x_2)$, where X is the random variable describing distribution I, is indicated by the shaded region where the lines slope up to the right. If X is the random variable describing distribution II, then the $Pr(x_1 < X < x_2)$ is given by the shaded region where the lines slope down to the right. Obviously the two shaded regions are different in size; therefore, the probability associated with each distribution will be different.

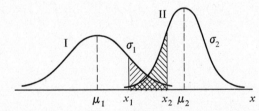

Figure 4.6 $Pr(x_1 < X < x_2)$ for different normal curves.

The difficulty encountered in solving integrals of normal density functions necessitates the tabulation of normal curve areas for quick reference. However, it would be a hopeless task to attempt to set up separate tables for every conceivable value of μ and σ. Fortunately we are able to transform all the observations of any normal random variable X to a new set of observations of a normal random variable Z with mean zero and variance 1. This can be done by means of the transformation

$$Z = \frac{X - \mu}{\sigma}.$$

Whenever X assumes a value x, the corresponding value of Z is given by $z = (x - \mu)/\sigma$. Therefore, if X falls between the values $x = x_1$ and $x = x_2$, the random variable Z will fall between the corresponding values

$z_1 = (x_1 - \mu)/\sigma$ and $z_2 = (x_2 - \mu)/\sigma$. Consequently, we may write

$$Pr(x_1 < X < x_2) = \frac{1}{\sqrt{2\pi}\sigma} \int_{x_1}^{x_2} e^{-(1/2)\,[(x-\mu)/\sigma]^2}\,dx$$

$$= \frac{1}{\sqrt{2\pi}} \int_{z_1}^{z_2} e^{-z^2/2}\,dz$$

$$= \int_{z_1}^{z_2} n(z;0,1)\,dz$$

$$= Pr(z_1 < Z < z_2),$$

where Z is seen to be a normal random variable with mean zero and variance 1.

> **DEFINITION** 4.1 *The distribution of a normal random variable with mean zero and variance 1 is called a* standard normal distribution.

The original and transformed distributions are illustrated in Figure 4.7. Since all the values of X falling between x_1 and x_2 have corresponding z values between z_1 and z_2, the area under the X curve between the ordinates $x = x_1$ and $x = x_2$ in Figure 4.7 equals the area under the Z curve between the transformed ordinates $z = z_1$ and $z = z_2$.

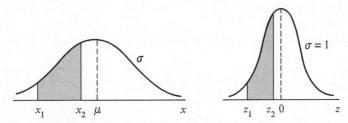

Figure 4.7 The original and transformed normal distributions.

We have now reduced the required number of tables of normal curve areas to one, namely, that of the standard normal distribution. Table IV (see Statistical Tables) gives the area under the standard normal curve corresponding to $Pr(Z < z)$ for values of z from -3.4 to 3.4. To illustrate the use of this table, let us find the probability that Z is less than 1.74. First, we locate a value of z equal to 1.7 in the left column, then move across the row to the column under 0.04, where we read 0.9591. Therefore, $Pr(Z < 1.74) = 0.9591$.

Example 4.1 Given a normal distribution with $\mu = 50$ and $\sigma = 10$, find the probability that X assumes a value between 45 and 62.

Solution. The z values corresponding to $x_1 = 45$ and $x_2 = 62$ are

$$z_1 = \frac{45 - 50}{10} = -0.5$$

$$z_2 = \frac{62 - 50}{10} = 1.2.$$

Therefore

$$Pr(45 < X < 62) = Pr(-0.5 < Z < 1.2).$$

The $Pr(-0.5 < Z < 1.2)$ is given by the area of the shaded region in Figure 4.8. This area may be found by subtracting the area to the left

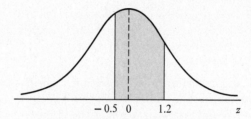

Figure 4.8 Area for Example 4.1.

of the ordinate $z = -0.5$ from the entire area to the left of $z = 1.2$. Using Table IV we have

$$
\begin{aligned}
Pr(45 < X < 62) &= Pr(-0.5 < Z < 1.2) \\
&= Pr(Z < 1.2) - Pr(Z < -0.5) \\
&= 0.8849 - 0.3085 \\
&= 0.5764.
\end{aligned}
$$

According to Chebyshev's theorem, the probability that a random variable assumes a value within two standard deviations of the mean is at least 3/4. If the random variable has a normal distribution, the z values corresponding to $x_1 = \mu - 2\sigma$ and $x_2 = \mu + 2\sigma$ are easily computed to be

$$z_1 = \frac{(\mu - 2\sigma) - \mu}{\sigma} = -2$$

and

$$z_2 = \frac{(\mu + 2\sigma) - \mu}{\sigma} = 2.$$

Hence
$$Pr(\mu - 2\sigma < X < \mu + 2\sigma) = Pr(-2 < Z < 2)$$
$$= Pr(Z < 2) - Pr(Z < -2)$$
$$= 0.9772 - 0.0228$$
$$= 0.9544,$$

which is a much stronger statement than that given by Chebyshev's theorem.

Some of the many problems where the normal distribution is applicable are treated in the following examples. The use of the normal curve to approximate binomial probabilities will be considered in Section 4.3.

Example 4.2 A certain type of storage battery lasts on the average 3.0 years, with a standard deviation of 0.5 years. Assuming the battery lives are normally distributed, find the probability that a given battery will last less than 2.3 years.

Solution. First construct a diagram such as Figure 4.9, showing the given distribution of battery lives and the desired area. To find the

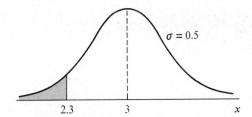

Figure 4.9 Area for Example 4.2.

$Pr(X < 2.3)$ we need to evaluate the area under the normal curve to the left of 2.3. This is accomplished by finding the area to the left of the corresponding z value. Hence we find

$$z = \frac{2.3 - 3}{0.5} = -1.4,$$

and then using Table IV we have

$$Pr(X < 2.3) = Pr(Z < -1.4)$$
$$= 0.0808.$$

Example 4.3 An electrical firm manufactures light bulbs that have a length of life that is normally distributed with mean equal to 800 hours and a standard deviation of 40 hours. Find the probability that a bulb burns between 778 and 834 hours.

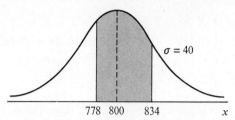

Figure 4.10 Area for Example 4.3.

Solution. The distribution of light bulbs is illustrated in Figure 4.10. the z values corresponding to $x_1 = 778$ and $x_2 = 834$ are

$$z_1 = \frac{778 - 800}{40} = -0.55$$

$$z_2 = \frac{834 - 800}{40} = 0.85.$$

Hence

$$Pr(778 < X < 834) = Pr(-0.55 < Z < 0.85)$$
$$= Pr(Z < 0.85) - Pr(Z < -0.55)$$
$$= 0.8023 - 0.2912$$
$$= 0.5111.$$

Example 4.4 Gauges are used to reject all components in which a certain dimension is not within the specification $1.50 \pm d$. It is known that this measurement is normally distributed with mean 1.50 and standard deviation 0.2. Determine the value d such that the specifications "cover" 95% of the measurements.

Solution. Examples 4.2 and 4.3 were solved by going first from a value of x to a z value and then computing the desired area. In this problem we reverse the process and begin with a known area or probability, find the z value, and then determine x from the formula $x = \sigma z + \mu$. We know from Table IV that

$$Pr(-1.96 < Z < 1.96) = 0.95.$$

Thus

$$1.50 + d = (0.2)(1.96) + 1.50,$$

from which we obtain

$$d = (0.2)(1.96) = 0.392.$$

An illustration of the specifications is given in Figure 4.11.

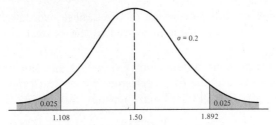

Figure 4.11 Specifications for Example 4.4.

Example 4.5 A certain machine makes electrical resistors having a mean resistance of 40 ohms and a standard deviation of 2 ohms. Assuming that the resistance follows a normal distribution and can be measured to any degree of accuracy, what percentage of resistors will have a resistance that exceeds 43 ohms?

Solution. A percentage is found by multiplying the relative frequency by 100%. Since the relative frequency for an interval is equal to the probability of falling in the interval, we must find the area to the right of $x = 43$ in Figure 4.12. This can be done by transforming $x = 43$ to the

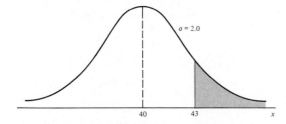

Figure 4.12 Area for Example 4.5.

corresponding z value, obtaining the area to the left of z from Table IV, and then subtracting this area from 1. We find

$$z = \frac{43 - 40}{2} = 1.5.$$

Hence

$$Pr(X > 43) = Pr(Z > 1.5)$$
$$= 1 - Pr(Z < 1.5)$$
$$= 1 - 0.9332$$
$$= 0.0668.$$

therefore 6.68% of the resistors will have a resistance exceeding 43 ohms.

Example 4.6 Find the percentage of resistances exceeding 43 ohms in Example 4.5 if the resistance is measured to the nearest ohm.

Solution. This problem differs from Example 4.5 in that we now assign a measurement of 43 ohms to all resistors whose resistances are greater than 42.5 and less than 43.5. We are actually approximating a discrete

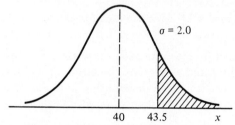

Figure 4.13 Area for Example 4.6.

distribution by means of a continuous normal distribution. The required area is the region shaded to the right of 43.5 in Figure 4.13. We now find

$$z = \frac{43.5 - 40}{2} = 1.75.$$

Hence

$$Pr(X > 43.5) = Pr(Z > 1.75)$$
$$= 1 - Pr(Z < 1.75)$$
$$= 1 - 0.9599$$
$$= 0.0401.$$

Therefore 4.01% of the resistances exceed 43 ohms when measured to the nearest ohm. The difference 6.68% − 4.01% = 2.67% between this answer and that of Example 4.5 represents all of those resistors having a resistance greater than 43 and less than 43.5 that are now being recorded as 43 ohms.

Example 4.7 The quality grade-point averages of 300 college freshmen follow approximately a normal distribution with a mean of 2.1 and a standard deviation of 1.2. How many of these freshmen would you expect to have a score between 2.5 and 3.5 inclusive, if the grade-point averages are computed to the nearest tenth?

Solution. Since the scores are recorded to the nearest tenth we require the area between $x_1 = 2.45$ and $x_2 = 3.55$, as indicated in Figure 4.14.

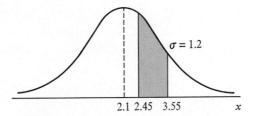

Figure 4.14 Area for Example 4.7.

The corresponding z values are

$$z_1 = \frac{2.45 - 2.1}{1.2} = 0.29$$

$$z_2 = \frac{3.55 - 2.1}{1.2} = 1.21.$$

Therefore

$$Pr(2.45 < X < 3.55) = Pr(0.29 < Z < 1.21)$$
$$= Pr(Z < 1.21) - Pr(Z < 0.29)$$
$$= 0.8869 - 0.6141$$
$$= 0.2728.$$

Hence 27.28%, or approximately 82 of the 300 freshmen, should have a score between 2.5 and 3.5 inclusive.

4.3 Normal Approximation to the Binomial

Probabilities associated with binomial experiments are readily obtainable from the formula $b(x;n,p)$ of the binomial distribution or from Table II (see Statistical Tables) when n is small. If n is not listed in any available table, we must compute the binomial probabilities by approximation procedures. In Section 3.5 we illustrated how the Poisson distribution can be used to approximate binomial probabilities when n is large and p is close to zero or 1. Both the binomial and Poisson distributions are discrete. The first application of a continuous probability distribution to approximate probabilities over a discrete sample space was demonstrated in Section 4.2, Examples 4.6 and 4.7, where the normal curve was used. We shall now state a theorem that allows us to use areas under the normal curve to approximate binomial probabilities when n is sufficiently large.

THEOREM 4.1 *If X is a binomial random variable with mean $\mu = np$ and variance $\sigma^2 = npq$, then the limiting form of the distribution of*

$$Z = \frac{X - np}{\sqrt{npq}},$$

as $n \to \infty$, is the standardized normal distribution $n(z;0,1)$.

It turns out that the proper normal distribution provides a very accurate approximation to the binomial distribution when n is large and p is close to $1/2$. In fact, even when n is small and p is not extremely close to zero or 1 the approximation is fairly good.

To investigate the normal approximation to the binomial distribution we first draw the histogram for $b(x;15,0.4)$ and then superimpose the particular normal curve having the same mean and variance as the binomial variable X. Hence we draw a normal curve with

$$\mu = np = (15)(0.4) = 6$$

and

$$\sigma^2 = npq = (15)(0.4)(0.6) = 3.6.$$

The histogram of $b(x;15,0.4)$ and the corresponding superimposed normal curve, which is completely determined by its mean and variance, are illustrated in Figure 4.15.

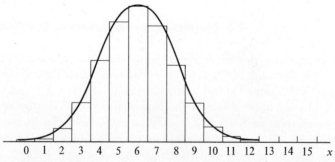

Figure 4.15 Normal curve approximation of $b(x;15,0.4)$.

The exact probability of the binomial random variable X assuming a given value x is equal to the area of the rectangle whose base is centered at x. For example, the exact probability that X assumes the value 4 is equal to the area of the rectangle with base centered at $x = 4$. Using the formula for the binomial distribution we find this area to be

$$b(4;15,0.4) = 0.1268.$$

This same probability is approximately equal to the area of the shaded region under the normal curve between the two ordinates $x_1 = 3.5$ and

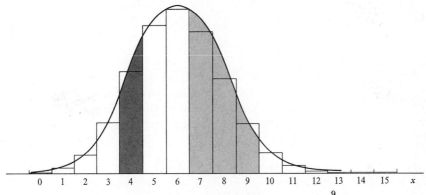

Figure 4.16 Normal approximation for $b(4;15,0.4)$ and $\sum_{x=7}^{9} b(x;15,0.4)$.

$x_2 = 4.5$ in Figure 4.16. Converting to z values we have

$$z_1 = \frac{3.5 - 6}{1.9} = -1.316$$

$$z_2 = \frac{4.5 - 6}{1.9} = -0.789.$$

If X is a binomial random variable and Z a standard normal variable, then

$$
\begin{aligned}
Pr(X = 4) &= b(4;15,0.4) \\
&\simeq Pr(-1.316 < Z < -0.789) \\
&= Pr(Z < -0.789) - Pr(Z < -1.316) \\
&= 0.2151 - 0.0941 \\
&= 0.1210.
\end{aligned}
$$

This agrees very closely with the exact value of 0.1268.

The normal approximation is most useful in calculating binomial sums for large values of n, which, without tables of binomial sums, is an impossible task. Referring to Figure 4.16, we might be interested in the probability that X assumes a value from 7 to 9 inclusive. The exact probability is given by

$$Pr(7 \leq X \leq 9) = \sum_{x=7}^{9} b(x;15,0.4)$$

$$= \sum_{x=0}^{9} b(x;15,0.4) - \sum_{x=0}^{6} b(x;15,0.4)$$

$$= 0.9662 - 0.6098$$

$$= 0.3564,$$

which is equal to the sum of the areas of the rectangles with bases centered at $x = 7$, 8, and 9. For the normal approximation we find the area of the shaded region under the curve between the ordinates $x_1 = 6.5$ and $x_2 = 9.5$ in Figure 4.16. The corresponding z values are

$$z_1 = \frac{6.5 - 6}{1.9} = 0.263$$

$$z_2 = \frac{9.5 - 6}{1.9} = 1.842.$$

Now

$$Pr(7 < X < 9) \simeq Pr(0.263 < Z < 1.842)$$

$$= Pr(Z < 1.842) - Pr(Z < 0.263)$$

$$= 0.9673 - 0.6037$$

$$= 0.3636.$$

Once again the normal curve approximation provides a value that agrees very closely to the exact value of 0.3564. The degree of accuracy, which depends on how well the curve fits the histogram, will increase as n

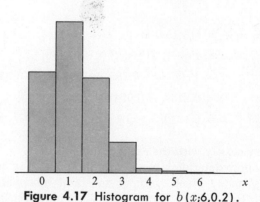

Figure 4.17 Histogram for $b(x;6,0.2)$.

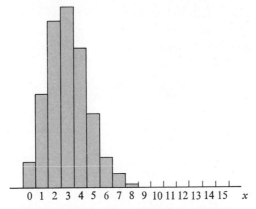

Figure 4.18 Histogram for $b(x;15,0.2)$.

increases. This is particularly true when p is not very close to $1/2$ and the histogram is no longer symmetric. Figures 4.17 and 4.18 show the histograms for $b(x;6,0.2)$ and $b(x;15,0.2)$, respectively. It is evident that a normal curve would fit the histogram when $n = 15$ considerably better than when $n = 6$.

In summary, we use the normal approximation to evaluate binomial probabilities whenever p is not close to zero or 1. The approximation is excellent when n is large and fairly good for small values of n if p is reasonably close to $1/2$. One possible guide to determine when the normal approximation may be used is provided by calculating np and nq. If both np and nq are greater than 5, the approximation will be good.

Example 4.8 A process yields 10% defective items. If 100 items are randomly selected from the process, what is the probability that the number of defectives exceeds 13?

Solution. The number of defectives has the binomial distribution with parameters $n = 100$ and $p = 0.1$. Since the sample size is large, we should obtain fairly accurate results using the normal curve approximation with

$$\mu = np = (100)(0.1) = 10$$
$$\sigma = \sqrt{npq} = \sqrt{(100)(0.1)(0.9)} = 3.0.$$

To obtain the desired probability we have to find the area to the right of $x = 13.5$. The z value corresponding to 13.5 is

$$z = \frac{13.5 - 10}{3} = 1.167$$

and the probability that the number of defectives exceeds 13 is given by the area of the shaded region in Figure 4.19. Hence, if X represents the number of defectives, then

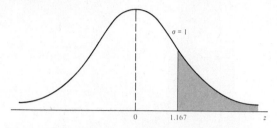

Figure 4.19 Area for Example 4.8.

$$Pr(X > 13) = \sum_{x=14}^{100} b(x;100,0.1)$$
$$\simeq Pr(Z > 1.167)$$
$$= 1 - Pr(Z < 1.167)$$
$$= 1 - 0.8784$$
$$= 0.1216.$$

Example 4.9 A multiple-choice quiz has 200 questions each with four possible answers of which only one is the correct answer. What is the probability that sheer guesswork yields from 25 to 30 correct answers for 80 of the 200 problems about which the student has no knowledge?

Solution. The probability of a correct answer for each of the 80 questions is $p = 1/4$. If X represents the number of correct answers due to guesswork, then

$$Pr(25 \leq X \leq 30) = \sum_{x=25}^{30} b(x;80,1/4).$$

Using the normal curve approximation with

$$\mu = np = (80)(1/4) = 20$$
$$\sigma = \sqrt{npq} = \sqrt{(80)(1/4)(3/4)} = 3.87,$$

we need the area between $x_1 = 24.5$ and $x_2 = 30.5$. The corresponding z values are

$$z_1 = \frac{24.5 - 20}{3.87} = 1.163$$

$$z_2 = \frac{30.5 - 20}{3.87} = 2.713.$$

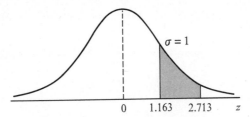

Figure 4.20 Area for Example 4.9.

The probability of correctly guessing from 25 to 30 questions is given by the area of the shaded region in Figure 4.20. From Table IV we find

$$Pr(25 \leq X \leq 30) = \sum_{x=25}^{30} b(x;80,1/4)$$
$$\simeq Pr(1.163 < Z < 2.713)$$
$$= Pr(Z < 2.713) - Pr(Z < 1.163)$$
$$= 0.9967 - 0.8776$$
$$= 0.1191.$$

4.4 Gamma, Exponential, and Chi-square Distributions

Although the normal distribution can be used to solve many problems in engineering and science, there are still numerous situations that require a different type of density function. Three such density functions, the *gamma, exponential,* and *chi-square distributions,* will be discussed in this section. Additional densities will be presented in Section 4.5 and in Chapter 5, Sections 5.7 and 5.8.

The gamma distribution derives its name from the well-known *gamma function* studied in many areas of mathematics. Before we proceed to the gamma distribution let us review this function and some of its important properties.

DEFINITION 4.2 *The* gamma function *is defined by*

$$\Gamma(\alpha) = \int_0^\infty x^{\alpha-1}e^{-x}\,dx$$

for $\alpha > 0$.

Integrating by parts with $u = x^{\alpha-1}$ and $dv = e^{-x}\,dx$, we obtain

$$\Gamma(\alpha) = -e^{-x}x^{\alpha-1}\Big|_0^\infty + \int_0^\infty e^{-x}(\alpha - 1)x^{\alpha-2}\,dx$$
$$= (\alpha - 1)\int_0^\infty e^{-x}x^{\alpha-2}\,dx,$$

which yields the recursion formula

$$\Gamma(\alpha) = (\alpha - 1)\Gamma(\alpha - 1).$$

Repeated application of the recursion formula gives

$$\Gamma(\alpha) = (\alpha - 1)(\alpha - 2)\Gamma(\alpha - 2)$$
$$= (\alpha - 1)(\alpha - 2)(\alpha - 3)\Gamma(\alpha - 3),$$

and so forth. Note that when $\alpha = n$, where n is a positive integer,

$$\Gamma(n) = (n - 1)(n - 2) \cdots \Gamma(1).$$

However, by Definition 4.2,

$$\Gamma(1) = \int_0^\infty e^{-x}\,dx = 1$$

and hence

$$\Gamma(n) = (n - 1)!.$$

One important property of the gamma function, left for the student to verify (See Exercise 25), is that $\Gamma(1/2) = \sqrt{\pi}$.

We shall now include the gamma function in our definition of the gamma distribution.

GAMMA DISTRIBUTION *The continuous random variable X has a gamma distribution, with parameters α and β, if its density function is given by*

$$f(x) = \frac{1}{\beta^\alpha \Gamma(\alpha)}\,x^{\alpha-1}e^{-x/\beta}, \qquad x > 0$$
$$= 0, \qquad\qquad\qquad elsewhere,$$

where $\alpha > 0$ and $\beta > 0$.

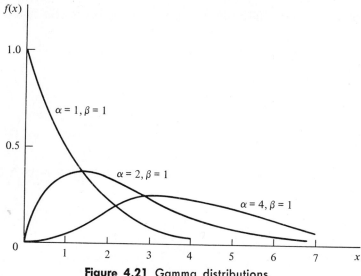

Figure 4.21 Gamma distributions.

Graphs of several gamma distributions are shown in Figure 4.21 for certain specified values of the parameters α and β. The special gamma distribution for which $\alpha = 1$ is called the *exponential distribution*.

EXPONENTIAL DISTRIBUTION *The continuous random variable X has an* exponential distribution, *with parameter β, if its density function is given by*

$$f(x) = \frac{1}{\beta} e^{-x/\beta}, \qquad x > 0$$

$$= 0, \qquad\qquad elsewhere,$$

where $\beta > 0$.

The exponential distribution has many applications in the field of statistics, particularly in the areas of *reliability theory* and *waiting times* or *queueing problems*. One interesting application of the exponential distribution to reliability theory is given in the following example.

Example 4.10 Suppose a system contains a certain type of component whose time in years to failure is given by the random variable T, distributed exponentially with parameter $\beta = 5$. If five of these components are installed in different systems, what is the probability that at least two are still functioning at the end of 8 years?

Solution. The probability that a given component is still functioning after 8 years is given by

$$Pr(T > 8) = \frac{1}{5} \int_8^\infty e^{-t/5}\, dt$$

$$= e^{-8/5}$$

$$\simeq 0.2.$$

Let X represent the number of components functioning after 8 years. Then, using the binomial distribution,

$$Pr(X \geq 2) = \sum_{x=2}^{5} b(x;5,0.2)$$

$$= 1 - \sum_{x=0}^{1} b(x;5,0.2)$$

$$= 1 - 0.7373$$

$$= 0.2627.$$

A second special case of the gamma distribution is obtained by letting $\alpha = \nu/2$ and $\beta = 2$, where ν is a positive integer. The probability density so obtained is called the *chi-square distribution* with ν *degrees of freedom.*

CHI-SQUARE DISTRIBUTION *The continuous random variable X has a* chi-square distribution, *with ν degrees of freedom, if its density function is given by*

$$f(x) = \frac{1}{2^{\nu/2}\Gamma(\nu/2)}\, x^{\nu/2-1} e^{-x/2}, \qquad x > 0$$

$$= 0, \qquad\qquad\qquad\qquad elsewhere,$$

where ν is a positive integer.

The chi-square distribution is one of our main tools in the area of hypothesis testing, a subject to be studied in later chapters. We shall now derive the mean and variance of these three density functions.

THEOREM 4.2 *The mean and variance of the gamma distribution are*

$$\mu = \alpha\beta \quad and \quad \sigma^2 = \alpha\beta^2.$$

Proof. The rth moment about the origin for the gamma distribution is

$$\mu'_r = E(X^r) = \frac{1}{\beta^\alpha \Gamma(\alpha)} \int_0^\infty x^{r+\alpha-1} e^{-x/\beta} \, dx.$$

Now, let $y = x/\beta$ to give

$$\mu'_r = \frac{\beta^r}{\Gamma(\alpha)} \int_0^\infty y^{r+\alpha-1} e^{-y} \, dy$$

$$= \frac{\beta^r \Gamma(\alpha + r)}{\Gamma(\alpha)}.$$

Then

$$\mu = \mu'_1 = \frac{\beta \Gamma(\alpha + 1)}{\Gamma(\alpha)} = \alpha\beta$$

and

$$\sigma^2 = \mu'_2 - \mu^2 = \frac{\beta^2 \Gamma(\alpha + 2)}{\Gamma(\alpha)} - \alpha^2\beta^2$$

$$= \alpha\beta^2.$$

COROLLARY 1 *The mean and variance of the exponential distribution are*

$$\mu = \beta \quad and \quad \sigma^2 = \beta^2.$$

COROLLARY 2 *The mean and variance of the chi-square distribution are*

$$\mu = \nu \quad and \quad \sigma^2 = 2\nu.$$

4.5 Weibull Distribution

Modern technology has enabled us to design many complicated systems whose operation, or perhaps safety, depend on the reliability of the various components making up the systems. For example, a fuse may burn out, a steel column may buckle, or a heat-sensing device may fail. Identical components subjected to identical environmental conditions will fail at different and unpredictable times. The *time to failure* or *life length* of a component, measured from some specified time until it fails, is represented by the continuous random variable T with probability density function $f(t)$. One distribution that has been used extensively in recent years to deal with such problems as reliability and life testing is the *Weibull distribution*.

WEIBULL DISTRIBUTION *The continuous random variable T has a* Weibull distribution, *with parameters α and β, if its density function is given by*

$$f(t) = \alpha\beta t^{\beta-1}e^{-\alpha t^\beta}, \qquad t > 0$$
$$= 0, \qquad\qquad elsewhere,$$

where $\alpha > 0$ and $\beta > 0$.

The graph of the Weibull distribution for various values of the parameters α and β are illustrated in Figure 4.22. We see that the curves change in shape considerably for different values of the parameters, particularly the parameter β. If we let $\beta = 1$, the Weibull distribution reduces to the exponential distribution. For values of $\beta > 1$, the curves become somewhat bell-shaped and resemble the normal curves, but display some skewness.

The mean and variance of the Weibull distribution are stated in the following theorem. The reader is asked to provide the proof in Exercise 28.

THEOREM 4.3 *The mean and variance of the Weibull distribution are*

$$\mu = \alpha^{-1/\beta}\Gamma\left(1 + \frac{1}{\beta}\right),$$
$$\sigma^2 = \alpha^{-2/\beta}\left\{\Gamma\left(1 + \frac{2}{\beta}\right) - \left[\Gamma\left(1 + \frac{1}{\beta}\right)\right]^2\right\}.$$

To apply the Weibull distribution to reliability theory, let us first define the reliability of a component or product as the *probability that it will function properly for at least a specified time under specified experi-*

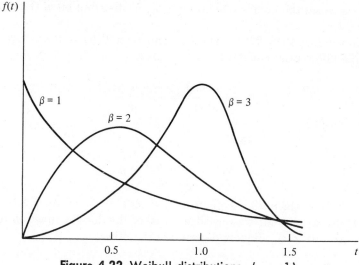

Figure 4.22 Weibull distributions. $(\alpha = 1)$

mental conditions. Therefore if $R(t)$ is defined to be the reliability of the given component at time t, we may write

$$
\begin{aligned}
R(t) &= Pr(T > t) \\
&= \int_t^\infty f(t) \, dt \\
&= 1 - F(t),
\end{aligned}
$$

where $F(t)$ is the cumulative distribution of T. The conditional probability that a component will fail in the interval from $T = t$ to $T = t + \Delta t$, given that it survived to time t, is given by

$$
\frac{F(t + \Delta t) - F(t)}{R(t)}.
$$

Dividing this ratio by Δt and taking the limit as $\Delta t \to 0$, we get the *failure rate*, denoted by $Z(t)$. Hence

$$
\begin{aligned}
Z(t) &= \lim_{\Delta t \to 0} \frac{F(t + \Delta t) - F(t)}{\Delta t} \frac{1}{R(t)} \\
&= \frac{F'(t)}{R(t)} = \frac{f(t)}{R(t)} = \frac{f(t)}{1 - F(t)},
\end{aligned}
$$

which expresses the failure rate in terms of the distribution of the time to failure.

From the fact that $R(t) = 1 - F(t)$ and then $R'(t) = -F'(t)$, we can write the differential equation

$$Z(t) = \frac{-R'(t)}{R(t)} = \frac{-d[\ln R(t)]}{dt}$$

and then solving,

$$\ln R(t) = -\int Z(t)\, dt$$

or

$$R(t) = e^{-\int Z(t)\, dt} + c,$$

where c satisfies the initial assumption that $R(0) = 1$ or $F(0) = 1 - R(0) = 0$. Thus, we see that a knowledge of either the density function $f(t)$ or the failure rate $Z(t)$ uniquely determines the other.

Example 4.11 Show that the failure-rate function is given by

$$Z(t) = \alpha\beta t^{\beta-1}, \qquad t > 0,$$

if and only if the time to failure distribution is the Weibull distribution

$$f(t) = \alpha\beta t^{\beta-1} e^{-\alpha t^\beta}, \qquad t > 0.$$

Solution. Assume $Z(t) = \alpha\beta t^{\beta-1}$, $t > 0$. Then we can write

$$f(t) = Z(t)R(t),$$

where

$$R(t) = e^{-\int Z(t)\, dt} = e^{-\int \alpha\beta t^{\beta-1}\, dt} = e^{-\alpha t^\beta} + c.$$

From the condition that $R(0) = 1$, we find $c = 0$. Hence

$$R(t) = e^{-\alpha t^\beta}$$

and

$$f(t) = \alpha\beta t^{\beta-1} e^{-\alpha t^\beta}, \qquad t > 0.$$

Now, if we assume

$$f(t) = \alpha\beta t^{\beta-1} e^{-\alpha t^\beta}, \qquad t > 0,$$

then $Z(t)$ is determined by writing

$$Z(t) = \frac{f(t)}{R(t)},$$

where

$$R(t) = 1 - F(t) = 1 - \int_0^t \alpha\beta x^{\beta-1} e^{-\alpha x^\beta} \, dx$$

$$= 1 + \int_0^t de^{-\alpha x^\beta}$$

$$= e^{-\alpha t^\beta}.$$

Then

$$Z(t) = \frac{\alpha\beta t^{\beta-1} e^{-\alpha t^\beta}}{e^{-\alpha t^\beta}} = \alpha\beta t^{\beta-1}, \qquad t > 0.$$

In Example 4.11, the failure rate is seen to decrease with time if $\beta < 1$, increases with time if $\beta > 1$, and is constant if $\beta = 1$. In view of the fact that the Weibull distribution with $\beta = 1$ reduces to the exponential distribution, the assumption of constant failure rate is often referred to as the *exponential assumption*.

EXERCISES

1. Given a normal distribution with $\mu = 40$ and $\sigma = 6$, find
   (a) The area below 32.
   (b) The area above 27.
   (c) The area between 42 and 51.
   (d) The point that has 45% of the area below it.
   (e) The point that has 13% of the area above it.
2. Given a normal distribution with $\mu = 200$ and $\sigma^2 = 100$, find
   (a) The area below 214.
   (b) The area above 179.
   (c) The area between 188 and 206.
   (d) The point that has 80% of the area below it.
   (e) The two points containing the middle 75% of the area.
3. Given the normally distributed variable X with mean 18 and standard deviation 2.5, find
   (a) $Pr(X < 15)$.
   (b) The value of k such that $Pr(X < k) = 0.2578$.
   (c) $Pr(17 < X < 21)$.
   (d) The value of k such that $Pr(X > k) = 0.1539$.
4. A soft drink machine is regulated so that it discharges an average of 7 ounces per cup. If the amount of drink is normally distributed with standard deviation equal to 0.5 ounce,
   (a) What fraction of the cups will contain more than 7.8 ounces?
   (b) What is the probability that a cup contains between 6.7 and 7.3 ounces?
   (c) How many cups will likely overflow if 8-ounce cups are used for the next 1000 drinks?
   (d) Below what value do we get the smallest 25% of the drinks?

5. The finished inside diameter of a piston ring is normally distributed with a mean of 4.00 inches and a standard deviation of 0.01 inch.
   (a) What proportion of rings will have inside diameters exceeding 4.025 inches?
   (b) What is the probability that a piston ring will have an inside diameter between 3.99 and 4.01 inches?
   (c) Below what value of inside diameter will 15% of the piston rings fall?

6. If a set of grades on a statistics examination are approximately normally distributed with a mean of 74 and a standard deviation of 7.9, find
   (a) The lowest passing grade if the lowest 10% of the students are given F's.
   (b) The highest B if the top 5% of the students are given A's.

7. The heights of 1000 students are normally distributed with a mean of 68.5 inches and a standard deviation of 2.7 inches. Assuming the heights are recorded to the nearest half inch, how many of these students would you expect to have heights
   (a) Less than 63.0 inches?
   (b) Between 67.5 and 71.0 inches inclusive?
   (c) Equal to 69.0 inches?
   (d) Greater than or equal to 74.0 inches?

8. In a mathematics examination the average grade was 82 and the standard deviation was 5. All students with grades from 88 to 94 received a grade of B. If the grades are approximately normally distributed and eight students received a B grade, how many students took the examination?

9. A company pays its employees an average wage of $3.25 an hour with a standard deviation of 60 cents. If the wages are approximately normally distributed,
   (a) What percent of the workers receive wages between $2.75 and $3.69 an hour inclusive?
   (b) The highest 5% of the hourly wages are greater than what amount?

10. The tensile strength of a certain metal component is normally distributed with a mean of 10,000 pounds per square inch and a standard deviation of 100 pounds per square inch. Measurements are recorded to the nearest 50 pounds per square inch.
    (a) What proportion of these components exceed 10,150 pounds per square inch in tensile strength?
    (b) If specifications require that all components have tensile strength between 9,800 and 10,200 pounds per square inch inclusive, what proportion of pieces would we expect to scrap?

11. If a set of observations are normally distributed, what percent of these differ from the mean by
    (a) More than 1.3σ?
    (b) Less than 0.52σ?

12. The I.Q.'s of 600 applicants to a certain college are approximately normally distributed with a mean of 115 and a standard deviation of 12. If the college requires an I.Q. of at least 95, how many of these students will be rejected on this basis regardless of their other qualifications?

13. The average rainfall, recorded to the nearest hundredth of an inch, in Roanoke, Virginia, for the month of March is 3.63 inches. Assuming a normal distribution with a standard deviation of 1.03 inches, find the probability that next March Roanoke receives
    (a) Less than 0.72 inches of rain.
    (b) More than 2 inches but not over 3 inches.
    (c) More than 5.3 inches.

14. The average life of a certain type of small motor is 10 years with a standard deviation of 2 years. The manufacturer replaces free all motors that fail while under guarantee. If he is willing to replace only 3% of the motors that fail, how long a guarantee should he offer? Assume the lives of the motors follow a normal distribution.

15. Find the error in approximating $\sum_{x=1}^{4} b(x;20,0.1)$ by the normal curve approximation.

16. A coin is tossed 400 times. Use the normal curve approximation to find the probability of obtaining
    (a) Between 185 and 210 heads inclusive.
    (b) Exactly 205 heads.
    (c) Less than 176 or more than 227 heads.

17. A pair of dice is rolled 180 times. What is the probability that a total of 7 occurs
    (a) At least 25 times.
    (b) Between 33 and 41 times inclusive.
    (c) Exactly 30 times?

18. The probability that a patient recovers from a delicate heart operation is 0.9. What is the probability that between 84 and 95 inclusive of the next 100 patients having this operation survive?

19. A drug manufacturer claims that a certain drug cures a blood disease on the average 80% of the time. To check the claim, government testers used the drug on a sample of 100 individuals and decide to accept the claim if 75 or more are cured.
    (a) What is the probability that the claim will be rejected when the cure probability is in fact 0.8?
    (b) What is the probability that the claim will be accepted by the government when the cure probability is as low as 0.7?

20. A survey of the residents in a United States city showed that 20% preferred a white telephone over any other color available. What is the probability that between 170 and 185 inclusive of the next 1000 telephones installed in this city will be white?

21. One sixth of the male freshmen entering a large state school are out-of-state students. If the students are assigned at random to the dormitories, 180 to a building, what is the probability that in a given dormitory at least one fifth of the students are from out of state?

22. A certain pharmaceutical company knows that, on the average, 5% of a certain type of pill has an ingredient that is below the minimum strength and

thus unacceptable. What is the probability that fewer than 10 in a sample of 200 pills will be unacceptable?

23. If a random variable X has the gamma distribution with $\alpha = 2$ and $\beta = 1$, find $Pr(1.8 < X < 2.4)$.

24. In a certain city, the daily consumption of water (in millions of gallons) follows approximately a gamma distribution with $\alpha = 2$ and $\beta = 3$. If the daily capacity of this city is 9 million gallons of water, what is the probability that on any given day the water supply is inadequate?

25. Use the gamma function, with $x = y^2/2$, to show that $\Gamma(1/2) = \sqrt{\pi}$.

26. The length of time for one individual to be served at a cafeteria is a random variable having an exponential distribution with a mean of 4 minutes. What is the probability that a person is served in less than 3 minutes on at least 4 of the next 6 days?

27. The life in years of a certain type of electrical switch has an exponential distribution with a failure rate of $\beta = 2$. If 100 of these switches are installed in different systems, what is the probability that at most 30 fail during the first year?

28. Derive the mean and variance of the Weibull distribution.

29. The lives of a certain automobile seal have the Weibull distribution with failure rate $Z(t) = 1/\sqrt{t}$. Find the probability that such a seal is still in use after 4 years.

5 FUNCTIONS OF RANDOM VARIABLES

5.1 Transformations of Variables

Frequently in statistics one encounters the need to derive the probability distribution of a function of one or more random variables. For example, suppose X is a discrete random variable with probability distribution $f(x)$ and suppose further that $Y = u(X)$ defines a one-to-one transformation between the values of X and Y. We wish to find the probability distribution of Y. It is important to note that the one-to-one transformation implies that each value x is related to one, and only one, value $y = u(x)$ and that each value y is related to one, and only one, value $x = w(y)$, where $w(y)$ is obtained by solving $y = u(x)$ for x in terms of y.

From our discussion of discrete probability distributions in Chapter 2 it is clear that the random variable Y assumes the value y when X assumes the value $w(y)$. Consequently, the probability distribution of Y is given by

$$g(y) = Pr(Y = y) = Pr[X = w(y)] = f[w(y)].$$

THEOREM 5.1 *Suppose that X is a* discrete *random variable with probability distribution $f(x)$. Let $Y = u(X)$ define a one-to-one transformation between the values of X and Y so that the equation $y = u(x)$ can be uniquely solved for x in terms of y, say $x = w(y)$. Then the probability distribution of Y is*

$$g(y) = f[w(y)].$$

Example 5.1 Let X be a geometric random variable with probability distribution $f(x) = \frac{3}{4}(\frac{1}{4})^{x-1}$, $x = 1, 2, 3, \ldots$. Find the probability distribution of the random variable $Y = X^2$.

Solution. Since the values of X are all positive, the transformation defines a one-to-one correspondence between the x and y values, namely, $y = x^2$ and $x = \sqrt{y}$. Hence

$$g(y) = f(\sqrt{y}) = \tfrac{3}{4}(\tfrac{1}{4})^{\sqrt{y}-1}, \qquad y = 1, 4, 9, \ldots$$
$$= 0, \qquad\qquad\qquad\quad \text{elsewhere.}$$

Consider a problem where X_1 and X_2 are two discrete random variables with joint probability distribution $f(x_1,x_2)$ and we wish to find the joint probability distribution $g(y_1,y_2)$ of the two new random variables $Y_1 = u_1(X_1,X_2)$ and $Y_2 = u_2(X_1,X_2)$, which define a one-to-one transformation between the set of points (x_1,x_2) and (y_1,y_2). Solving the equations $y_1 = u_1(x_1,x_2)$ and $y_2 = u_2(x_1,x_2)$ simultaneously, we obtain the unique inverse solutions $x_1 = w_1(y_1,y_2)$ and $x_2 = w_2(y_1,y_2)$. Hence the random variables Y_1 and Y_2 assume the values y_1 and y_2, respectively, when X_1 assumes the value $w_1(y_1,y_2)$ and X_2 assumes the value $w_2(y_1,y_2)$. The joint probability distribution of Y_1 and Y_2 is then

$$g(y_1,y_2) = Pr(Y_1 = y_1, Y_2 = y_2)$$
$$= Pr[X_1 = w_1(y_1,y_2), X_2 = w_2(y_1,y_2)]$$
$$= f[w_1(y_1,y_2),w_2(y_1,y_2)].$$

THEOREM 5.2 *Suppose that X_1 and X_2 are* discrete *random variables with joint probability distribution $f(x_1,x_2)$. Let $Y_1 = u_1(X_1,X_2)$ and $Y_2 = u_2(X_1,X_2)$ define a one-to-one transformation between the points (x_1,x_2) and (y_1,y_2) so that the equations $y_1 = u_1(x_1,x_2)$ and $y_2 = u_2(x_1,x_2)$ may be uniquely solved for x_1 and x_2 in terms of y_1 and y_2, say $x_1 = w_1(y_1,y_2)$ and $x_2 = w_2(y_1,y_2)$. Then the joint probability distribution of Y_1 and Y_2 is*

$$g(y_1,y_2) = f[w_1(y_1,y_2),w_2(y_1,y_2)].$$

Theorem 5.2 is extremely useful in finding the distribution of some random variable $Y_1 = u_1(X_1, X_2)$ where X_1 and X_2 are discrete random variables with joint probability distribution $f(x_1, x_2)$. We simply define a second function, say $Y_2 = u_2(X_1, X_2)$, maintaining a one-to-one correspondence between the points (x_1, x_2) and (y_1, y_2), and obtain the joint probability distribution $g(y_1, y_2)$. The distribution of Y_1 is just the marginal distribution of $g(y_1, y_2)$, found by summing over the y_2 values. Denoting the distribution of Y_1 by $h(y_1)$, we can then write

$$h(y_1) = \sum_{y_2} g(y_1, y_2).$$

Example 5.2 Let X_1 and X_2 be two independent random variables having Poisson distributions with parameters μ_1 and μ_2, respectively. Find the distribution of the random variable $Y_1 = X_1 + X_2$.

Solution. Since X_1 and X_2 are independent, we can write

$$f(x_1, x_2) = f(x_1)f(x_2)$$

$$= \frac{e^{-\mu_1}\mu_1^{x_1}}{x_1!} \frac{e^{-\mu_2}\mu_2^{x_2}}{x_2!}$$

$$= \frac{e^{-(\mu_1+\mu_2)}\mu_1^{x_1}\mu_2^{x_2}}{x_1! x_2!},$$

where $x_1 = 0, 1, 2, \ldots$ and $x_2 = 0, 1, 2, \ldots$. Let us now define a second random variable, say $Y_2 = X_2$. The inverse functions are given by $x_1 = y_1 - y_2$ and $x_2 = y_2$. Using Theorem 5.2 we find the joint probability distribution of Y_1 and Y_2 to be

$$g(y_1, y_2) = \frac{e^{-(\mu_1+\mu_2)}\mu_1^{y_1-y_2}\mu_2^{y_2}}{(y_1 - y_2)! y_2!},$$

where $y_1 = 0, 1, 2, \ldots$ and $y_2 = 0, 1, 2, \ldots, y_1$. Note that since $x_1 > 0$, the transformation $x_1 = y_1 - x_2$ implies that x_2 and hence y_2 must always be less than or equal to y_1. Consequently the marginal probability distribution of Y_1 is

$$h(y_1) = \sum_{y_2=0}^{y_1} g(y_1, y_2)$$

$$= e^{-(\mu_1+\mu_2)} \sum_{y_2=0}^{y_1} \frac{\mu_1^{y_1-y_2}\mu_2^{y_2}}{(y_1 - y_2)! y_2!}$$

$$= \frac{e^{-(\mu_1+\mu_2)}}{y_1!} \sum_{y_2=0}^{y_1} \frac{y_1!}{y_2!(y_1 - y_2)!} \mu_1^{y_1-y_2}\mu_2^{y_2}$$

$$= \frac{e^{-(\mu_1+\mu_2)}}{y_1!} \sum_{y_2=0}^{y_1} \binom{y_1}{y_2} \mu_1^{y_1-y_2}\mu_2^{y_2}.$$

Recognizing this sum as the binomial expansion of $(\mu_1 + \mu_2)^{y_1}$, we obtain

$$h(y_1) = \frac{e^{-(\mu_1+\mu_2)}(\mu_1 + \mu_2)^{y_1}}{y_1!}, \qquad y_1 = 0, 1, 2, \ldots,$$

from which we conclude that the sum of the two independent random variables having Poisson distributions, with parameters μ_1 and μ_2, has a Poisson distribution with parameter $\mu_1 + \mu_2$.

To find the probability distribution of the random variable $Y = u(X)$ when X is a continuous random variable and the transformation is one to one, we shall need Theorem 5.3.

THEOREM 5.3 *Suppose that* X *is a* continuous *random variable with probability distribution* $f(x)$. *Let* $Y = u(X)$ *define a one-to-one correspondence between the values of* X *and* Y *so that the equation* $y = u(x)$ *can be uniquely solved for* x *in terms of* y, *say* $x = w(y)$. *Then the probability distribution of* Y *is*

$$g(y) = f[w(y)]|J|,$$

where $J = w'(y)$ *and is called the Jacobian of the transformation.*

Proof. (1) Suppose $y = u(x)$ is an increasing function as in Figure 5.1. Then we see that whenever Y falls between a and b the random variable X

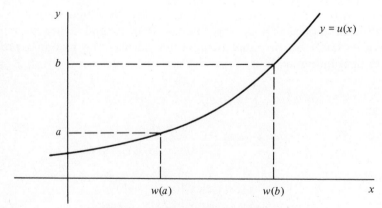

Figure 5.1 Increasing function.

must fall between $w(a)$ and $w(b)$. Hence,

$$Pr(a < Y < b) = Pr[w(a) < X < w(b)]$$
$$= \int_{w(a)}^{w(b)} f(x)\, dx.$$

Changing the variable of integration from x to y by the relation $x = w(y)$ we obtain $dx = w'(y)\, dy$ and hence

$$Pr(a < Y < b) = \int_{a}^{b} f[w(y)]w'(y)\, dy.$$

Since the integral gives the desired probability for every $a < b$ within the permissible set of y values, then the probability distribution of Y is

$$g(y) = f[w(y)]w'(y) = f[w(y)]J.$$

Recognizing $J = w'(y)$ as the reciprocal of slope of the curve of an increasing function $y = u\ (x)$, it is then obvious that $J = |J|$. Hence,

$$g(y) = f[w(y)]|J|.$$

(2) Suppose $y = u(x)$ is a decreasing function as in Figure 5.2. Then we write

$$Pr(a < Y < b) = Pr[w(b) < X < w(a)]$$
$$= \int_{w(b)}^{w(a)} f(x)\, dx.$$

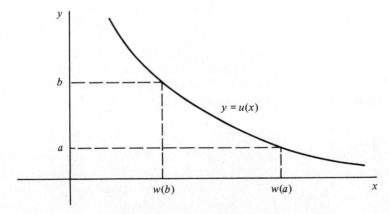

Figure 5.2 Decreasing function.

Again changing the variable of integration to y, we obtain

$$Pr(a < Y < b) = \int_b^a f[w(y)]w'(y)\,dy$$

$$= -\int_a^b f[w(y)]w'(y)\,dy,$$

from which we conclude that

$$g(y) = -f[w(y)]w'(y) = -f[w(y)]J.$$

In this case the slope of the curve is negative and $J = -|J|$. Hence

$$g(y) = f[w(y)]|J|,$$

as before.

Example 5.3 Let X be a continuous random variable with probability distribution

$$f(x) = \frac{x}{12}, \qquad 1 < x < 5$$

$$= 0, \qquad \text{elsewhere.}$$

Find the probability distribution of the random variable $Y = 2X - 3$.

Solution. The inverse solution of $y = 2x - 3$ yields $x = (y + 3)/2$, from which we obtain $J = dx/dy = 1/2$. Therefore, using Theorem 5.3, we find the density function of Y to be

$$g(y) = \left[\frac{(y+3)/2}{12}\right]\left(\frac{1}{2}\right) = \frac{(y+3)}{48}, \qquad -1 < y < 7$$

$$= 0, \qquad\qquad\qquad\qquad \text{elsewhere.}$$

To find the joint probability distribution of the random variables $Y_1 = u_1(X_1,X_2)$ and $Y_2 = u_2(X_1,X_2)$ when X_1 and X_2 are continuous and the transformation is one to one, we need an additional theorem, analogous to Theorem 5.2, which we state without proof.

THEOREM 5.4 *Suppose that X_1 and X_2 are continuous random variables with joint probability distribution $f(x_1,x_2)$. Let $Y_1 = u_1(X_1,X_2)$ and $Y_2 = u_2(X_1,X_2)$ define a one-to-one transformation between the points (x_1,x_2) and (y_1,y_2) so that the equations $y_1 = u_1(x_1,x_2)$ and $y_2 = u_2(x_1,x_2)$ may be uniquely solved for x_1 and x_2 in terms of y_1 and y_2, say $x_1 =$*

$w_1(y_1,y_2)$ and $x_2 = w_2(y_1,y_2)$. Then the joint probability distribution of Y_1 and Y_2 is

$$g(y_1,y_2) = f[w_1(y_1,y_2),w_2(y_1,y_2)]|J|,$$

where the Jacobian is the 2×2 determinant

$$J = \begin{vmatrix} \partial x_1/\partial y_1 & \partial x_1/\partial y_2 \\ \partial x_2/\partial y_1 & \partial x_2/\partial y_2 \end{vmatrix}$$

and $\partial x_1/\partial y_1$ is simply the derivative of $x_1 = w_1(y_1,y_2)$ with respect to y_1 with y_2 held constant, referred to in calculus as the partial derivative of x_1 with respect to y_1. The other partial derivatives are defined in a similar manner.

Example 5.4 Let X_1 and X_2 be two continuous random variables with joint probability distribution

$$f(x_1,x_2) = 4x_1x_2, \qquad 0 < x_1 < 1, 0 < x_2 < 1$$
$$= 0, \qquad \text{elsewhere.}$$

Find the joint probability distribution of $Y_1 = X_1^2$ and $Y_2 = X_1X_2$.

Solution. The inverse solutions of $y_1 = x_1^2$ and $y_2 = x_1x_2$ are $x_1 = \sqrt{y_1}$ and $x_2 = y_2/\sqrt{y_1}$, from which we obtain

$$J = \begin{vmatrix} 1/2\sqrt{y_1} & -y_2/2y_1^{3/2} \\ 0 & 1/\sqrt{y_1} \end{vmatrix} = \frac{1}{2y_1}.$$

The transformation is one to one, mapping the points $\{(x_1,x_2) \mid 0 < x_1 < 1, \ 0 < x_2 < 1\}$ into the set $\{(y_1,y_2) \mid y_2^2 < y_1 < 1, \ 0 < y_2 < 1\}$. From Theorem 5.4, the joint probability distribution of Y_1 and Y_2 is

$$g(y_1,y_2) = 4(\sqrt{y_1}) \frac{y_2}{\sqrt{y_1}} \frac{1}{2y_1}.$$
$$= \frac{2y_2}{y_1}, \qquad y_2^2 < y_1 < 1, 0 < y_2 < 1$$
$$= 0, \qquad \text{elsewhere.}$$

Problems frequently arise where we wish to find the probability distribution of the random variable $Y = u(X)$ when X is a continuous random variable and the transformation is not one to one. That is, to each value x there corresponds exactly one value y, but to each y value there corresponds more than one x value. For example, suppose $f(x)$ is positive over the interval $-1 < x < 2$ and zero elsewhere. Consider the transformation $y = x^2$. In this case $x = \pm\sqrt{y}$ for $0 < y < 1$ and $x = \sqrt{y}$ for $1 < y < 4$. For the interval $1 < y < 4$, the probability distribution of Y is found as before, using Theorem 5.3. That is,

$$g(y) = f[w(y)]|J| = \frac{f(\sqrt{y})}{2\sqrt{y}}, \qquad 1 < y < 4.$$

However, when $0 < y < 1$, we may partition the interval $-1 < x < 1$ to obtain the two inverse functions

$$\begin{aligned} x &= -\sqrt{y}, & -1 < x < 0 \\ &= \sqrt{y}, & 0 < x < 1. \end{aligned}$$

Then to every y value there corresponds a single x value for each partition.

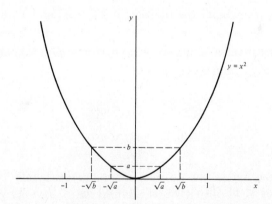

Figure 5.3 Decreasing and increasing function.

From Figure 5.3 we see that

$$Pr(a < Y < b) = Pr(-\sqrt{b} < X < -\sqrt{a}) + Pr(\sqrt{a} < X < \sqrt{b})$$
$$= \int_{-\sqrt{b}}^{-\sqrt{a}} f(x) \, dx + \int_{\sqrt{a}}^{\sqrt{b}} f(x) \, dx.$$

Changing the variable of integration from x to y, we obtain

$$Pr(a < Y < b) = \int_b^a f(-\sqrt{y})J_1\, dy + \int_a^b f(\sqrt{y})J_2\, dy$$
$$= -\int_a^b f(-\sqrt{y})J_1\, dy + \int_a^b f(\sqrt{y})J_2\, dy,$$

where

$$J_1 = \frac{d(-\sqrt{y})}{dy} = \frac{-1}{2\sqrt{y}} = -|J_1|$$

and

$$J_2 = \frac{d(\sqrt{y})}{dy} = \frac{1}{2\sqrt{y}} = |J_2|.$$

Hence, we can write

$$Pr(a < Y < b) = \int_a^b [f(-\sqrt{y})|J_1| + f(\sqrt{y})|J_2|]\, dy,$$

and then

$$g(y) = f(-\sqrt{y})|J_1| + f(\sqrt{y})|J_2|$$
$$= [f(-\sqrt{y}) + f(\sqrt{y})]/2\sqrt{y}, \qquad 0 < y < 1.$$

The probability distribution of Y for $0 < y < 4$ may now be written

$$g(y) = \frac{[f(-\sqrt{y}) + f(\sqrt{y})]}{2\sqrt{y}}, \qquad 0 < y < 1$$
$$= \frac{f(\sqrt{y})}{2\sqrt{y}}, \qquad 1 < y < 4$$
$$= 0, \qquad \text{elsewhere.}$$

The above procedure for finding $g(y)$ when $0 < y < 1$ is generalized in Theorem 5.5 for k inverse functions. For transformations not one to one of functions of several variables the reader is referred to *Introduction to Mathematical Statistics* by Hogg and Craig.

THEOREM 5.5 *Suppose that X is a* continuous *random variable with probability distribution $f(x)$. Let $Y = u(X)$ define a transformation between the values of X and Y that is not one to one. If the interval over which X is defined can be partitioned into k mutually disjoint sets such* *that each of the inverse functions* $x_1 = w_1(y)$, $x_2 = w_2(y)$, ..., $x_k = w_k(y)$ *of* $y = u(x)$ *defines a one-to-one correspondence, then the probability distribution of* Y *is*

$$g(y) = \sum_{i=1}^{k} f[w_i(y)]|J_i|,$$

where $J_i = w_i'(y)$, $i = 1, 2, ..., k$.

Example 5.5 Show that $Y = (X - \mu)^2/\sigma^2$ has a chi-square distribution with 1 degree of freedom when X has a normal distribution with mean μ and variance σ^2.

Solution. Let $Z = (X - \mu)/\sigma$, where the random variable Z has the standard normal distribution

$$f(z) = \frac{1}{\sqrt{2\pi}} e^{-z^2/2}, \qquad -\infty < z < \infty.$$

We shall now find the distribution of the random variable $Y = Z^2$. The inverse solutions of $y = z^2$ are $z = \pm\sqrt{y}$. Designating $z_1 = -\sqrt{y}$ and $z_2 = \sqrt{y}$, $J_1 = -1/2\sqrt{y}$ and $J_2 = 1/2\sqrt{y}$. Hence by Theorem 5.5, we have

$$g(y) = \frac{1}{\sqrt{2\pi}} e^{-y/2} \left| \frac{-1}{2\sqrt{y}} \right| + \frac{1}{\sqrt{2\pi}} e^{-y/2} \left| \frac{1}{2\sqrt{y}} \right|$$

$$= \frac{1}{2^{1/2}\sqrt{\pi}} y^{1/2-1} e^{-y/2}, \qquad y > 0.$$

Since $g(y)$ is a density function it follows that

$$1 = \frac{1}{2^{1/2}\sqrt{\pi}} \int_0^\infty y^{1/2-1} e^{-y/2} \, dy$$

$$= \frac{\Gamma(1/2)}{\sqrt{\pi}} \int_0^\infty \frac{1}{2^{1/2}\Gamma(1/2)} y^{1/2-1} e^{-y/2} \, dy$$

$$= \frac{\Gamma(1/2)}{\sqrt{\pi}},$$

the integral being the area under a gamma probability curve with parameters $\alpha = 1/2$ and $\beta = 2$. Therefore, $\sqrt{\pi} = \Gamma(1/2)$ and the probability distribution of Y is given by

$$g(y) = \frac{1}{2^{1/2}\Gamma(1/2)}\, y^{1/2-1}e^{-y/2}, \qquad y > 0,$$

which is seen to be a chi-square distribution with 1 degree of freedom.

5.2 Moment-Generating Functions

Although the method of transforming variables provides an effective way of finding the distribution of a function of several random variables, there is an alternative and often preferred procedure when the function in question is the sum of independent random variables. We shall refer to this method as the *moment-generating function* technique.

DEFINITION 5.1 *The* moment-generating function *of the random variable* X *is given by* $E(e^{tX})$ *and is denoted by* $M_X(t)$. *Hence*

$$M_X(t) = E(e^{tX})$$

$$= \sum_{i=1}^{n} e^{tx_i}f(x_i) \qquad \textit{if } X \textit{ is discrete}$$

$$= \int_{-\infty}^{\infty} e^{tx}f(x)\, dx \quad \textit{if } X \textit{ is continuous.}$$

Moment-generating functions will exist only if the sum or integral of Definition 5.1 converges. If a moment-generating function of a random variable X does exist, it can be used to generate all the moments of that variable. The method is described in Theorem 5.6.

THEOREM 5.6 *Let* X *be a random variable with moment-generating function* $M_X(t)$. *Then*

$$\frac{d^r M_X(t)}{dt^r}\bigg|_{t=0} = \mu_r'.$$

Proof. Assuming we can differentiate inside summation and integral signs,

$$\frac{d^r M_X(t)}{dt^r} = \sum_{i=1}^{n} x_i^r e^{tx_i} f(x_i) \qquad \text{(discrete case)}$$

$$= \int_{-\infty}^{\infty} x^r e^{tx} f(x) \, dx \qquad \text{(continuous case)}.$$

Setting $t = 0$, we see that both cases reduce to $E(X^r) = \mu_r'$.

Example 5.6 Find the moment-generating function of the binomial random variable X and then use it to verify that $\mu = np$ and $\sigma^2 = npq$.

Solution. From Definition 5.1 we have

$$M_X(t) = \sum_{x=0}^{n} e^{tx} \binom{n}{x} p^x q^{n-x}$$

$$= \sum_{x=0}^{n} \binom{n}{x} (pe^t)^x q^{n-x}.$$

Recognizing this last sum as the binomial expansion of $(pe^t + q)^n$, we obtain

$$M_X(t) = (pe^t + q)^n.$$

Now

$$\frac{dM_X(t)}{dt} = n(pe^t + q)^{n-1} pe^t$$

and

$$\frac{d^2 M_X(t)}{dt^2} = np[e^t(n-1)(pe^t + q)^{n-2} pe^t + (pe^t + q)^{n-1} e^t].$$

Setting $t = 0$, we get

$$\mu_1' = np \quad \text{and} \quad \mu_2' = np[(n-1)p + 1].$$

Therefore

$$\mu = \mu_1' = np$$

and

$$\sigma^2 = \mu_2' - \mu^2 = np(1-p) = npq,$$

which agrees with our previous results obtained in Chapter 3.

Example 5.7 Show that the moment-generating function of the random variable X having a normal probability distribution with mean μ and variance σ^2 is given by $M_X(t) = e^{\mu t + \sigma^2 t^2 / 2}$.

Solution. From Definition 5.1, the moment-generating function of the normal random variable X is

$$M_X(t) = \int_{-\infty}^{\infty} e^{tx} \frac{1}{\sqrt{2\pi}\,\sigma} e^{-(1/2)\,[(x-\mu)/\sigma]^2}\,dx$$

$$= \int_{-\infty}^{\infty} \frac{1}{\sqrt{2\pi}\,\sigma} e^{-[x^2-2(\mu+t\sigma^2)x+\mu^2]/2\sigma^2}\,dx.$$

Completing the square in the exponent, we can write

$$x^2 - 2(\mu + t\sigma^2)x + \mu^2 = [x - (\mu + t\sigma^2)]^2 - 2\mu t\sigma^2 - t^2\sigma^4$$

and then

$$M_X(t) = \int_{-\infty}^{\infty} \frac{1}{\sqrt{2\pi}\,\sigma} e^{-\{[x-(\mu+t\sigma^2)]^2-2\mu t\sigma^2-t^2\sigma^4\}/2\sigma^2}\,dx$$

$$= e^{\mu t+\sigma^2 t^2/2} \int_{-\infty}^{\infty} \frac{1}{\sqrt{2\pi}\,\sigma} e^{-(1/2)\{[x-(\mu+t\sigma^2)]/\sigma\}^2}\,dx.$$

Let $w = [x - (\mu + t\sigma^2)]/\sigma$; then $dx = \sigma\,dw$ and

$$M_X(t) = e^{\mu t+\sigma^2 t^2/2} \int_{-\infty}^{\infty} \frac{1}{\sqrt{2\pi}} e^{-w^2/2}\,dw$$

$$= e^{\mu t+\sigma^2 t^2/2},$$

since the last integral represents the area under a standard normal density curve and hence equals 1.

Example 5.8 Show that the moment-generating function of the random variable X having a chi-square distribution with ν degrees of freedom is $M_X(t) = (1 - 2t)^{-\nu/2}$.

Solution. The chi-square distribution was obtained as a special case of the gamma distribution by setting $\alpha = \nu/2$ and $\beta = 2$. Substituting for $f(x)$ in Definition 5.1 we obtain

$$M_X(t) = \int_0^{\infty} e^{tx} \frac{1}{2^{\nu/2}\Gamma(\nu/2)} x^{\nu/2-1} e^{-x/2}\,dx$$

$$= \frac{1}{2^{\nu/2}\Gamma(\nu/2)} \int_0^{\infty} x^{\nu/2-1} e^{-x(1-2t)/2}\,dx.$$

Writing $y = x(1 - 2t)/2$ and $dx = [2/(1 - 2t)]\,dy$, we get

$$M_X(t) = \frac{1}{2^{\nu/2}\Gamma(\nu/2)} \int_0^\infty \left[\frac{2y}{(1 - 2t)}\right]^{\nu/2-1} e^{-y}\, \frac{2}{(1 - 2t)}\, dy$$

$$= \frac{1}{\Gamma(\nu/2)(1 - 2t)^{\nu/2}} \int_0^\infty y^{\nu/2-1} e^{-y}\, dy$$

$$= (1 - 2t)^{-\nu/2},$$

since the last integral equals $\Gamma(\nu/2)$.

The properties of moment-generating functions discussed in the following four theorems will be of particular importance in determining the distributions of linear combinations of independent random variables. In keeping with the mathematical scope of this text we state Theorem 5.7 without proof.

THEOREM 5.7 (UNIQUENESS THEOREM) *Let X and Y be two random variables with moment-generating functions $M_X(t)$ and $M_Y(t)$, respectively. If $M_X(t) = M_Y(t)$ for all values of t, then X and Y have the same probability distribution.*

THEOREM 5.8 $M_{X+a}(t) = e^{at}M_X(t)$.

Proof.

$$M_{X+a}(t) = E[e^{t(X+a)}]$$

$$= e^{at}E(e^{tX}) = e^{at}M_X(t).$$

THEOREM 5.9 $M_{aX}(t) = M_X(at)$.

Proof.

$$M_{aX}(t) = E[e^{t(aX)}] = E[e^{(at)X}]$$

$$= M_X(at).$$

THEOREM 5.10 *If X_1 and X_2 are independent random variables with moment-generating functions $M_{X_1}(t)$ and $M_{X_2}(t)$, respectively, and $Y = X_1 + X_2$, then*

$$M_Y(t) = M_{X_1+X_2}(t) = M_{X_1}(t)M_{X_2}(t).$$

Proof.

$$M_Y(t) = E(e^{tY}) = E[e^{t(X_1+X_2)}]$$
$$= \int_{-\infty}^{\infty} \int_{-\infty}^{\infty} e^{t(x_1+x_2)} f(x_1,x_2) \, dx_1 \, dx_2.$$

Since the variables are independent, we have $f(x_1,x_2) = g(x_1)h(x_2)$ and then

$$M_Y(t) = \int_{-\infty}^{\infty} e^{tx_1} g(x_1) \, dx_1 \int_{-\infty}^{\infty} e^{tx_2} h(x_2) \, dx_2$$
$$= M_{X_1}(t)M_{X_2}(t).$$

The proof for the discrete case is obtained in a similar manner by replacing integrals with summations.

One might use Theorems 5.7 and 5.10 along with the result of Exercise 14 as an alternative method to Example 5.2 in finding the distribution of the sum of two independent Poisson random variables. For example, if X_1 and X_2 are independent Poisson variables with moment-generating functions given by

$$M_{X_1}(t) = e^{\mu_1(e^t-1)} \quad \text{and} \quad M_{X_2}(t) = e^{\mu_2(e^t-1)},$$

respectively, then according to Theorem 5.10, the moment-generating function of the random variable $Y_1 = X_1 + X_2$ is

$$M_{Y_1}(t) = M_{X_1}(t)M_{X_2}(t)$$
$$= e^{\mu_1(e^t-1)} e^{\mu_2(e^t-1)}$$
$$= e^{(\mu_1+\mu_2)(e^t-1)},$$

which we immediately identify as the moment-generating function of a random variable having a Poisson distribution with the parameter $\mu_1 + \mu_2$. Hence, according to Theorem 5.7, we again conclude that the sum of two independent random variables having Poisson distributions, with parameters μ_1 and μ_2, has a Poisson distribution with parameter $\mu_1 + \mu_2$.

In applied statistics one frequently needs to know the probability distribution of a linear combination of independent normal random variables. Let us obtain the distribution of the random variable $Y = a_1X_1 + a_2X_2$ when X_1 is a normal variable with mean μ_1 and variance σ_1^2 and X_2 is also a normal variable but independent of X_1, with mean μ_2 and variance σ_2^2. First, by Theorem 5.10, we find

$$M_Y(t) = M_{a_1X_1}(t)M_{a_2X_2}(t),$$

and then using Theorem 5.9

$$M_Y(t) = M_{X_1}(a_1t)M_{X_2}(a_2t).$$

Substituting a_1t for t in the moment-generating function of the normal distribution derived in Example 5.7 and then a_2t for t, we have

$$M_Y(t) = e^{a_1\mu_1 t + a_1{}^2\sigma_1{}^2 t^2/2}e^{a_2\mu_2 t + a_2{}^2\sigma_2{}^2 t^2/2}$$
$$= e^{(a_1\mu_1 + a_2\mu_2)t + (a_1{}^2\sigma_1{}^2 + a_2{}^2\sigma_2{}^2)t^2/2},$$

which we recognize as the moment-generating function of a distribution that is normal with mean $a_1\mu_1 + a_2\mu_2$ and variance $a_1^2\sigma_1^2 + a_2^2\sigma_2^2$.

Generalizing to the case of n independent normal variables we state the following result.

THEOREM 5.11 *If X_1, X_2, ..., X_n are independent random variables having normal distributions with means μ_1, μ_2, ..., μ_n and variances $\sigma_1^2, \sigma_2^2, ..., \sigma_n^2$, respectively, then the random variable*

$$Y = a_1X_1 + a_2X_2 + \cdots + a_nX_n$$

has a normal distribution with mean

$$\mu_Y = a_1\mu_1 + a_2\mu_2 + \cdots + a_n\mu_n$$

and variance

$$\sigma_Y^2 = a_1^2\sigma_1^2 + a_2^2\sigma_2^2 + \cdots + a_n^2\sigma_n^2.$$

It is now evident that the Poisson distribution and the normal distribution possess a reproductive property in that the sums of independent random variables having either of these distributions is a random variable that also has the same type of distribution. This reproductive property is also possessed by the chi-square distribution.

THEOREM 5.12 *If X_1, X_2, $\ldots$, X_n are mutually independent random variables that have, respectively, chi-square distributions with $\nu_1, \nu_2, \ldots, \nu_n$ degrees of freedom, then the random variable*

$$Y = X_1 + X_2 + \cdots + X_n$$

has a chi-square distribution with $\nu = \nu_1 + \nu_2 + \cdots + \nu_n$ degrees of freedom.

Proof. By Theorem 5.10,

$$M_Y(t) = M_{X_1}(t) M_{X_2}(t) \cdots M_{X_n}(t).$$

From Example 5.8,

$$M_{X_i}(t) = (1 - 2t)^{-\nu_i/2}, \qquad i = 1, 2, \ldots, n.$$

Therefore

$$M_Y(t) = (1 - 2t)^{-\nu_1/2}(1 - 2t)^{-\nu_2/2} \cdots (1 - 2t)^{-\nu_n/2}$$
$$= (1 - 2t)^{-(\nu_1+\nu_2+\cdots+\nu_n)/2},$$

which we recognize as the moment-generating function of a chi-square distribution with $\nu = \nu_1 + \nu_2 + \cdots + \nu_n$ degrees of freedom.

COROLLARY *If X_1, X_2, $\ldots$, X_n are independent random variables having identical normal distributions with mean μ and variance σ^2, then the random variable*

$$Y = \sum_{i=1}^{n} \left(\frac{X_i - \mu}{\sigma} \right)^2$$

has a chi-square distribution with $\nu = n$ degrees of freedom.

The preceding corollary is an immediate consequence of Example 5.5, which states that each of the n independent random variables $[(X_i - \mu)/\sigma]^2$, $i = 1, 2, \ldots, n$, has a chi-square distribution with 1 degree of freedom.

5.3 Random Sampling

The outcome of a statistical experiment may be recorded either as a numerical value or as a descriptive representation. When a pair of dice are tossed and the total is the outcome of interest, we record a numerical value. However, if the students in a certain school are given blood tests and the type of blood is of interest, then a descriptive representation might be the most useful. A person's blood can be classified in eight ways. It must be AB, A, B, or O, with a plus or minus sign, depending on the presence or absence of the Rh antigen.

The statistician works primarily with numerical observations. For the experiment involving the blood types he will probably let numbers 1 to 8 represent each blood type and then record the appropriate number for each student. In the classification of blood types we can have only as many observations as there are students in the school. The experiment, therefore, results in a finite number of observations. In the die-tossing experiment we are interested in recording the total that occurs. Therefore, if we toss the dice indefinitely, we obtain an infinite set of values, each representing the result of a single toss of a pair of dice.

The totality of observations with which we are concerned, whether finite or infinite, constitute what we call a *population*. In past years the word *population* referred to observations obtained from statistical studies involving people. Today the statistician uses the term to refer to observations relevant to anything of interest, whether it be groups of people, animals, or objects.

DEFINITION 5.2 *A* population *consists of the totality of the observations with which we are concerned.*

The number of observations in the population is defined to be the *size* of the population. If there are 600 students in the school that are classified according to blood type, we say we have a population of size 600. The die-tossing experiment generates a population whose size is infinite. The numbers on the cards in a deck, the heights of residents in a certain city, and the lengths of fish in a particular lake are examples of populations with finite size. In each case the total number of observations is a finite number. The observations obtained by measuring the atmospheric pressure every day from the past on into the future or all measurements on the depth of a lake from any conceivable position are examples of populations whose sizes are infinite. Some finite populations are so large that in theory we assume them to be infinite. This is true if you consider the population of lives of a certain type of storage battery being manufactured for mass distribution throughout the country.

Each observation in a population is a value of a random variable X having some probability distribution $f(x)$. If one is inspecting items coming off an assembly line for defects, then each observation in the population might be a value zero or 1 of the binomial random variable X with probability distribution

$$b(x;1,p) = p^x q^{1-x}, \qquad x = 0, 1,$$

where zero indicates a nondefective item and 1 indicates a defective item. Of course it is assumed that p, the probability of any item being defective, remains constant from trial to trial. In the blood-type experiment the random variable X represents the type of blood by assuming a value from 1 to 8. Each student is given one of the values of the discrete random variable. The lives of the storage batteries are values assumed by a continuous random variable having perhaps a normal distribution. When we speak hereafter about a "binomial population," a "normal population," or, in general, the "population $f(x)$," we shall mean a population whose observations are values of a random variable having a binomial distribution, a normal distribution, or the probability distribution $f(x)$. Hence the mean and variance of a random variable or probability distribution are also referred to as the mean and variance of the corresponding population.

The statistician is interested in arriving at conclusions concerning unknown population parameters. In a normal population, for example, the parameters μ and σ^2 may be unknown and are to be estimated from the information provided by a sample selected from the population. This takes us into the theory of sampling. If our inferences are to be accurate, we must understand the relation of a sample to its population. Certainly the sample should be representative of the population. It should be a *random sample* in the sense that the observations are made independently and at random.

In selecting a random sample of size n from a population $f(x)$ let us define the random variable X_i, $i = 1, 2, \ldots, n$, to represent the ith measurement or sample value that we observe. The random variables $X_1, X_2, \ldots, X_n$ will then constitute a random sample from the population $f(x)$ with numerical values $x_1, x_2, \ldots, x_n$ if the measurements are obtained by repeating the experiment n independent times under essentially the same conditions. Owing to the identical conditions under which the elements of the sample are selected, it is reasonable to assume that the n random variables $X_1, X_2, \ldots, X_n$ are independent and that each has the same probability distribution $f(x)$. That is, the probability distributions of X_1, X_2, $\ldots$, X_n are, respectively, $f(x_1)$, $f(x_2)$, $\ldots$, $f(x_n)$ and their joint probability distribution is

$$f(x_1, x_2, \ldots, x_n) = f(x_1)f(x_2) \cdots f(x_n).$$

The concept of a random sample is defined formally in the following definition.

DEFINITION 5.3 *Let X_1, X_2, $\ldots$, X_n be n independent random variables each having the same probability distribution $f(x)$. We then define X_1, X_2, $\ldots$, X_n to be a* random sample *of size n from the population $f(x)$ and write its joint probability distribution as*

$$f(x_1, x_2, \ldots, x_n) = f(x_1)f(x_2) \cdots f(x_n).$$

If one makes a random selection of $n = 8$ storage batteries from a manufacturing process, which has maintained the same specifications, and records the length of life for each battery with the first measurement x_1 being a value of X_1, the second measurement x_2 a value of X_2, and so forth, then x_1, x_2, $\ldots$, x_8 are the values of the random sample X_1, X_2, $\ldots$, X_8. Assuming the population of battery lives to be normal, the possible values of any X_i, $i = 1, 2, \ldots, 8$, will be precisely the same as those in the original population and hence X_i has the same identical normal distribution as X.

5.4 Sampling Theory

Our main purpose in selecting random samples is to elicit information about the unknown population parameters. Suppose we wish to arrive at a conclusion concerning the proportion of people in the United States who prefer a certain brand of coffee. It would be impossible to question every American and compute the parameter representing the true proportion. Instead, a large random sample is selected and the proportion of this sample favoring the brand of coffee in question is calculated. This value is now used to make some inference concerning the true proportion.

A value computed from a sample is called a *statistic*. Since many random samples are possible from the same population, we would expect the statistic to vary somewhat from sample to sample. Hence a statistic is a *random variable*.

DEFINITION 5.4 *A* statistic *is a random variable that depends only on the observed random sample.*

A statistic is usually represented by ordinary Latin letters. The sample proportion in the preceding illustration is a statistic that is commonly represented by $\hat{P}$. The value of the random variable $\hat{P}$ for the given sample is denoted by $\hat{p}$. To use $\hat{p}$ to estimate, with some degree of accuracy, the true proportion p of people in the United States who prefer the given brand of coffee we must first know more about the probability distribution of the statistic $\hat{P}$.

In Chapter 2 we introduced the two parameters μ and σ^2, which measure the center and the variability of a probability distribution. These are constant population parameters and are in no way affected or influenced by the observations of a random sample. We shall, however, define some important statistics that describe corresponding measures of a random sample. The most commonly used statistics for measuring the center of a set of data, arranged in order of magnitude, are the *mean, median,* and *mode.* The most important of these and the one we shall consider first is the mean.

DEFINITION 5.5 *If X_1, X_2, ..., X_n represent a random sample of size n, then the* sample mean *is defined by the statistic*

$$\bar{X} = \frac{\sum_{i=1}^{n} X_i}{n}.$$

Note that the statistic $\bar{X}$ assumes the value $\bar{x} = \sum_{i=1}^{n} x_i/n$ when X_1 assumes the value x_1, X_2 assumes the value x_2, and so forth.

Example 5.9 Find the mean of the random sample whose observations are 20, 27, and 25.

Solution. The observed value $\bar{x}$ of the statistic $\bar{X}$ is

$$\bar{x} = \frac{20 + 27 + 25}{3} = 24.$$

The second most useful statistic for measuring the center of a set of data is the median. We shall designate the median by the symbol $\tilde{X}$.

DEFINITION 5.6 *If* X_1, X_2, ..., X_n *represent a random sample of size n, arranged in increasing order of magnitude, then the* sample median *is defined by the statistic*

$$\tilde{X} = X_{(n+1)/2} \qquad \text{if n is odd}$$

$$= \frac{X_{n/2} + X_{(n/2)+1}}{2} \quad \text{if n is even.}$$

Example 5.10 Find the median for the random sample whose observations are 8, 3, 9, 5, 6, 8, and 5.

Solution. Arranging the observations in order of magnitude, 3, 5, 5, 6, 8, 8, 9, gives $\tilde{x} = 6$.

Example 5.11 Find the median for the random sample whose observations are 10, 8, 4, and 7.

Solution. Arranging the observations in order of magnitude, 4, 7, 8, 10, the median is the arithmetic mean of the two middle values. Therefore $\tilde{x} = (7 + 8)/2 = 7.5$.

The third and final statistic for measuring the center of a random sample that we shall discuss is the mode, designated by the statistic M.

DEFINITION 5.7 *If* X_1, X_2, ..., X_n, *not necessarily all different, represent a random sample of size n, then the* mode M *is that value of the sample that occurs most often or with the greatest frequency. The mode may not exist, and when it does it is not necessarily unique.*

Example 5.12 The mode of the random sample whose observations are 2, 4, 4, 5, 6, 6, 6, 7, 7, and 8 is $m = 6$.

Example 5.13 The observations 3, 4, 4, 4, 4, 6, 7, 7, 8, 8, 8, 8, and 9 have two modes, 4 and 8, since both 4 and 8 occur with the greatest frequency. The distribution of the sample is said to be *bimodal*.

When the mode could be either of two adjacent numbers arranged in order of magnitude, we take the arithmetic mean of the two numbers as the mode. Therefore, the modes of the observations 3, 5, 5, 5, 6, 6, 6, 7, 9, 9, and 9 are $(5 + 6)/2 = 5.5$ and 9.

In summary, let us consider the relative merits of the mean, median, and mode. The mean is the most commonly used measure of central tendency in statistics. It is easy to calculate and it employs all available information. The distributions of sample means are well known, and consequently the methods used in statistical inference are based on the sample mean. The only real disadvantage to the mean is that it may be affected adversely by extreme values. That is, if most contributions to a charity are less than $5, then a very large contribution, say $10,000, would produce an average donation that is considerably higher than the majority of gifts.

The median has the advantage of being easy to compute. It is not influenced by extreme values and would give a truer average in the case of the charitable contributions. In dealing with samples selected from populations, the sample means will not vary as much from sample to sample as will the medians. Therefore, if we are attempting to estimate the center of a population based on a sample value, the mean is more stable than the median. Hence a sample mean is likely to be closer to the population mean than the sample median would be to the population median.

The mode is the least used measure of the three. For small sets of data its value is almost useless, if in fact it exists at all. Only in the case of a large mass of data does it have a significant meaning. Its only advantage is that it requires no calculation.

The three statistics defined above do not by themselves give an adequate description of the distribution of our data. We need to know how the observations spread out from the average. It is quite possible to have two sets of observations with the same mean or median that differ considerably in the variability or their measurements about the average.

Consider the following measurements, in ounces, for two samples of orange juice bottled by two different companies A and B:

Sample A	31	32	30	33	34
Sample B	34	32	28	29	37

Both samples have the same mean, namely, 32. It is quite obvious that company A bottles orange juice with a more uniform content than company B. We say the variability or the dispersion of the observations from the average is less for sample A than for sample B. Therefore in buying orange juice we would feel more confident that the bottle we select will be closer to the advertised average if we buy from company A.

The most important statistics for measuring the variability of a random sample are the *range* and the *variance*. The simplest of these to compute is the range.

> DEFINITION 5.8 *The* range *of a random sample* X_1, X_2, $\ldots$, X_n, *arranged in increasing order of magnitude, is defined by the statistic* $X_n - X_1$.

Example 5.14 The range of the set of observations 10, 12, 12, 18, 19, 22, and 24 is $24 - 10 = 14$.

In the case of the companies bottling orange juice, the range for company A is 4 compared to a range of 9 for company B, indicating a greater spread in the values for company B.

The range is a poor measure of variability, particularly if the size of the sample is large. It considers only the extreme values and tells us nothing about the distribution of values in between. Consider, for example, the following two sets of data both with a range of 12:

$$3, \quad 4, \quad 5, \quad 6, \quad 8, \quad 9, \quad 10, \quad 12, \quad 15$$
$$3, \quad 8, \quad 8, \quad 9, \quad 9, \quad 9, \quad 10, \quad 10, \quad 15.$$

In the first set the mean and median are both 8, but the numbers vary over the entire interval from 3 to 15. In the second set the mean and median are both 9, but most of the values are closer to the average. Although the range fails to measure this variability between the upper and lower observations, it does have some useful applications. In industry the range for measurements on items coming off an assembly line might be specified in advance. As long as all measurements fall within the specified range the process is said to be in control.

To overcome the disadvantage of the range we shall consider a measure of variability, namely, the *sample variance*, that considers the position of each observation relative to the sample mean.

> DEFINITION 5.9 *If* X_1, X_2, $\ldots$, X_n *represent a random sample of size* n, *then the* sample variance *is defined by the statistic*
>
> $$S^2 = \frac{\sum_{i=1}^{n} (X_i - \bar{X})^2}{n - 1}.$$

The computed value of S^2 for a given sample is denoted by s^2. Note that S^2 is essentially defined to be the average of the squares of the deviations of the observations from their mean. The reason for using $n-1$ as a divisor rather than the more obvious choice n will become apparent in Chapter 6.

THEOREM 5.13 *If S^2 is the variance of a random sample of size n, we may write*

$$S^2 = \frac{n \sum_{i=1}^{n} X_i^2 - \left(\sum_{i=1}^{n} X_i \right)^2}{n(n-1)}.$$

Proof. By definition

$$S^2 = \frac{\sum_{i=1}^{n} (X_i - \bar{X})^2}{n-1}$$

$$= \frac{\sum_{i=1}^{n} (X_i^2 - 2\bar{X}X_i + \bar{X}^2)}{n-1}$$

$$= \frac{\sum_{i=1}^{n} X_i^2 - 2\bar{X} \sum_{i=1}^{n} X_i + n\bar{X}^2}{n-1}.$$

Replacing $\bar{X}$ by $\sum_{i=1}^{n} X_i / n$, and multiplying numerator and denominator by n, we obtain the more useful computational formula

$$S^2 = \frac{n \sum_{i=1}^{n} X_i^2 - \left(\sum_{i=1}^{n} X_i \right)^2}{n(n-1)}.$$

The *sample standard deviation*, denoted by S, is defined to be the positive square root of the sample variance.

Example 5.15 Find the variance of the sample whose observations are 3, 4, 5, 6, 6, and 7.

Solution. We find that $\sum_{i=1}^{6} x_i^2 = 171$, $\sum_{i=1}^{6} x_i = 31$, $n = 6$. Hence

$$s^2 = \frac{(6)(171) - (31)^2}{(6)(5)} = \frac{13}{6}.$$

The field of inductive statistics is basically concerned with generalizations and predictions. Generalizations from a statistic to a parameter can be made with confidence only if we understand the fluctuating behavior of our statistic when computed for different random samples from the same population. The distribution of the statistic in question will depend on the size of the population, the size of the samples, and the method of choosing the random samples. If the size of the population is large or infinite, the statistic has the same distribution whether we sample with or without replacement. On the other hand, sampling with replacement from a small finite population gives a slightly different distribution for the statistic than if we sample without replacement. Sampling with replacement from a finite population is equivalent to sampling from an infinite population since there is no limit on the possible size of the sample selected.

DEFINITION 5.10 *The probability distribution of a statistic is called a* sampling distribution.

DEFINITION 5.11 *The standard deviation of the sampling distribution of a statistic is called the* standard error *of the statistic.*

The probability distribution of $\bar{X}$ is called the *sampling distribution of the mean,* and the standard error of the mean is the standard deviation of the sampling distribution of $\bar{X}$. Every sample of size n selected from a specified population provides a value s of the statistic S, the sample standard deviation. The standard error of the sample standard deviation is then the standard deviation of the statistic S.

In the remainder of this chapter we shall study several important sampling distributions of frequently used statistics. The applications of these sampling distributions to problems of statistical inference will be considered in Chapters 6 and 7.

5.5 Sampling Distributions of Means

The first important sampling distribution to be considered is that of the mean $\bar{X}$. Suppose a random sample of n observations is taken from a normal population with mean μ and variance σ^2. Each observation X_i, $i = 1, 2, \ldots, n$, of the random sample will then have the same normal distribution as the population being sampled. Hence, by the reproductive property of the normal distribution established in Theorem 5.11, we conclude that

$$\bar{X} = \frac{X_1 + X_2 + \cdots + X_n}{n}$$

has a normal distribution with mean

$$\mu_{\bar{X}} = \frac{\mu + \mu + \cdots + \mu}{n} = \mu$$

and variance

$$\sigma_{\bar{X}}^2 = \frac{\sigma^2 + \sigma^2 + \cdots + \sigma^2}{n^2} = \frac{\sigma^2}{n}.$$

If we are sampling from a population with unknown distribution, either finite or infinite, the sampling distribution of $\bar{X}$ will still be approximately normal with mean μ and variance σ^2/n provided the sample size is large. This amazing result is an immediate consequence of the following theorem, called the *central limit theorem*. The proof is outlined in Exercise 30.

THEOREM 5.14 *If $\bar{X}$ is the mean of a random sample of size n taken from a population with mean μ and finite variance σ^2, then the limiting form of the distribution of*

$$Z = \frac{\bar{X} - \mu}{\sigma/\sqrt{n}},$$

as $n \to \infty$, is the standardized normal distribution $n(z;0,1)$.

The normal approximation for $\bar{X}$ will generally be good if $n \geq 30$ regardless of the shape of the population. If $n < 30$ the approximation is good only if the population is not too different from a normal population. If the population is known to be normal, the sampling distribution of $\bar{X}$ will follow a normal distribution exactly, no matter how small the size of the samples.

Example 5.16 An electrical firm manufactures light bulbs that have a length of life that is approximately normally distributed, with mean equal to 800 hours and a standard deviation of 40 hours. Find the probability that a random sample of 16 bulbs will have an average life of less than 775 hours.

Solution. The sampling distribution of $\bar{X}$ will be approximately normal, with $\mu_{\bar{X}} = 800$ and $\sigma_{\bar{X}} = 40/\sqrt{16} = 10$. The desired probability is

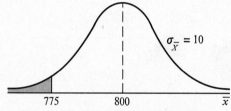

Figure 5.4 Area for Example 5.16.

given by the area of the shaded region in Figure 5.4. Corresponding to $\bar{x} = 775$, we find

$$z = \frac{775 - 800}{10} = -2.5,$$

and, therefore,

$$Pr(\bar{X} < 775) = Pr(Z < -2.5)$$
$$= 0.006.$$

Example 5.17 Given the discrete uniform population

$$f(x) = \tfrac{1}{4}, \qquad x = 0, 1, 2, 3$$
$$= 0, \qquad \text{elsewhere,}$$

find the probability that a random sample of size 36, selected with replacement, will yield a sample mean greater than 1.4 but less than 1.8 if the mean is measured to the nearest tenth.

Solution. Calculating the mean and variance of the uniform distribution by means of the formulas in Theorem 3.1, we find

$$\mu = \frac{0 + 1 + 2 + 3}{4} = \frac{3}{2}$$

and

$$\sigma^2 = \frac{(0 - 3/2)^2 + (1 - 3/2)^2 + (2 - 3/2)^2 + (3 - 3/2)^2}{4}$$

$$= \frac{5}{4}.$$

The sampling distribution of $\bar{X}$ may be approximated by the normal distribution with mean $\mu_{\bar{X}} = 3/2$ and variance $\sigma_{\bar{X}}^2 = \sigma^2/n = 5/144$. Taking the square root we find the standard deviation to be $\sigma_{\bar{X}} = 0.186$. The probability that $\bar{X}$ is greater than 1.4 but less than 1.8 is given by the area of the shaded region in Figure 5.5. The z values corresponding

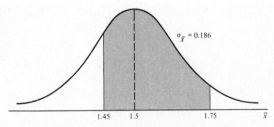

Figure 5.5 Area for Example 5.17.

to $\bar{x}_1 = 1.45$ and $\bar{x}_2 = 1.75$ are

$$z_1 = \frac{1.45 - 1.5}{0.186} = -0.269$$

$$z_2 = \frac{1.75 - 1.5}{0.186} = 1.344.$$

Therefore

$$Pr(1.4 < X < 1.8) \simeq Pr(-0.269 < Z < 1.344)$$

$$= Pr(Z < 1.344) - Pr(Z < -0.269)$$

$$= 0.9105 - 0.3932$$

$$= 0.5173.$$

Suppose we now have two populations, the first with mean μ_1 and variance σ_1^2, and the second with mean μ_2 and variance σ_2^2. Let the statistic $\bar{X}_1$ represent the mean of a random sample of size n_1 selected from the first population, and the statistic $\bar{X}_2$ represent the mean of a random sample selected from the second population, independent of the sample from the first population. What can we say about the sampling distribution of the difference $\bar{X}_1 - \bar{X}_2$ for repeated samples of size n_1 and n_2? According to Theorem 5.14 the variables $\bar{X}_1$ and $\bar{X}_2$ are both approximately normally distributed with means μ_1 and μ_2 and variances σ_1^2/n_1 and σ_2^2/n_2, respectively. This approximation improves as n_1 and n_2 increase. By choosing independent samples from the two populations the variables $\bar{X}_1$ and $\bar{X}_2$ will be independent, and then using Theorem 5.11, with $a_1 = 1$ and $a_2 = -1$, we conclude that $\bar{X}_1 - \bar{X}_2$ is approximately normally distributed with mean

$$\mu_{\bar{X}_1 - \bar{X}_2} = \mu_{\bar{X}_1} - \mu_{\bar{X}_2} = \mu_1 - \mu_2$$

and variance

$$\sigma_{\bar{X}_1 - \bar{X}_2}^2 = \sigma_{\bar{X}_1}^2 + \sigma_{\bar{X}_2}^2 = \frac{\sigma_1^2}{n_1} + \frac{\sigma_2^2}{n_2}.$$

THEOREM 5.15 *If independent samples of size n_1 and n_2 are drawn at random from two populations, discrete or continuous, with means μ_1 and μ_2 and variances σ_1^2 and σ_2^2, respectively, then the sampling distribution of the differences of means, $\bar{X}_1 - \bar{X}_2$, is approximately normally distributed with mean and variance given by*

$$\mu_{\bar{X}_1 - \bar{X}_2} = \mu_1 - \mu_2$$

$$\sigma_{\bar{X}_1 - \bar{X}_2}^2 = \frac{\sigma_1^2}{n_1} + \frac{\sigma_2^2}{n_2}.$$

Hence

$$Z = \frac{(\bar{X}_1 - \bar{X}_2) - (\mu_1 - \mu_2)}{\sqrt{(\sigma_1^2/n_1) + (\sigma_2^2/n_2)}}$$

is approximately a standard normal variable.

If both n_1 and n_2 are greater than or equal to 30, the normal approximation for the distribution of $\bar{X}_1 - \bar{X}_2$ is very good.

Example 5.18 The television picture tubes of manufacturer A have a mean lifetime of 6.5 years and a standard deviation of 0.9 year, while those of manufacturer B have a mean lifetime of 6.0 years and a standard deviation of 0.8 year. What is the probability that a random sample of 36 tubes from manufacturer A will have a mean lifetime that is at least 1 year more than the mean lifetime of a sample of 49 tubes from manufacturer B?

Solution. We are given the following information:

Population 1	Population 2
$\mu_1 = 6.5$	$\mu_2 = 6.0$
$\sigma_1 = 0.9$	$\sigma_2 = 0.8$
$n_1 = 36$	$n_2 = 49$

Using Theorem 5.15, the sampling distribution of $\bar{X}_1 - \bar{X}_2$ will have a mean and standard deviation given by

$$\mu_{\bar{X}_1 - \bar{X}_2} = 6.5 - 6.0 = 0.5$$

$$\sigma_{\bar{X}_1 - \bar{X}_2} = \sqrt{\frac{0.81}{36} + \frac{0.64}{49}} = 0.189.$$

The probability that the mean of 36 tubes from manufacturer A will be at least 1 year longer than the mean of 49 tubes from manufacturer B is given by the area of the shaded region in Figure 5.6. Corresponding to

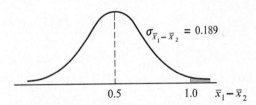

$$\sigma_{\bar{X}_1 - \bar{X}_2} = 0.189$$

$$0.5 \qquad 1.0 \qquad \bar{x}_1 - \bar{x}_2$$

Figure 5.6 Area for Example 5.18.

the value $\bar{x}_1 - \bar{x}_2 = 1.0$, we find

$$z = \frac{1.0 - 0.5}{0.189} = 2.646,$$

and hence

$$Pr(\bar{X}_1 - \bar{X}_2 \geq 1.0) = Pr(Z > 2.646)$$
$$= 1 - Pr(Z < 2.646)$$
$$= 1 - 0.9959$$
$$= 0.0041.$$

5.6 Sampling Distribution of $(n - 1)S^2/\sigma^2$

If a random sample of size n is drawn from a normal population with mean μ and variance σ^2, and the sample variance s^2 is computed, we obtain a value of the statistic S^2. The sampling distribution of S^2 has little practical application in statistics. Instead we shall consider the distribution of the random variable $(n - 1)S^2/\sigma^2$.

By the addition and subtraction of the sample mean $\bar{X}$, it is easy to see that

$$\sum_{i=1}^{n} (X_i - \mu)^2 = \sum_{i=1}^{n} [(X_i - \bar{X}) + (\bar{X} - \mu)]^2$$

$$= \sum_{i=1}^{n} (X_i - \bar{X})^2 + \sum_{i=1}^{n} (\bar{X} - \mu)^2 + 2(\bar{X} - \mu) \sum_{i=1}^{n} (X_i - \bar{X})$$

$$= \sum_{i=1}^{n} (X_i - \bar{X})^2 + n(\bar{X} - \mu)^2.$$

Dividing each term of the equality by σ^2 and substituting $(n - 1)S^2$ for $\sum_{i=1}^{n} (X_i - \bar{X})^2$, we obtain

$$\frac{\sum_{i=1}^{n} (X_i - \mu)^2}{\sigma^2} = \frac{(n - 1)S^2}{\sigma^2} + \frac{(\bar{X} - \mu)^2}{\sigma^2/n}.$$

Now, according to the corollary of Theorem 5.12 we know that $\sum_{i=1}^{n} (X_i - \mu)^2/\sigma^2$ is a chi-square random variable with n degrees of freedom. The second term on the right of the equality is the square of a standard normal variable since $\bar{X}$ is a normal random variable with mean $\mu_{\bar{X}} = \mu$ and variance $\sigma_{\bar{X}}^2 = \sigma^2/n$. Therefore we may conclude from Example 5.5 that $(\bar{X} - \mu)^2/(\sigma^2/n)$ is a chi-square random variable with 1 degree of freedom. Using advanced techniques beyond the scope of this text one can also show that the two chi-square variables $\sum_{i=1}^{n} (X_i - \mu)^2/\sigma^2$ and $(\bar{X} - \mu)^2/(\sigma^2/n)$ are independent. Owing to the reproductive property of independent chi-square random variables, established in Theorem 5.12, it would seem reasonable to assume that $(n - 1)S^2/\sigma^2$ is also a chi-square random variable with $\nu = n - 1$ degrees of freedom. We state this result, without formal proof, in the following theorem.

THEOREM 5.16 *If S^2 is the variance of a random sample of size n taken from a normal population having the variance σ^2, then the random variable*

$$X^2 = \frac{(n-1)S^2}{\sigma^2}$$

has a chi-square distribution with $\nu = n - 1$ degrees of freedom.

The values of the random variable X^2 are calculated from each sample by the formula

$$\chi^2 = \frac{(n-1)s^2}{\sigma^2}.$$

The probability that a random sample produces a χ^2 value greater than some specified value is equal to the area under the curve to the right of this value. It is customary to let χ^2_α represent the χ^2 value above which we find an area of α. This is illustrated by the shaded region in Figure 5.7.

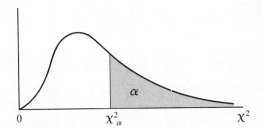

Figure 5.7 Tabulated values of the chi-square distribution.

Table VI (see Statistical Tables) gives values of χ^2_α for various values of α and ν. The areas α are the column headings, the degrees of freedom ν are given in the left column, and the table entries are the χ^2 values. Hence, the χ^2 value with 7 degrees of freedom, leaving an area of 0.05 to the right, is $\chi^2_{0.05} = 14.067$. Owing to lack of symmetry we must also use the tables to find $\chi^2_{0.95} = 2.167$ for $\nu = 7$.

Exactly 95% of a chi-square distribution with $n - 1$ degrees of freedom lies between $\chi^2_{0.975}$ and $\chi^2_{0.025}$. A χ^2 value falling to the right of $\chi^2_{0.025}$ is not likely to occur unless our assumed value of σ^2 is too small. Likewise, a χ^2 value falling to the left of $\chi^2_{0.975}$ is unlikely unless our assumed value of σ^2 is too large. In other words, it is possible to have a χ^2 value to the left of

$\chi_{0.975}^2$ or to the right of $\chi_{0.025}^2$ when σ^2 is correct, but if this should occur it is more probable that the assumed value of σ^2 is in error.

Example 5.19 A manufacturer of car batteries guarantees that his batteries will last, on the average, 3 years with a standard deviation of 1 year. If five of these batteries have lifetimes of 1.9, 2.4, 3.0, 3.5, and 4.2 years, is the manufacturer still convinced that his batteries have a standard deviation of 1 year?

Solution. We first find the sample variance:

$$s^2 = \frac{(5)(48.26) - (15)^2}{(5)(4)} = 0.815.$$

Then

$$\chi^2 = \frac{(4)(0.815)}{1} = 3.26$$

is a value from a chi-square distribution with 4 degrees of freedom. Since 95% of the χ^2 values with 4 degrees of freedom fall between 0.484 and 11.143, the computed value with $\sigma^2 = 1$ is reasonable, and therefore the manufacturer has no reason to suspect that the standard deviation is other than 1 year.

5.7 *t* Distribution

Most of the time we are not fortunate enough to know the variance of the population from which we select our random samples. For samples of size $n \geq 30$, a good estimate of σ^2 is provided by calculating a value for S^2. What then happens to our statistic $(\bar{X} - \mu)/(\sigma/\sqrt{n})$ of Theorem 5.14 if we replace σ^2 by S^2? As long as S^2 provides a good estimate of σ^2 and does not vary from sample to sample, which is usually the case for $n \geq 30$, the distribution of the statistic $(\bar{X} - \mu)/(S/\sqrt{n})$ is still approximately distributed as a standard normal variable Z.

If the sample size is small $(n < 30)$, the values of S^2 fluctuate considerably from sample to sample (see Exercise 36) and the distribution of the random variable $(\bar{X} - \mu)/(S/\sqrt{n})$ is no longer a standard normal distribution. We are now dealing with the distribution of a statistic that we shall call T, where

$$T = \frac{\bar{X} - \mu}{S/\sqrt{n}}.$$

In deriving the sampling distribution of T we shall assume our random sample was selected from a normal population. We can then write

$$T = \frac{(\bar{X} - \mu)/(\sigma/\sqrt{n})}{\sqrt{S^2/\sigma^2}} = \frac{Z}{\sqrt{V/(n-1)}},$$

where

$$Z = \frac{\bar{X} - \mu}{\sigma/\sqrt{n}}$$

has the standard normal distribution, and

$$V = \frac{(n-1)S^2}{\sigma^2}$$

has a chi-square distribution with $\nu = n - 1$ degrees of freedom. In sampling from normal populations one can show that $\bar{X}$ and S^2 are independent and consequently so are Z and V. We are now in a position to derive the distribution of T.

THEOREM 5.17 *Let Z be a standard normal random variable and V a chi-square random variable with ν degrees of freedom. If Z and V are independent, then the distribution of the random variable T, where*

$$T = \frac{Z}{\sqrt{V/\nu}},$$

is given by

$$h(t) = \frac{\Gamma[(\nu+1)/2]}{\Gamma(\nu/2)\sqrt{\pi\nu}}\left(1 + \frac{t^2}{\nu}\right)^{-(\nu+1)/2}, \qquad -\infty < t < \infty.$$

This is known as the t *distribution with ν degrees of freedom.*

Proof. Since Z and V are independent random variables, their joint probability distribution is given by the product of the distribution of Z and of V. That is,

$$f(z,v) = \frac{1}{\sqrt{2\pi}}\, e^{-z^2/2}\, \frac{1}{2^{\nu/2}\Gamma(\nu/2)}\, v^{\nu/2-1}e^{-v/2}, \qquad -\infty < z < \infty, 0 < v < \infty$$

$$= 0, \qquad\qquad\qquad\qquad \text{elsewhere.}$$

Let us define a second random variable $U = V$. The inverse solutions of $t = z/\sqrt{v/\nu}$ and $u = v$ are $z = t\sqrt{u}/\sqrt{\nu}$ and $v = u$, from which we obtain

$$J = \begin{vmatrix} \sqrt{u}/\sqrt{\nu} & t/2\sqrt{u\nu} \\ 0 & 1 \end{vmatrix} = \frac{\sqrt{u}}{\sqrt{\nu}}.$$

The transformation is one to one, mapping the points $\{(z,v) \mid -\infty < z < \infty, 0 < v < \infty\}$ into the set $\{(t,u) \mid -\infty < t < \infty, 0 < u < \infty\}$. Using Theorem 5.4 we find the joint probability distribution of T and U to be

$$g(t,u) = \frac{1}{\sqrt{2\pi}\, 2^{\nu/2}\Gamma(\nu/2)}\, u^{\nu/2-1} e^{-\{(u/2)[1+(t^2/\nu)]\}} \frac{\sqrt{u}}{\sqrt{\nu}}, \qquad -\infty < t < \infty,$$

$$0 < u < \infty$$

$$= 0, \qquad\qquad\qquad\qquad\qquad\qquad\qquad \text{elsewhere.}$$

Integrating out u, the distribution of T is given by

$$h(t) = \int_0^\infty g(t,u)\, du$$

$$= \int_0^\infty \frac{1}{\sqrt{2\pi\nu}\, 2^{\nu/2}\Gamma(\nu/2)}\, u^{[(\nu+1)/2]-1} e^{-\{(u/2)[1+(t^2/\nu)]\}}\, du.$$

Let us substitute $z = u(1 + t^2/\nu)/2$ and $du = dz/(1 + t^2/\nu)$ to give

$$h(t) = \frac{1}{\sqrt{2\pi\nu}\, 2^{\nu/2}\Gamma(\nu/2)} \int_0^\infty \left(\frac{2z}{1+t^2/\nu}\right)^{[(\nu+1)/2]-1} e^{-z} \left(\frac{2}{1+t^2/\nu}\right) dz$$

$$= \frac{1}{\Gamma(\nu/2)\sqrt{\pi\nu}} \left(1 + \frac{t^2}{\nu}\right)^{-[(\nu+1)/2]} \int_0^\infty z^{[(\nu+1)/2]-1} e^{-z}\, dz$$

$$= \frac{\Gamma[(\nu+1)/2]}{\Gamma(\nu/2)\sqrt{\pi\nu}} \left(1 + \frac{t^2}{\nu}\right)^{-(\nu+1)/2}, \qquad -\infty < t < \infty$$

$$= 0, \qquad\qquad\qquad\qquad\qquad\qquad\qquad \text{elsewhere.}$$

The probability distribution of T was first published in 1908 in a paper by W. S. Gosset. At the time, Gosset was employed by an Irish brewery that disallowed publication of research by members of its staff. To circumvent this restriction he published his work secretly under the name "Student." Consequently, the distribution of T is usually called the *Student t distribution*, or simply the *t distribution*. In deriving the equation of this distribution, Gosset assumed the samples were selected from a normal population. Although this would seem to be a very restrictive assumption, it can be shown that nonnormal populations possessing bell-shaped distributions will still provide values of T that approximate the t distribution very closely.

The distribution of T is similar to the distribution of Z in that they both are symmetric about a mean of zero. Both distributions are bell-shaped but the t distribution is more variable, owing to the fact that the T values depend on the fluctuations of two quantities, $\bar{X}$ and S^2, whereas the Z values depend only on the changes of $\bar{X}$ from sample to sample. The distribution of T differs from that of Z in that the variance of T depends on the sample size n and is always greater than 1. Only when the sample size $n \to \infty$ will the two distributions become the same. In Figure 5.8 we show the relationship between a standard normal distribution ($\nu = \infty$) and t distributions with 2 and 5 degrees of freedom.

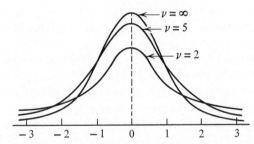

Figure 5.8 *t* distribution curves for $\nu = 2, 5,$ and ∞.

The probability that a random sample produces a value $t = (\bar{x} - \mu)/(s/\sqrt{n})$ falling between any two specified values is equal to the area under the curve of the t distribution between the two ordinates corresponding to the specified values. It would be a tedious task to attempt to set up separate tables giving the areas between every conceivable pair of ordinates for all values of $n \leq 30$. Table VI (see Statistical Tables) gives only those t values above which we find a specified area α, where α is 0.1, 0.05, 0.025, 0.01, or 0.005. This table is set up differently than the table of normal curve areas in that the areas are now the column headings and the entries are the t values. The left column gives the degrees of freedom. It is customary to let t_α represent the t value above which we find an area equal to α. Hence, the t value with 10 degrees of freedom leaving an area of 0.025 to the right is $t = 2.228$. Since the t distribution is symmetric about a mean of zero, we have $t_{1-\alpha} = -t_\alpha$; that is, the t value leaving an area of $1 - \alpha$ to the right and therefore an area of α to the left is equal to the negative t value that leaves an area of α in the right tail of the distribution (see Figure 5.9). For a t distribution with 10 degrees of freedom we have $t_{0.975} = -t_{0.025} = -2.228$. This means that the t value of a random sample of size 11, selected from a normal population, will fall between -2.228 and 2.228 with probability equal to 0.95.

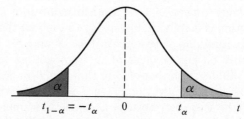

Figure 5.9 Symmetry property of the t distribution.

Exactly 95% of a t distribution with $n - 1$ degrees of freedom lies between $-t_{0.025}$ and $t_{0.025}$. Therefore a t value falling below $-t_{0.025}$ or above $t_{0.025}$ would tend to make one believe that either a rare event has taken place or our assumption about μ is in error. The importance of μ will determine the length of the interval for an acceptable t value. In other words, if you do not mind having the true mean slightly different from what you claim it to be, you might choose a wide interval from $-t_{0.01}$ to $t_{0.01}$ in which the t value should fall. A t value falling at either end of the interval, but within the interval, would lead us to believe that our assumed value for μ is correct although it is very probable that some other value close to μ is the true value. If μ is to be known with a high degree of accuracy, a short interval such as $-t_{0.05}$ to $t_{0.05}$ should be used. In this case a t value falling outside the interval would lead you to believe that the assumed value of μ is in error when it is entirely possible that it is correct. The problems connected with the establishment of proper intervals in testing hypotheses concerning the parameter μ will be treated in Chapter 7.

Example 5.20 A manufacturer of light bulbs claims that his bulbs will burn on the average 500 hours. To maintain this average he tests 25 bulbs each month. If the computed t value falls between $-t_{0.05}$ and $t_{0.05}$, he is satisfied with his claim. What conclusion should he draw from a sample that has a mean $\bar{x} = 518$ hours and a standard deviation $s = 40$ hours? Assume the distribution of burning times to be approximately normal.

Solution. From Table V we find $t_{0.05} = 1.711$ for 24 degrees of freedom. Therefore the manufacturer is satisfied with his claim if a sample of 25 bulbs yields a t value between -1.711 and 1.711. If $\mu = 500$, then

$$t = \frac{518 - 500}{40/\sqrt{25}} = 2.25,$$

a value well above 1.711. The probability of obtaining a t value, with $\nu = 24$, equal to or greater than 2.25 is approximately 0.02. If $\mu > 500$, the value of t computed from the sample would be more reasonable. Hence the manufacturer is likely to conclude that his bulbs are a better product than he thought.

5.8 F Distribution

One of the most important distributions in applied statistics is the F *distribution*. The statistic F is defined to be the ratio of two independent chi-square random variables, each divided by their degrees of freedom. Hence, we can write

$$F = \frac{U/\nu_1}{V/\nu_2},$$

where U and V are independent random variables having chi-square distributions with ν_1 and ν_2 degrees of freedom, respectively. We shall now derive the sampling distribution of F.

THEOREM 5.18 *Let U and V be two independent random variables having chi-square distributions with ν_1 and ν_2 degrees of freedom, respectively. Then the distribution of the random variable*

$$F = \frac{U/\nu_1}{V/\nu_2}$$

is given by

$$h(f) = \frac{\Gamma[(\nu_1 + \nu_2)/2](\nu_1/\nu_2)^{\nu_1/2}}{\Gamma(\nu_1/2)\Gamma(\nu_2/2)} \frac{f^{\nu_1/2-1}}{(1 + \nu_1 f/\nu_2)^{(\nu_1+\nu_2)/2}}, \qquad 0 < f < \infty$$

$$= 0, \qquad\qquad\qquad\qquad\qquad\qquad\qquad\qquad\qquad\qquad\qquad elsewhere.$$

This is known as the F *distribution with ν_1 and ν_2 degrees of freedom.*

Proof. The joint probability distribution of the independent random variables U and V is given by

$$\phi(u,v) = r(u)s(v),$$

where $r(u)$ and $s(v)$ represent the distributions of U and V, respectively. Hence,

$$
\begin{aligned}
\phi(u,v) &= \frac{1}{2^{\nu_1/2}\Gamma(\nu_1/2)}\, u^{\nu_1/2-1}e^{-u/2}\, \frac{1}{2^{\nu_2/2}\Gamma(\nu_2/2)}\, v^{\nu_2/2-1}e^{-v/2} \\
&= \frac{1}{2^{(\nu_1+\nu_2)/2}\Gamma(\nu_1/2)\Gamma(\nu_2/2)}\, u^{\nu_1/2-1}v^{\nu_2/2-1}e^{-(u+v)/2}, \quad \begin{array}{l} 0 < u < \infty, \\ 0 < v < \infty, \end{array} \\
&= 0, \qquad\qquad\qquad\qquad\qquad\qquad\qquad\qquad\qquad\qquad \text{elsewhere.}
\end{aligned}
$$

Let us define a second random variable $W = V$. The inverse solutions of $f = (u/\nu_1)/(v/\nu_2)$ and $w = v$ are $u = (\nu_1/\nu_2)fw$ and $v = w$, from which we obtain

$$
J = \begin{vmatrix} (\nu_1/\nu_2)w & (\nu_1/\nu_2)f \\ 0 & 1 \end{vmatrix} = \frac{\nu_1}{\nu_2}\, w.
$$

The transformation is one to one, mapping the points $\{(u,v)\mid 0 < u < \infty, 0 < v < \infty\}$ into the set $\{(f,w)\mid 0 < f < w, 0 < w < \infty\}$. Using Theorem 5.4, the joint probability distribution of F and W is

$$
\begin{aligned}
g(f,w) &= \frac{1}{2^{(\nu_1+\nu_2)/2}\Gamma(\nu_1/2)\Gamma(\nu_2/2)}\left(\frac{\nu_1 fw}{\nu_2}\right)^{\nu_1/2-1} w^{\nu_2/2-1}e^{-(w/2)\,[(\nu_1 f/\nu_2)+1]}\,\frac{\nu_1 w}{\nu_2}, \\
&\qquad 0 < f < \infty, \, 0 < w < \infty \\
&= 0, \qquad \text{elsewhere.}
\end{aligned}
$$

The distribution of F is then given by the marginal distribution

$$
\begin{aligned}
h(f) &= \int_0^\infty g(f,w)\, dw \\
&= \frac{(\nu_1/\nu_2)^{\nu_1/2}f^{\nu_1/2-1}}{2^{(\nu_1+\nu_2)/2}\Gamma(\nu_1/2)\Gamma(\nu_2/2)}\int_0^\infty w^{[(\nu_1+\nu_2)/2]-1}e^{-(w/2)\,[(\nu_1 f/\nu_2)+1]}\, dw.
\end{aligned}
$$

Substituting $z = (w/2)[(\nu_1 f/\nu_2) + 1]$ and $dw = [2/(\nu_1 f/\nu_2 + 1)]\, dz$, we obtain

$$
\begin{aligned}
h(f) &= \frac{(\nu_1/\nu_2)^{\nu_1/2}f^{\nu_1/2-1}}{2^{(\nu_1+\nu_2)/2}\Gamma(\nu_1/2)\Gamma(\nu_2/2)}\int_0^\infty \left(\frac{2z}{\nu_1 f/\nu_2 + 1}\right)^{(\nu_1+\nu_2)/2-1} e^{-z}\frac{2}{\nu_1 f/\nu_2 + 1}\, dz \\
&= \frac{(\nu_1/\nu_2)^{\nu_1/2}f^{\nu_1/2-1}}{\Gamma(\nu_1/2)\Gamma(\nu_2/2)(1 + \nu_1 f/\nu_2)^{(\nu_1+\nu_2)/2}}\int_0^\infty z^{(\nu_1+\nu_2)/2-1}e^{-z}\, dz \\
&= \frac{\Gamma[(\nu_1 + \nu_2)/2](\nu_1/\nu_2)^{\nu_1/2}}{\Gamma(\nu_1/2)\Gamma(\nu_2/2)}\,\frac{f^{\nu_1/2-1}}{(1 + \nu_1 f/\nu_2)^{(\nu_1+\nu_2)/2}}, \qquad 0 < f < \infty \\
&= 0, \qquad\qquad\qquad\qquad\qquad\qquad\qquad\qquad\qquad\quad \text{elsewhere.}
\end{aligned}
$$

The number of degrees of freedom associated with the chi-square random variable appearing in the numerator of F is always stated first, followed by the number of degrees of freedom associated with the chi-square random variable appearing in the denominator. Thus the curve of the F distribution depends not only on the two parameters ν_1 and ν_2 but also on the order in which we state them. Once these two values are given we can identify the curve. Typical F curves are shown in Figure 5.10.

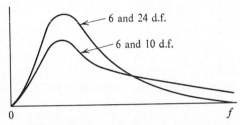

Figure 5.10 Typical F distributions.

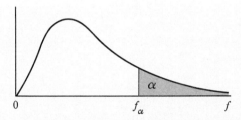

Figure 5.11 Tabulated values of the F distribution.

Let us define f_α to be the particular value f of the random variable F above which we find an area equal to α. This is illustrated by the shaded region in Figure 5.11. Table VII (see Statistical Tables) gives values of f_α only for $\alpha = 0.05$ and $\alpha = 0.01$ for various combinations of the degrees of freedom ν_1 and ν_2. Hence, the f value with 6 and 10 degrees of freedom, leaving an area of 0.05 to the right, is $f_{0.05} = 3.22$. By means of the following theorem, Table VII can be used to find values of $f_{0.95}$ and $f_{0.99}$. The proof is left for the reader.

THEOREM 5.19 *Writing $f_\alpha(\nu_1,\nu_2)$ for f_α with ν_1 and ν_2 degrees of freedom,*

$$f_{1-\alpha}(\nu_1,\nu_2) = \frac{1}{f_\alpha(\nu_2,\nu_1)}.$$

Thus the f value with 6 and 10 degrees of freedom, leaving an area of 0.95 to the right, is

$$f_{0.95}(6,10) = \frac{1}{f_{0.05}(10,6)} = \frac{1}{4.06} = 0.246.$$

Suppose random samples of size n_1 and n_2 are selected from two normal populations with variances σ_1^2 and σ_2^2, respectively. From Theorem 5.16, we know that

$$X_1^2 = (n_1 - 1)S_1^2/\sigma_1^2$$

and

$$X_2^2 = (n_2 - 1)S_2^2/\sigma_2^2$$

are random variables having chi-square distributions with $\nu_1 = n_1 - 1$ and $\nu_2 = n_2 - 1$ degrees of freedom. Furthermore, since the samples are selected at random, we are dealing with independent random variables and then using Theorem 5.18, with $X_1^2 = U$ and $X_2^2 = V$, we obtain the following result.

THEOREM 5.20 *If S_1^2 and S_2^2 are the variances of independent random samples of size n_1 and n_2 taken from normal populations with variances σ_1^2 and σ_2^2, respectively, then*

$$F = \frac{S_1^2/\sigma_1^2}{S_2^2/\sigma_2^2} = \frac{\sigma_2^2 S_1^2}{\sigma_1^2 S_2^2}$$

has an F distribution with $\nu_1 = n_1 - 1$ and $\nu_2 = n_2 - 1$ degrees of freedom.

In Chapters 6 and 7 we shall use Theorem 5.20 to make inferences concerning the variances of two normal populations. The F distribution is applied primarily, however, in the analysis-of-variance procedures of Chapters 11, 12, and 13, where we wish to test the equality of several means simultaneously.

EXERCISES

1. Let X be a random variable with probability distribution

$$f(x) = \tfrac{1}{3}, \qquad x = 1, 2, 3$$
$$= 0, \qquad \text{elsewhere.}$$

Find the probability distribution of the random variable $Y = 2X - 1$.

2. Let X be a binomial random variable with probability distribution

$$f(x) = \binom{3}{x}\left(\frac{2}{5}\right)^x\left(\frac{3}{5}\right)^{3-x}, \qquad x = 0, 1, 2, 3$$

$$= 0, \qquad\qquad\qquad \text{elsewhere.}$$

Find the probability distribution of the random variable $Y = X^2$.

3. Let X_1 and X_2 be discrete random variables with the multinomial distribution

$$f(x_1, x_2) = \binom{2}{x_1, x_2, 2 - x_1 - x_2}\left(\frac{1}{4}\right)^{x_1}\left(\frac{1}{3}\right)^{x_2}\left(\frac{5}{12}\right)^{2-x_1-x_2}$$

for $x_1 = 0, 1, 2$; $x_2 = 0, 1, 2$; $x_1 + x_2 \le 2$; and zero elsewhere. Find the joint probability distribution of $Y_1 = X_1 + X_2$ and $Y_2 = X_1 - X_2$.

4. Let X_1 and X_2 be discrete random variables with joint probability distribution

$$f(x_1, x_2) = \frac{x_1 x_2}{18}, \qquad x_1 = 1, 2, x_2 = 1, 2, 3$$

$$= 0, \qquad \text{elsewhere.}$$

Find the probability distribution of the random variable $Y = X_1 X_2$.

5. Let X have the probability distribution

$$f(x) = 1, \qquad 0 < x < 1$$

$$= 0, \qquad \text{elsewhere.}$$

Show that the random variable $Y = -2 \ln X$ has a chi-square distribution with 2 degrees of freedom.

6. Given the random variable X with probability distribution

$$f(x) = 2x, \qquad 0 < x < 1$$

$$= 0, \qquad \text{elsewhere,}$$

find the probability distribution of Y, where $Y = 8X^3$.

7. The speed of a molecule in a uniform gas at equilibrium is a random variable V whose probability distribution is given by

$$f(v) = kv^2 e^{-bv^2}, \qquad v > 0$$

$$= 0, \qquad \text{elsewhere,}$$

where k is an appropriate constant and b depends on the absolute temperature and mass of the molecule. Find the probability distribution of the kinetic energy of the molecule W, where $W = mV^2/2$.

8. Let X_1 and X_2 be independent random variables each having the probability distribution

$$f(x) = e^{-x}, \quad x > 0$$
$$= 0, \quad \text{elsewhere.}$$

Show that the random variables Y_1 and Y_2 are independent when $Y_1 = X_1 + X_2$ and $Y_2 = X_1/(X_1 + X_2)$.

9. A current of I amperes flowing through a resistance of R ohms varies according to the probability distribution

$$f(i) = 6i(1 - i), \quad 0 < i < 1$$
$$= 0, \quad \text{elsewhere.}$$

If the resistance varies independently of the current according to the probability distribution

$$g(r) = 2r, \quad 0 < r < 1$$
$$= 0, \quad \text{elsewhere,}$$

find the probability distribution for the power $W = I^2 R$ watts.

10. Let X be a random variable with probability distribution

$$f(x) = \frac{(1 + x)}{2}, \quad -1 < x < 1$$
$$= 0, \quad \text{elsewhere.}$$

Find the probability distribution of the random variable $Y = X^2$.

11. Let X have the probability distribution

$$f(x) = \frac{2(x + 1)}{9}, \quad -1 < x < 2$$
$$= 0, \quad \text{elsewhere.}$$

Find the probability distribution of the random variable $Y = X^2$.

12. Given the discrete uniform distribution

$$f(x) = \frac{1}{n}, \quad x = 1, 2, 3, \ldots, n$$
$$= 0, \quad \text{elsewhere,}$$

show that the moment-generating function of X is

$$M_X(t) = \frac{e^t(1 - e^{nt})}{n(1 - e^t)}.$$

13. Show that the moment-generating function of the geometric random variable is

$$M_X(t) = \frac{pe^t}{1 - qe^t},$$

   and then use $M_X(t)$ to find the mean and variance of the geometric distribution.

14. Show that the moment-generating function of the random variable X having a Poisson distribution with parameter μ is $M_X(t) = e^{\mu(e^t-1)}$. Using $M_X(t)$, find the mean and variance of the Poisson distribution.

15. The moment-generating function of a certain Poisson random variable X is given by

$$M_X(t) = e^{4(e^t-1)}.$$

   Find $Pr(\mu - 2\sigma < X < \mu + 2\sigma)$.

16. Using the moment-generating function of Example 5.8, show that the mean and variance of the chi-square distribution with ν degrees of freedom are, respectively, ν and 2ν.

17. In a random sample of 18 students at Roanoke College the following numbers of days absent were recorded for the previous semester: 1, 3, 4, 0, 4, 2, 3, 1, 2, 3, 0, 4, 1, 1, 1, 5, 1, and 0. Find the mean, median, and mode.

18. The numbers of trout caught by eight fishermen on the first day of the trout season are 7, 4, 6, 7, 4, 4, 8, and 7. If these eight values represent the catch of a random sample of fishermen at Smith Mountain Lake, define a suitable population. If the values represent the catch of a random sample of fishermen at various lakes and streams in Montgomery County, define a suitable population. Find the mean, median, and mode for the data.

19. Find the mean, median, and mode for the sample 18, 10, 11, 98, 22, 15, 11, 25, and 17. Which value appears to be the best measure of central tendency? Give reasons for your preference.

20. Calculate the range and standard deviation for the data of Exercise 17.

21. The grade-point averages of 15 college seniors selected at random from the graduating class are as follows:

$$
\begin{array}{ccc}
2.3 & 3.4 & 2.9 \\
2.6 & 2.1 & 2.4 \\
3.1 & 2.7 & 2.6 \\
1.9 & 2.0 & 3.6 \\
2.1 & 1.8 & 2.1.
\end{array}
$$

   Calculate the standard deviation.

22. Show that the sample variance is unchanged if a constant is added to or subtracted from each value in the sample.

23. If each observation in a sample is multiplied by k, show that the sample variance becomes k^2 times its original value.

24. (a) Calculate the variance of the sample 3, 5, 8, 7, 5, and 7.
  (b) Without calculating, state the variance of the sample 6, 10, 16, 14, 10, and 14.
  (c) Without calculating, state the variance of the sample 25, 27, 30, 29, 27, and 29.

25. If all possible samples of size 16 are drawn from a normal population with mean equal to 50 and standard deviation equal to 5, what is the probability that a sample mean $\bar{X}$ will fall in the interval from $\mu_{\bar{X}} - 1.9\sigma_{\bar{X}}$ to $\mu_{\bar{X}} - 0.4\sigma_{\bar{X}}$? Assume the sample means can be measured to any degree of accuracy.

26. If the size of a sample is 36 and the standard error of the mean is 2, what must the size of the sample become if the standard error is to be reduced to 1.2?

27. A soft-drink machine is regulated so that the amount of drink dispensed is approximately normally distributed with a mean of 7 ounces per cup and a standard deviation equal to 0.5 ounce. Periodically the machine is checked by taking a sample of nine drinks and computing the average content. If the mean, $\bar{X}$, of the nine drinks falls within the interval $\mu_{\bar{X}} \pm 2\sigma_{\bar{X}}$, the machine is thought to be operating satisfactorily; otherwise, adjustments must be made. What action should one take if a sample of nine drinks has a mean content of 7.4 ounces?

28. The heights of 1000 students are approximately normally distributed with a mean of 68.5 inches and a standard deviation of 2.7 inches. If 200 random samples of size 25 are drawn from this population and the means recorded to the nearest tenth of an inch, determine
  (a) The expected mean and standard deviation of the sampling distribution of the mean.
  (b) The number of sample means that fall between 67.9 and 69.2 inclusive.
  (c) The number of sample means falling below 67.0.

29. By expanding e^{tx} in a Maclaurin series and integrating term by term, show that

$$M_X(t) = \int_{-\infty}^{\infty} e^{tx} f(x)\, dx$$

$$= 1 + \mu t + \mu_2' \frac{t^2}{2!} + \cdots + \mu_r' \frac{t^r}{r!} + \cdots.$$

30. *Central Limit Theorem.*
  (a) Using Theorems 5.8, 5.9, and 5.10 show that

$$M_{\frac{\bar{X}-\mu}{\sigma/\sqrt{n}}}(t) = e^{-\mu\sqrt{n}t/\sigma}\left[M_X\left(\frac{t}{\sigma\sqrt{n}}\right)\right]^n,$$

where $\bar{X}$ is the mean of a random sample of size n from a population $f(x)$ with mean μ and variance σ^2, and hence

$$\ln M_{\frac{\bar{X}-\mu}{\sigma/\sqrt{n}}}(t) = -\mu\sqrt{n}t/\sigma + n \ln M_X\left(\frac{t}{\sigma\sqrt{n}}\right).$$

(b) Use the result of Exercise 29 to expand $M_X(t/\sigma_{\bar{x}}\sqrt{n})$ as an infinite series in powers of t. We can then write $M_X(t/\sigma\sqrt{n}) = 1 + v$, where v is an infinite series.

(c) Assuming n sufficiently large, expand $\ln(1 + v)$ in a Maclaurin series and then show that

$$\lim_{n\to\infty} \ln M_{\frac{\bar{X}-\mu}{\sigma/\sqrt{n}}}(t) = \frac{t^2}{2}$$

and hence

$$\lim_{n\to\infty} M_{\frac{\bar{X}-\mu}{\sigma/\sqrt{n}}}(t) = e^{t^2/2}.$$

31. A random sample of size 25 is taken from a normal population having a mean of 80 and a standard deviation of 5. A second random sample of size 36 is taken from a different normal population having a mean of 75 and a standard deviation of 3. Find the probability that the sample mean computed from the 25 measurements will exceed the sample mean computed from the 36 measurements by at least 3.4 but less than 5.9. Assume the means to be measured to the nearest tenth.

32. The mean score for freshmen on an aptitude test, at a certain college, is 540, with a standard deviation of 50. What is the probability that two groups of students selected at random, consisting of 32 and 50 students, respectively, will differ in their mean scores by
(a) More than 20 points?
(b) An amount between 5 and 10 points?
Assume the means to be measured to any degree of accuracy.

33. For a chi-square distribution find
(a) $\chi^2_{0.01}$ with $\nu = 18$.
(b) $\chi^2_{0.975}$ with $\nu = 29$.
(c) χ^2_α such that $Pr(X^2 < \chi^2_\alpha) = 0.99$ with $\nu = 4$.

34. Find the probability that a random sample of 25 observations, from a normal population with variance $\sigma^2 = 6$, will have a variance s^2
(a) Greater than 9.1.
(b) Between 3.462 and 10.745.
Assume the sample variances to be continuous measurements.

35. A placement test has been given for the past 5 years to college freshmen with a mean $\mu = 74$ and a variance $\sigma^2 = 8$. Would a school consider these values valid today if 20 students obtained a mean $\bar{x} = 72$ and a variance $s^2 = 16$ on this test?

36. Show that the variance of S^2 decreases as n becomes large. [Hint: First find the variance of $(n-1)S^2/\sigma^2$.]

37. (a) Find $t_{0.025}$ when $\nu = 17$.
(b) Find $t_{0.99}$ when $\nu = 10$.
(c) Find t_α such that $Pr(-t_\alpha < T < t_\alpha) = 0.90$ when $\nu = 23$.

38. A normal population with unknown variance has a mean of 20. Is one likely to obtain a random sample of size 9 from this population with a mean of 24 and a standard deviation of 4.1? If not, what conclusion would you draw?

39. If a cigarette manufacturer claims that his cigarettes have an average nicotine content of 18.3 milligrams, is it likely that we could select a random sample of eight cigarettes and find the nicotine contents to be 20, 17, 21, 19, 22, 21, 20, and 16 milligrams?

40. For an F distribution find
   (a) $f_{0.05}$ with $\nu_1 = 7$ and $\nu_2 = 15$.
   (b) $f_{0.05}$ with $\nu_1 = 15$ and $\nu_2 = 7$.
   (c) $f_{0.01}$ with $\nu_1 = 24$ and $\nu_2 = 19$.
   (d) $f_{0.95}$ with $\nu_1 = 19$ and $\nu_2 = 24$.
   (e) $f_{0.99}$ with $\nu_1 = 28$ and $\nu_2 = 12$.

41. If S_1^2 and S_2^2 represent the variances of independent random samples of size $n_1 = 25$ and $n_2 = 31$, taken from normal populations with variances $\sigma_1^2 = 10$ and $\sigma_2^2 = 15$, respectively, find the $Pr(S_1^2/S_2^2 > 1.26)$.

42. If S_1^2 and S_2^2 represent the variances of independent random samples of size $n_1 = 8$ and $n_2 = 12$, taken from normal populations with equal variances, find the $Pr(S_1^2/S_2^2 < 4.89)$.

6 ESTIMATION THEORY

6.1 Introduction

The theory of *statistical inference* may be defined to be those methods by which one makes inferences or generalizations about a population. The trend of today is to distinguish between the *classical* method of estimating a population parameter, whereby inferences are based strictly on information obtained from a random sample selected from the population, and the *Bayesian* method, which utilizes prior subjective knowledge about the probability distribution of the unknown parameters in conjunction with the information provided by the sample data. Throughout most of this chapter we shall obtain classical estimates of unknown population parameters such as the mean, proportion, and the standard deviation by computing statistics from random samples and applying the theory of sampling distributions from Chapter 5. For completeness, the Bayesian approach to statistical decision theory is presented in Sections 6.9 and 6.10.

Statistical inference may be divided into two major areas: *estimation* and *tests of hypotheses*. We shall treat these two areas separately, dealing with the theory of estimation in this chapter and the theory of hypothesis testing in Chapter 7. To distinguish clearly between the two areas, consider the following examples. A candidate for public office may wish to estimate the true proportion of voters favoring him by obtaining the opinions from a random sample of 100 eligible voters. The fraction of voters in the sample favoring the candidate could be used as an estimate of the true proportion of the population of voters. A knowledge of the sampling distribution of a proportion enables one to establish the degree of accuracy of our estimate. This problem falls in the area of estimation.

Now consider the case where a housewife is interested in finding out whether brand A floor wax is more scuff-resistant than brand B floor wax. She might hypothesize that brand A is better than brand B and, after proper testing, accept or reject this hypothesis. In this example we do not attempt to estimate a parameter, but instead we try to arrive at a correct decision about a prestated hypothesis. Once again we are dependent on sampling theory to provide us with some measure of accuracy for our decision.

6.2 Classical Methods of Estimation

An estimate of a population parameter may be given as a *point estimate* or as an *interval estimate*. A point estimate of some population parameter θ is a single value $\hat{\theta}$ of a statistic $\hat{\Theta}$. For example, the value $\bar{x}$ of the statistic $\bar{X}$, computed from a sample of size n, is a point estimate of the population parameter μ.

The statistic that one uses to obtain a point estimate is called an *estimator* or a *decision function*. Hence, the decision function S, which is a function of the random sample, is an estimator of σ and the estimate s is the "action" taken. Different samples will generally lead to different actions or estimates.

DEFINITION 6.1 *The set of all possible actions that can be taken in an estimation problem is called the* action space *or* decision space.

An estimator is not expected to estimate the population parameter without error. We do not expect $\bar{X}$ to estimate μ exactly, but we certainly hope that it is not too far off. For a particular sample it is possible to obtain a closer estimate of μ by using the sample median $\tilde{X}$ as an estimator. Consider, for instance, a sample consisting of the values 2, 5, and 11 from a population whose mean is 4 but supposedly unknown. We would estimate μ to be $\bar{x} = 6$, using the sample mean as our estimate, or $\tilde{x} = 5$, using the sample median as our estimate. In this case the estimator $\tilde{X}$ produces an estimate closer to the true parameter than that of the estimator $\bar{X}$. On the other hand, if our random sample contains the values 2, 6, and 7, then $\bar{x} = 5$ and $\tilde{x} = 6$, so that $\bar{X}$ is now the better estimator.

Not knowing the true value of μ, we must decide in advance whether to use $\bar{X}$ or $\tilde{X}$ as our estimator.

What are the desirable properties of a "good" decision function that would influence us to choose one estimator rather than another? Let $\hat{\Theta}$ be an estimator whose value $\hat{\theta}$ is a point estimate of some unknown population parameter θ. Certainly we would like the sampling distribution of $\hat{\Theta}$ to have a mean equal to the parameter estimated. An estimator possessing this property is said to be *unbiased*.

DEFINITION 6.2 *A statistic $\hat{\Theta}$ is said to be an* unbiased *estimator of the parameter θ if $\mu_{\hat{\Theta}} = E(\hat{\Theta}) = \theta$.*

Example 6.1 Show that S^2 is an unbiased estimator of the parameter σ^2.

Solution. Let us write

$$\sum_{i=1}^{n} (X_i - \bar{X})^2 = \sum_{i=1}^{n} [(X_i - \mu) - (\bar{X} - \mu)]^2$$

$$= \sum_{i=1}^{n} (X_i - \mu)^2 - 2(\bar{X} - \mu) \sum_{i=1}^{n} (X_i - \mu) + n(\bar{X} - \mu)^2$$

$$= \sum_{i=1}^{n} (X_i - \mu)^2 - n(\bar{X} - \mu)^2.$$

Now

$$E(S^2) = E\left[\frac{\sum_{i=1}^{n} (X_i - \bar{X})^2}{n - 1}\right]$$

$$= \frac{1}{n - 1}\left[\sum_{i=1}^{n} E(X_i - \mu)^2 - nE(\bar{X} - \mu)^2\right]$$

$$= \frac{1}{n - 1}\left(\sum_{i=1}^{n} \sigma_{X_i}^2 - n\sigma_{\bar{X}}^2\right).$$

However,

$$\sigma_{X_i}^2 = \sigma^2 \qquad \text{for } i = 1, 2, \ldots, n$$

and

$$\sigma_{\bar{X}}^2 = \frac{\sigma^2}{n}.$$

Therefore

$$E(S^2) = \frac{1}{n - 1}\left(n\sigma^2 - n\frac{\sigma^2}{n}\right) = \sigma^2.$$

Although S^2 is an unbiased estimator of σ^2, S on the other hand is a biased estimator of σ with the bias becoming insignificant for large samples.

If $\hat{\Theta}_1$ and $\hat{\Theta}_2$ are two unbiased estimators of the same population parameter θ, we would choose the estimator whose sampling distribution has the smallest variance. Hence, if $\sigma_{\hat{\Theta}_1}^2 < \sigma_{\hat{\Theta}_2}^2$, we say that $\hat{\Theta}_1$ is a *more efficient* estimator than $\hat{\Theta}_2$.

> DEFINITION 6.3 *If we consider all possible unbiased estimators of some parameter θ, the one with the smallest variance is called the* most efficient *estimator of θ.*

In Figure 6.1 we illustrate the sampling distributions of three different estimators $\hat{\Theta}_1$, $\hat{\Theta}_2$, and $\hat{\Theta}_3$, all estimating θ. It is clear that only $\hat{\Theta}_1$ and $\hat{\Theta}_2$ are unbiased since their distributions are centered at θ. The estimator $\hat{\Theta}_1$ has a smaller variance than $\hat{\Theta}_2$ and is therefore more efficient. Hence, our choice for an estimator of θ, among the three considered, would be $\hat{\Theta}_1$.

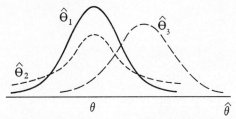

Figure 6.1 Sampling distributions of different estimators of θ.

For normal populations one can show that both $\bar{X}$ and $\tilde{X}$ are unbiased estimators of the population mean μ, but the variance of $\bar{X}$ is smaller than the variance of $\tilde{X}$. Thus both estimates $\bar{x}$ and $\tilde{x}$ will, on the average, equal the population mean μ, but $\bar{x}$ is likely to be closer to μ for a given sample.

An interval estimate of the parameter θ is an interval of the form $a < \theta < b$, where a and b depend on the point estimate $\hat{\theta}$ for the particular sample chosen and also on the sampling distribution of the statistic $\hat{\Theta}$. Thus a random sample of SAT verbal scores for students of the entering freshman class might produce an interval from 530 to 550 within which we expect to find the true average of all SAT verbal scores for the freshman class. The values of the end points, 530 and 550, will depend on the computed sample mean $\bar{x}$ and the sampling distribution of $\bar{X}$. As the sample size increases, we know that $\sigma_{\bar{X}}^2 = \sigma^2/n$ decreases and consequently our estimate is likely to be closer to the parameter μ, resulting in a shorter interval. Thus the interval estimate indicates, by its length, the accuracy of the point estimate.

Different samples yield different $\hat{\theta}$ values, and therefore produce different interval estimates of the population parameter θ. Some of these intervals will contain θ and others will not. The sampling distribution of $\hat{\theta}$ will enable us to find a and b for all possible samples such that any specified fraction of these intervals will contain θ. Therefore, if a and b are computed so that 0.95 of all possible intervals, in repeated sampling, would contain θ, then we have a probability equal to 0.95 of selecting one of the samples that will produce an interval containing θ. This interval, computed from the selected random sample, is called a 95% *confidence interval*. In other words, we are 95% confident that our computed interval does in fact contain the population parameter θ. Generally speaking, the distribution of $\hat{\theta}$ enables us to compute the end points, a and b, so that any specified fraction $1 - \alpha$, $0 < \alpha < 1$, of the intervals computed from all possible samples contain the parameter θ. The interval computed from a particular sample is then called a $(1 - \alpha)100\%$ *confidence interval*. The fraction $1 - \alpha$ is called the *confidence coefficient*, and the end points, a and b, are called the *confidence limits* or *fiducial limits*.

The longer the confidence interval, the more confident we can be that the given interval contains the unknown parameter. Of course one might prefer to be 95% confident that the average life of a certain television tube is between 6 and 7 years than to be 99% confident that it is between 3 and 10 years. Ideally, we prefer a short interval with a high degree of confidence.

6.3 Estimating the Mean

A point estimator of the population mean μ is given by the statistic $\bar{X}$. The sampling distribution of $\bar{X}$ is centered at μ and in most applications the variance is smaller than that of any other estimator. Thus the sample mean $\bar{x}$ will be used as a point estimate for the population mean μ. Recall that $\sigma_{\bar{X}}^2 = \sigma^2/n$, so that a large sample will yield a value of $\bar{X}$ that comes from a sampling distribution with a small variance. Hence $\bar{X}$ is likely to estimate μ very closely when n is large.

Let us now consider the interval estimate of μ. If our sample is selected from a normal population or, failing this, if n is sufficiently large, we can establish a confidence interval for μ by considering the sampling distribution of $\bar{X}$. According to the central limit theorem, we can expect the sampling distribution of $\bar{X}$ to be approximately normally distributed with mean $\mu_{\bar{X}} = \mu$ and deviation standard $\sigma_{\bar{X}} = \sigma/\sqrt{n}$. Writing $z_{\alpha/2}$ for the z value above which we find an area of $\alpha/2$, we can see from Figure 6.2 that

$$Pr(-z_{\alpha/2} < Z < z_{\alpha/2}) = 1 - \alpha,$$

where

$$Z = \frac{\bar{X} - \mu}{\sigma/\sqrt{n}}.$$

Hence

$$Pr\left(-z_{\alpha/2} < \frac{\bar{X} - \mu}{\sigma/\sqrt{n}} < z_{\alpha/2}\right) = 1 - \alpha.$$

Multiplying each term in the inequality by $\sigma/\sqrt{n}$, and then subtracting

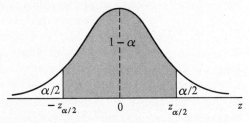

Figure 6.2 $Pr(-z_{\alpha/2} < Z < z_{\alpha/2}) = 1 - \alpha.$

$\bar{X}$ from each term and multiplying by -1 (reversing the sense of the inequalities), we obtain

$$Pr\left(\bar{X} - z_{\alpha/2}\frac{\sigma}{\sqrt{n}} < \mu < \bar{X} + z_{\alpha/2}\frac{\sigma}{\sqrt{n}}\right) = 1 - \alpha.$$

A random sample of size n is selected from a population whose variance σ^2 is known and the mean $\bar{x}$ is computed to give the $(1 - \alpha)100\%$ confidence interval:

$$\bar{x} - z_{\alpha/2}\frac{\sigma}{\sqrt{n}} < \mu < \bar{x} + z_{\alpha/2}\frac{\sigma}{\sqrt{n}}.$$

CONFIDENCE INTERVAL FOR μ; σ KNOWN *A $(1 - \alpha)100\%$ confidence interval for μ is*

$$\bar{x} - z_{\alpha/2}\frac{\sigma}{\sqrt{n}} < \mu < \bar{x} + z_{\alpha/2}\frac{\sigma}{\sqrt{n}},$$

where $\bar{x}$ is the mean of a sample of size n from a population with known variance σ^2 and $z_{\alpha/2}$ is the value of the standard normal distribution leaving an area of $\alpha/2$ to the right.

For small samples selected from nonnormal populations, we cannot expect our degree of confidence to be accurate. However, for samples of size $n \geq 30$, regardless of the shape of most populations, sampling theory guarantees good results.

To compute a $(1 - \alpha)100\%$ confidence interval for μ we have assumed that σ is known. Since this is generally not the case, we shall replace σ by the sample standard deviation s, provided $n \geq 30$.

Example 6.2 The mean and standard deviation for the quality point averages of a random sample of 36 college seniors are calculated to be 2.6 and 0.3, respectively. Find the 95% and 99% confidence intervals for the mean of the entire senior class.

Solution. The point estimate of μ is $\bar{x} = 2.6$. Since the sample size is large, the standard deviation σ can be approximated by $s = 0.3$. The z value, leaving an area of 0.025 to the right and therefore an area of 0.975 to the left, is $z_{0.025} = 1.96$ (Table IV). Hence, the 95% confidence interval is

$$2.6 - (1.96)(0.3/\sqrt{36}) < \mu < 2.6 + (1.96)(0.3/\sqrt{36}),$$

which reduces to

$$2.50 < \mu < 2.70.$$

To find a 99% confidence interval we find the z value leaving an area of 0.005 to the right and 0.995 to the left. Therefore, using Table IV again, $z_{0.005} = 2.575$, and the 99% confidence interval is

$$2.6 - (2.575)(0.3/\sqrt{36}) < \mu < 2.6 + (2.575)(0.3/\sqrt{36}),$$

or simply

$$2.47 < \mu < 2.73.$$

We now see that a longer interval is required to estimate μ with a higher degree of accuracy.

The $(1 - \alpha)100\%$ confidence interval provides an estimate of the accuracy of our point estimate. If μ is actually the center value of the interval, then $\bar{x}$ estimates μ without error. Most of the time, however, $\bar{x}$ will not be exactly equal to μ and the point estimate is in error. The size of this error will be the difference between μ and $\bar{x}$, and we can be $(1 - \alpha)100\%$ confident that this difference will be less than $z_{\alpha/2}\sigma/\sqrt{n}$.

We can readily see this if we draw a diagram of the confidence interval as in Figure 6.3.

> **THEOREM 6.1** *If $\bar{x}$ is used as an estimate of μ, we can be $(1 - \alpha)100\%$ confident that the error will be less than $z_{\alpha/2}\sigma/\sqrt{n}$.*

In Example 6.2 we are 95% confident that the sample mean $\bar{x} = 2.6$ differs from the true mean μ by an amount less than 0.1 and 99% confident that the difference is less than 0.13.

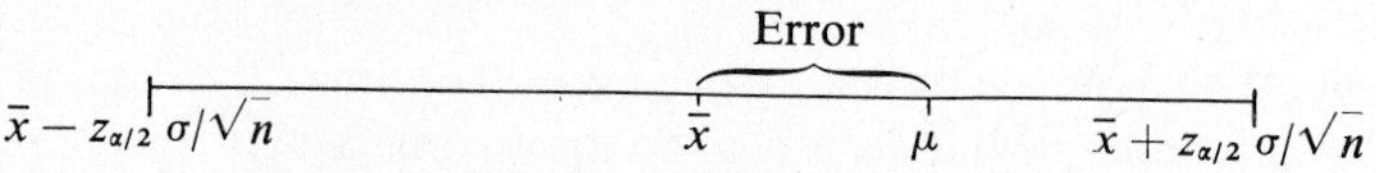

Figure 6.3 Error in estimating μ by $\bar{x}$.

Frequently we wish to know how large a sample is necessary to ensure that the error in estimating μ will be less than a specified amount e. By Theorem 6.1, this means we must choose n such that $z_{\alpha/2}\sigma/\sqrt{n} = e$.

> **THEOREM 6.2** *If $\bar{x}$ is used as an estimate of μ, we can be $(1 - \alpha)100\%$ confident that the error will be less than a specified amount e when the sample size is*
> $$n = \left[\frac{z_{\alpha/2}\sigma}{e} \right]^2.$$

Strictly speaking, the formula in Theorem 6.2 is applicable only if we know the variance of the population from which we are to select our sample. Lacking this information, a preliminary sample of size $n \geq 30$ could be taken to provide an estimate of σ; then, using Theorem 6.2, we could determine approximately how many observations are needed to provide the desired degree of accuracy.

Example 6.3 How large a sample is required in Example 6.2 if we want to be 95% confident that our estimate of μ is off by less than 0.05.

Solution. The sample standard deviation $s = 0.3$ obtained from the preliminary sample of size 36 will be used for σ. Then, by Theorem 6.2,

$$n = \left[\frac{(1.96)(0.3)}{0.05} \right]^2 = 138.3.$$

Therefore we can be 95% confident that a random sample of size 139 will provide an estimate $\bar{x}$ differing from μ by an amount less than 0.05.

Frequently we are attempting to estimate the mean of a population when the variance is unknown and it is impossible to obtain a sample of size $n \geq 30$. Cost can often be a factor that limits our sample size. As long as our population is approximately bell-shaped, confidence intervals can be computed when σ^2 is unknown and the sample size is small by using the sampling distribution of T, where

$$T = \frac{\bar{X} - \mu}{S/\sqrt{n}}.$$

The procedure is the same as for large samples except that we use the t distribution in place of the standard normal.

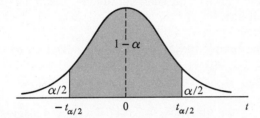

Figure 6.4 $Pr(-t_{\alpha/2} < T < t_{\alpha/2}) = 1 - \alpha.$

Referring to Figure 6.4, we can assert that

$$Pr(-t_{\alpha/2} < T < t_{\alpha/2}) = 1 - \alpha,$$

where $t_{\alpha/2}$ is the t value with $n - 1$ degrees of freedom, above which we find an area of $\alpha/2$. Owing to symmetry, an equal area of $\alpha/2$ will fall to the left of $-t_{\alpha/2}$. Substituting for T, we write

$$Pr\left(-t_{\alpha/2} < \frac{\bar{X} - \mu}{S/\sqrt{n}} < t_{\alpha/2}\right) = 1 - \alpha.$$

Multiplying each term in the inequality by $S/\sqrt{n}$, and then subtracting $\bar{X}$ from each term and multiplying by -1, we obtain

$$Pr(\bar{X} - t_{\alpha/2}S/\sqrt{n} < \mu < \bar{X} + t_{\alpha/2}S/\sqrt{n}) = 1 - \alpha.$$

For our particular random sample of size n, the mean $\bar{x}$ and standard deviation s are computed and the $(1 - \alpha)100\%$ confidence interval is given by

$$\bar{x} - t_{\alpha/2}s/\sqrt{n} < \mu < \bar{x} + t_{\alpha/2}s/\sqrt{n}.$$

CONFIDENCE INTERVAL FOR μ; σ UNKNOWN AND $n < 30$ A $(1 - \alpha)100\%$ *confidence interval for μ is*

$$\bar{x} - t_{\alpha/2}s/\sqrt{n} < \mu < \bar{x} + t_{\alpha/2}s/\sqrt{n},$$

where $\bar{x}$ and s are the mean and standard deviation, respectively, of a sample of size $n < 30$ from an approximate normal population, and $t_{\alpha/2}$ is the value of the t distribution, with $\nu = n - 1$ degrees of freedom, leaving an area of $\alpha/2$ to the right.

Example 6.4 The weights of seven similar containers of sulfuric acid are 9.8, 10.2, 10.4, 9.8, 10.0, 10.2, and 9.6 ounces. Find a 95% confidence interval for the mean of all such containers, assuming an approximate normal distribution.

Solution. The sample mean and standard deviation for the given data are

$$\bar{x} = 10.0 \quad \text{and} \quad s = 0.283.$$

Using Table V we find $t_{0.025} = 2.447$ for $\nu = 6$ degrees of freedom. Hence the 95% confidence interval for μ is

$$10.0 - (2.447)(0.283/\sqrt{7}) < \mu < 10.0 + (2.447)(0.283/\sqrt{7}),$$

which reduces to

$$9.74 < \mu < 10.26.$$

6.4 Estimating the Difference Between Two Means

If we have two populations with means μ_1 and μ_2 and variances σ_1^2 and σ_2^2, respectively, a point estimator of the difference between μ_1 and μ_2 is given by the statistic $\bar{X}_1 - \bar{X}_2$. Therefore, to obtain a point estimate of $\mu_1 - \mu_2$, we shall select two independent random samples, one from each population, of size n_1 and n_2, and compute the difference, $\bar{x}_1 - \bar{x}_2$, of the sample means.

If our independent samples are selected from normal populations, or failing this, if n_1 and n_2 are both greater than 30, we can establish a confidence interval for $\mu_1 - \mu_2$ by considering the sampling distribution of $\bar{X}_1 - \bar{X}_2$.

According to Theorem 5.15 we can expect the sampling distribution of $\bar{X}_1 - \bar{X}_2$ to be approximately normally distributed with mean $\mu_{\bar{X}_1 - \bar{X}_2} =$

$\mu_1 - \mu_2$ and standard deviation $\sigma_{\bar{X}_1 - \bar{X}_2} = \sqrt{(\sigma_1^2/n_1) + (\sigma_2^2/n_2)}$. Therefore we can assert with a probability of $1 - \alpha$ that the standard normal variable

$$Z = \frac{(\bar{X}_1 - \bar{X}_2) - (\mu_1 - \mu_2)}{\sqrt{(\sigma_1^2/n_1) + (\sigma_2^2/n_2)}}$$

will fall between $-z_{\alpha/2}$ and $z_{\alpha/2}$. Referring once again to Figure 6.2, we write

$$Pr(-z_{\alpha/2} < Z < z_{\alpha/2}) = 1 - \alpha.$$

Substituting for Z, we state equivalently that

$$Pr\left[-z_{\alpha/2} < \frac{(\bar{X}_1 - \bar{X}_2) - (\mu_1 - \mu_2)}{\sqrt{(\sigma_1^2/n_1) + (\sigma_2^2/n_2)}} < z_{\alpha/2}\right] = 1 - \alpha.$$

Multiplying each term of the inequality by $\sqrt{(\sigma_1^2/n_1) + (\sigma_2^2/n_2)}$, and then subtracting $\bar{X}_1 - \bar{X}_2$ from each term and multiplying by -1, we have

$$Pr\left[(\bar{X}_1 - \bar{X}_2) - z_{\alpha/2}\sqrt{\frac{\sigma_1^2}{n_1} + \frac{\sigma_2^2}{n_2}} < \mu_1 - \mu_2 < (\bar{X}_1 - \bar{X}_2) \right.$$
$$\left. + z_{\alpha/2}\sqrt{\frac{\sigma_1^2}{n_1} + \frac{\sigma_2^2}{n_2}}\right] = 1 - \alpha.$$

For any two independent random samples of size n_1 and n_2 selected from two populations whose variances σ_1^2 and σ_2^2 are known, the difference of the sample means, $\bar{x}_1 - \bar{x}_2$, is computed and the $(1 - \alpha)100\%$ confidence interval is given by

$$(\bar{x}_1 - \bar{x}_2) - z_{\alpha/2}\sqrt{\frac{\sigma_1^2}{n_1} + \frac{\sigma_2^2}{n_2}} < \mu_1 - \mu_2 < (\bar{x}_1 - \bar{x}_2) + z_{\alpha/2}\sqrt{\frac{\sigma_1^2}{n_1} + \frac{\sigma_2^2}{n_2}}.$$

CONFIDENCE INTERVAL FOR $\mu_1 - \mu_2$; σ_1^2 AND σ_2^2 KNOWN $A\ (1 - \alpha)100\%$
confidence interval for $\mu_1 - \mu_2$ is

$$(\bar{x}_1 - \bar{x}_2) - z_{\alpha/2}\sqrt{\frac{\sigma_1^2}{n_1} + \frac{\sigma_2^2}{n_2}} < \mu_1 - \mu_2 < (\bar{x}_1 - \bar{x}_2) + z_{\alpha/2}\sqrt{\frac{\sigma_1^2}{n_1} + \frac{\sigma_2^2}{n_2}},$$

where $\bar{x}_1$ and $\bar{x}_2$ are means of independent random samples of size n_1 and n_2 from populations with known variances σ_1^2 and σ_2^2, respectively, and $z_{\alpha/2}$ is the value of the standard normal curve leaving an area of $\alpha/2$ to the right.

The degree of confidence is exact when samples are selected from normal populations. For nonnormal populations we obtain an approximate confidence interval that is very good when both n_1 and n_2 exceed 30. As before, if σ_1^2 and σ_2^2 are unknown and our samples are sufficiently large, we may replace σ_1^2 by s_1^2 and σ_2^2 by s_2^2 without appreciably affecting the confidence interval.

Example 6.5 A standardized chemistry test was given to 50 girls and 75 boys. The girls made an average grade of 76 with a standard deviation of 6, while the boys made an average grade of 82 with a standard deviation of 8. Find a 96% confidence interval for the difference $\mu_1 - \mu_2$, where μ_1 is the mean score of all boys and μ_2 is the mean score of all girls who might take this test.

Solution. The point estimate of $\mu_1 - \mu_2$ is $\bar{x}_1 - \bar{x}_2 = 82 - 76 = 6$. Since n_1 and n_2 are both large, we can substitute $s_1 = 8$ for σ_1 and $s_2 = 6$ for σ_2. Using $\alpha = 0.04$, we find $z_{0.02} = 2.054$ from Table IV. Hence, substitution in the formula

$$(\bar{x}_1 - \bar{x}_2) - z_{\alpha/2}\sqrt{\frac{\sigma_1^2}{n_1} + \frac{\sigma_2^2}{n_2}} < \mu_1 - \mu_2 < (\bar{x}_1 - \bar{x}_2) + z_{\alpha/2}\sqrt{\frac{\sigma_1^2}{n_1} + \frac{\sigma_2^2}{n_2}}$$

yields the 96% confidence interval

$$6 - 2.054\sqrt{\tfrac{64}{75} + \tfrac{36}{50}} < \mu_1 - \mu_2 < 6 + 2.054\sqrt{\tfrac{64}{75} + \tfrac{36}{50}},$$

or

$$3.42 < \mu_1 - \mu_2 < 8.58.$$

This procedure for estimating the difference between two means is applicable if σ_1^2 and σ_2^2 are known or can be estimated from large samples. If the sample sizes are small we must again resort to the t distribution to provide confidence intervals that are valid when the populations are approximately normally distributed.

Let us now assume σ_1^2 and σ_2^2 are unknown and that n_1 and n_2 are small (<30). If $\sigma_1^2 = \sigma_2^2 = \sigma^2$, we obtain a standard normal variable in the form

$$Z = \frac{(\bar{X}_1 - \bar{X}_2) - (\mu_1 - \mu_2)}{\sqrt{\sigma^2[(1/n_1) + (1/n_2)]}}.$$

According to Theorem 5.16 the random variables $(n_1 - 1)S_1^2/\sigma^2$ and $(n_2 - 1)S_2^2/\sigma^2$ have chi-square distributions with $n_1 - 1$ and $n_2 - 1$ degrees of freedom, respectively. Furthermore, they are independent chi-square variables since the random samples were selected independently.

Consequently, their sum

$$V = \frac{(n_1 - 1)S_1^2}{\sigma^2} + \frac{(n_2 - 1)S_2^2}{\sigma^2} = \frac{(n_1 - 1)S_1^2 + (n_2 - 1)S_2^2}{\sigma^2}$$

has a chi-square distribution with $\nu = n_1 + n_2 - 2$ degrees of freedom.

Replacing Z and V in Theorem 5.17 by the preceding expressions, we obtain the statistic

$$T = \frac{(\bar{X}_1 - \bar{X}_2) - (\mu_1 - \mu_2)}{\sqrt{\sigma^2[(1/n_1) + (1/n_2)]}} \bigg/ \sqrt{\frac{(n_1 - 1)S_1^2 + (n_2 - 1)S_2^2}{\sigma^2(n_1 + n_2 - 2)}},$$

which has the t distribution with $\nu = n_1 + n_2 - 2$ degrees of freedom.

A point estimate of the unknown common variance σ^2 can be obtained by pooling the sample variances. Denoting the pooled estimator by S_p^2, we write

$$S_p^2 = \frac{(n_1 - 1)S_1^2 + (n_2 - 1)S_2^2}{n_1 + n_2 - 2}.$$

Substituting S_p^2 in the T statistic, we obtain the less cumbersome form

$$T = \frac{(\bar{X}_1 - \bar{X}_2) - (\mu_1 - \mu_2)}{S_p \sqrt{(1/n_1) + (1/n_2)}}.$$

Using the statistic T, we have

$$Pr(-t_{\alpha/2} < T < t_{\alpha/2}) = 1 - \alpha,$$

where $t_{\alpha/2}$ is the t value with $n_1 + n_2 - 2$ degrees of freedom, above which we find an area of $\alpha/2$. Substituting for T in the inequality we write

$$Pr\left[-t_{\alpha/2} < \frac{(\bar{X}_1 - \bar{X}_2) - (\mu_1 - \mu_2)}{S_p \sqrt{(1/n_1) + (1/n_2)}} < t_{\alpha/2} \right] = 1 - \alpha.$$

Multiplying each term of the inequality by $S_p \sqrt{(1/n_1) + (1/n_2)}$, and then subtracting $\bar{X}_1 - \bar{X}_2$ from each term and multiplying by -1, we obtain

$$Pr\left[(\bar{X}_1 - \bar{X}_2) - t_{\alpha/2}S_p \sqrt{\frac{1}{n_1} + \frac{1}{n_2}} < \mu_1 - \mu_2 < (\bar{X}_1 - \bar{X}_2) \right.$$

$$\left. + t_{\alpha/2}S_p \sqrt{\frac{1}{n_1} + \frac{1}{n_2}} \right] = 1 - \alpha.$$

For any two independent random samples of size n_1 and n_2 selected from two normal populations, the difference of the sample means, $\bar{x}_1 - \bar{x}_2$, and the pooled standard deviation, s_p, are computed and the $(1 - \alpha)100\%$ confidence interval is given by

$$(\bar{x}_1 - \bar{x}_2) - t_{\alpha/2}s_p \sqrt{\frac{1}{n_1} + \frac{1}{n_2}} < \mu_1 - \mu_2 < (\bar{x}_1 - \bar{x}_2) + t_{\alpha/2}s_p \sqrt{\frac{1}{n_1} + \frac{1}{n_2}}.$$

SMALL SAMPLE CONFIDENCE INTERVAL FOR $\mu_1 - \mu_2$; $\sigma_1^2 = \sigma_2^2$ BUT UNKNOWN $A(1 - \alpha)100\%$ *confidence interval for* $\mu_1 - \mu_2$ *is*

$$(\bar{x}_1 - \bar{x}_2) - t_{\alpha/2}s_p \sqrt{\frac{1}{n_1} + \frac{1}{n_2}} < \mu_1 - \mu_2 < (\bar{x}_1 - \bar{x}_2)$$

$$+ t_{\alpha/2}s_p \sqrt{\frac{1}{n_1} + \frac{1}{n_2}},$$

where $\bar{x}_1$ *and* $\bar{x}_2$ *are the means of small independent samples of size* n_1 *and* n_2, *respectively, from approximate normal distributions,* s_p *is the pooled standard deviation, and* $t_{\alpha/2}$ *is the value of the t distribution with* $\nu = n_1 + n_2 - 2$ *degrees of freedom, leaving an area of* $\alpha/2$ *to the right.*

Example 6.6 In a batch chemical process, two catalysts are being compared for their effect on the output of the process reaction. A sample of 12 batches are prepared using catalyst 1 and a sample of 10 batches were obtained using catalyst 2. The 12 batches for which catalyst 1 was used gave an average yield of 85 with a sample standard deviation of 4, while the average for the second sample gave an average of 81 and a sample standard deviation of 5. Find a 90% confidence interval for the difference between the population means, assuming the populations are approximately normally distributed with equal variances.

Solution. Let μ_1 and μ_2 represent the population means of all yields using catalyst 1 and catalyst 2, respectively. We wish to find a 90% confidence interval for $\mu_1 - \mu_2$. Our point estimate of $\mu_1 - \mu_2$ is $\bar{x}_1 - \bar{x}_2 = 85 - 81 = 4$. The pooled estimate, s_p^2, of the common variance, σ^2, is

$$s_p^2 = \frac{(n_1 - 1)s_1^2 + (n_2 - 1)s_2^2}{n_1 + n_2 - 2}$$

$$= \frac{(11)(16) + (9)(25)}{12 + 10 - 2} = 20.05.$$

Taking the square root, $s_p = 4.478$. Using $\alpha = 0.1$, we find in Table V that $t_{0.05} = 1.725$ for $\nu = n_1 + n_2 - 2 = 20$ degrees of freedom. Therefore, substituting in the formula

$$(\bar{x}_1 - \bar{x}_2) - t_{\alpha/2}s_p \sqrt{\frac{1}{n_1} + \frac{1}{n_2}} < \mu_1 - \mu_2 < (\bar{x}_1 - \bar{x}_2) + t_{\alpha/2}s_p \sqrt{\frac{1}{n_1} + \frac{1}{n_2}}$$

we obtain the 90% confidence interval

$$4 - (1.725)(4.478) \sqrt{\tfrac{1}{12} + \tfrac{1}{10}} < \mu_1 - \mu_2 < 4$$
$$+ (1.725)(4.478) \sqrt{\tfrac{1}{12} + \tfrac{1}{10}},$$

which simplifies to

$$0.69 < \mu_1 - \mu_2 < 7.31.$$

Hence, we are 90% confident that the interval from 0.69 to 7.31 contains the true difference of the yields for the two catalysts. The fact that both confidence limits are positive indicates that catalyst 1 is superior to catalyst 2.

The procedure for constructing confidence intervals for $\mu_1 - \mu_2$ from small samples assumes the populations to be normal and the population variances to be equal. Slight departures from either of these assumptions do not seriously alter the degree of confidence for our interval. A procedure will be presented in Chapter 7 for testing the equality of two unknown population variances based on the information provided by the sample variances. If the population variances are considerably different, we still obtain good results when the populations are normal, provided $n_1 = n_2$. Therefore, in a planned experiment, one should make every effort to equalize the size of the samples.

Let us now consider the problem of finding an interval estimate of $\mu_1 - \mu_2$ for small samples when the unknown population variances are not likely to be equal, and it is impossible to select samples of equal size. The statistic most often used in this case is

$$T' = \frac{(\bar{X}_1 - \bar{X}_2) - (\mu_1 - \mu_2)}{\sqrt{(S_1^2/n_1) + (S_2^2/n_2)}},$$

which has approximately a t distribution with ν degrees of freedom, where

$$\nu = \frac{(s_1^2/n_1 + s_2^2/n_2)^2}{[(s_1^2/n_1)^2/(n_1 - 1)] + [(s_2^2/n_2)^2/(n_2 - 1)]}.$$

Since ν is seldom an integer, we round it off to the nearest whole number.

Using the statistic T', we write

$$Pr(-t_{\alpha/2} < T' < t_{\alpha/2}) \simeq 1 - \alpha,$$

where $t_{\alpha/2}$ is the value of the t distribution with ν degrees of freedom, above which we find an area of $\alpha/2$. Substituting for T' in the inequality, and following the exact steps as before, we state the final result.

SMALL SAMPLE CONFIDENCE INTERVAL FOR $\mu_1 - \mu_2$; $\sigma_1^2 \neq \sigma_2^2$ AND UNKNOWN *An approximate $(1 - \alpha)100\%$ confidence interval for $\mu_1 - \mu_2$ is*

$$(\bar{x}_1 - \bar{x}_2) - t_{\alpha/2}\sqrt{\frac{s_1^2}{n_1} + \frac{s_2^2}{n_2}} < \mu_1 - \mu_2 < (\bar{x}_1 - \bar{x}_2) + t_{\alpha/2}\sqrt{\frac{s_1^2}{n_1} + \frac{s_2^2}{n_2}},$$

where $\bar{x}_1$ and s_1^2, and $\bar{x}_2$ and s_2^2, are the means and variances of small independent samples of size n_1 and n_2, respectively, from approximate normal distributions, and $t_{\alpha/2}$ is the value of the t distribution with

$$\nu = \frac{(s_1^2/n_1 + s_2^2/n_2)^2}{[(s_1^2/n_1)^2/(n_1 - 1)] + [(s_2^2/n_2)^2/(n_2 - 1)]}$$

degrees of freedom, leaving an area of $\alpha/2$ to the right.

Example 6.7 Records for the past 15 years have shown the average rainfall in a certain region of the country for the month of May to be 1.94 inches with a standard deviation of 0.45 inch. A second region of the country has had an average rainfall in May of 1.04 inches of rain with a standard deviation of 0.26 inch during the past 10 years. Find a 95% confidence interval for the difference of the true average rainfalls in these two regions, assuming the observations came from normal populations with different variances.

Solution. For the first region we have $\bar{x}_1 = 1.94$, $s_1 = 0.45$, and $n_1 = 15$. For the second region $\bar{x}_2 = 1.04$, $s_2 = 0.26$, and $n_2 = 10$. We wish to find a 95% confidence interval for $\mu_1 - \mu_2$. Since the population variances are assumed to be unequal and our sample sizes are not the same, we can only find an approximate 95% confidence interval based on the t distribution with ν degrees of freedom, where

$$\nu = \frac{(s_1^2/n_1 + s_2^2/n_2)^2}{[(s_1^2/n_1)^2/(n_1 - 1)] + [(s_2^2/n_2)^2/(n_2 - 1)]}$$

$$= \frac{(0.2025/15 + 0.0676/10)^2}{[(0.2025/15)^2/14] + [(0.0676/10)^2/9]}$$

$$= 22.7 \simeq 23.$$

Our point estimate of $\mu_1 - \mu_2$ is $\bar{x}_1 - \bar{x}_2 = 1.94 - 1.04 = 0.90$. Using $\alpha = 0.05$, we find in Table V that $t_{0.025} = 2.069$ for $\nu = 23$ degrees of freedom. Therefore, substituting in the formula

$$(\bar{x}_1 - \bar{x}_2) - t_{\alpha/2} \sqrt{\frac{s_1^2}{n_1} + \frac{s_2^2}{n_2}} < \mu_1 - \mu_2 < (\bar{x}_1 - \bar{x}_2) + t_{\alpha/2} \sqrt{\frac{s_1^2}{n_1} + \frac{s_2^2}{n_2}}$$

we obtain the approximate 95% confidence interval

$$0.90 - 2.069 \sqrt{\frac{0.2025}{15} + \frac{0.0676}{10}} < \mu_1 - \mu_2 < 0.90$$
$$+ 2.069 \sqrt{\frac{0.2025}{15} + \frac{0.0676}{10}},$$

which simplifies to

$$0.61 < \mu_1 - \mu_2 < 1.19.$$

Hence we are 95% confident that the interval from 0.61 to 1.19 contains the true difference of the average rainfall for the two regions.

We conclude this section by considering estimation procedures for the difference of two means when the samples are not independent and the variances of the two populations are not necessarily equal. This will be true if the observations in the two samples occur in pairs so that the two observations are related. For instance, if we run a test on a new diet using 15 individuals, the weights before and after completion of the test form our two samples. Observations in the two samples made on the same individual are related and hence form a pair. To determine if the diet is effective, we must consider the differences d_i of paired observations. These differences are the values of a random sample $D_1, D_2, \ldots, D_n$ from a population that we shall assume to be normal with mean μ_D and unknown variance σ_D^2. We estimate σ_D^2 by s_d^2, the variance of the differences constituting the sample. Therefore s_d^2 is a value of the statistic S_d^2 that fluctuates from sample to sample. The point estimator of $\mu_1 - \mu_2 = \mu_D$ is given by $\bar{D}$.

A $(1 - \alpha)100\%$ confidence interval for μ_D can be established by writing

$$Pr(-t_{\alpha/2} < T < t_{\alpha/2}) = 1 - \alpha,$$

where

$$T = \frac{\bar{D} - \mu_D}{S_d/\sqrt{n}}$$

and $t_{\alpha/2}$, as before, is a value of the t distribution with $n - 1$ degrees of freedom.

It is now a routine procedure to replace T, by its definition, in the above inequality and carry out the mathematical steps that lead to the $(1 - \alpha)100\%$ confidence interval:

$$\bar{d} - t_{\alpha/2}\frac{s_d}{\sqrt{n}} < \mu_D < \bar{d} + t_{\alpha/2}\frac{s_d}{\sqrt{n}}.$$

CONFIDENCE INTERVAL FOR $\mu_1 - \mu_2 = \mu_D$ FOR PAIRED OBSERVATIONS
$A (1 - \alpha)100\%$ *confidence interval for* μ_D *is*

$$\bar{d} - t_{\alpha/2}\frac{s_d}{\sqrt{n}} < \mu_D < \bar{d} + t_{\alpha/2}\frac{s_d}{\sqrt{n}},$$

where $\bar{d}$ and s_d are the mean and standard deviation of the differences of n pairs of measurements and $t_{\alpha/2}$ is the value of the t distribution with $\nu = n - 1$ degrees of freedom, leaving an area of $\alpha/2$ to the right.

Example 6.8 Twenty college freshmen were divided into 10 pairs, each member of the pair having approximately the same I.Q. One of each pair was selected at random and assigned to a mathematics section using programmed materials only. The other member of each pair was assigned to a section in which the professor lectured. At the end of the semester each group was given the same examination and the following results were recorded:

Pair	Programmed materials	Lectures	d
1	76	81	−5
2	60	52	8
3	85	87	−2
4	58	70	−12
5	91	86	5
6	75	77	−2
7	82	90	−8
8	64	63	1
9	79	85	−6
10	88	83	5

Find a 98% confidence interval for the true difference in the two learning procedures.

Solution. We wish to find a 98% confidence interval for $\mu_1 - \mu_2$, where μ_1 and μ_2 represent the average grades of all students by the programmed and lecture method of presentation, respectively. Since the observations are paired, $\mu_1 - \mu_2 = \mu_D$. The point estimate of μ_D is given by $\bar{d} = -1.6$. The variance s_d^2 of the sample differences is

$$
\begin{aligned}
s_d^2 &= \frac{n\Sigma d_i^2 - (\Sigma d_i)^2}{n(n-1)} \\
&= \frac{(10)(392) - (-16)^2}{(10)(9)} = 40.7.
\end{aligned}
$$

Taking the square root, $s_d = 6.38$. Using $\alpha = 0.02$, we find in Table V that $t_{0.01} = 2.821$ for $\nu = n - 1 = 9$ degrees of freedom. Therefore, substituting in the formula

$$
\bar{d} - t_{\alpha/2} \frac{s_d}{\sqrt{n}} < \mu_D < \bar{d} + t_{\alpha/2} \frac{s_d}{\sqrt{n}}
$$

we obtain the 98% confidence interval

$$
-1.6 - (2.821) \frac{6.38}{\sqrt{10}} < \mu_D < -1.6 + (2.821) \frac{6.38}{\sqrt{10}}
$$

or simply

$$
-7.29 < \mu_D < 4.09.
$$

Hence we are 98% confident that the interval from -7.29 to 4.09 contains the true difference of the average grades for the two methods of instruction. Since this interval allows for the possibility of μ_D being equal to zero, we are unable to state that one method of instruction is better than the other even though this particular sample of differences shows the lecture procedure to be superior.

6.5 Estimating a Proportion

A point estimator of the proportion p in a binomial experiment is given by the statistic $\hat{P} = X/n$. Therefore, the sample proportion $\hat{p} = x/n$ will be used as the point estimate for the parameter p.

If the unknown proportion p is not expected to be too close to zero or 1, we can establish a confidence interval for p by considering the sampling distribution of $\hat{P}$, which, of course, is the same as that of the random variable X except for a change in scale. Hence, by Theorem 4.1, the distribution of $\hat{P}$ is approximately normally distributed with mean

$$\mu_{\hat{P}} = E(\hat{P}) = E\left(\frac{X}{n}\right) = \frac{np}{n} = p$$

and variance

$$\sigma_{\hat{P}}^2 = \sigma_{X/n}^2 = \frac{\sigma_X^2}{n^2} = \frac{npq}{n^2} = \frac{pq}{n}.$$

Therefore, we can assert that

$$Pr(-z_{\alpha/2} < Z < z_{\alpha/2}) = 1 - \alpha,$$

where

$$Z = \frac{\hat{P} - p}{\sqrt{pq/n}}$$

and $z_{\alpha/2}$ is the value of the standard normal curve above which we find an area of $\alpha/2$. Substituting for Z, we write

$$Pr\left(-z_{\alpha/2} < \frac{\hat{P} - p}{\sqrt{pq/n}} < z_{\alpha/2}\right) = 1 - \alpha.$$

Multiplying each term of the inequality by $\sqrt{pq/n}$, and then subtracting $\hat{P}$ and multiplying by -1, we obtain

$$Pr\left(\hat{P} - z_{\alpha/2}\sqrt{\frac{pq}{n}} < p < \hat{P} + z_{\alpha/2}\sqrt{\frac{pq}{n}}\right) = 1 - \alpha.$$

It is difficult to manipulate the inequalities so as to obtain a random interval whose end points are independent of p, the unknown parameter. When n is large, very little error is introduced by substituting the point estimate $\hat{p} = x/n$ for the p under the radical sign. Then we can write

$$Pr\left(\hat{P} - z_{\alpha/2}\sqrt{\frac{\hat{p}\hat{q}}{n}} < p < \hat{P} + z_{\alpha/2}\sqrt{\frac{\hat{p}\hat{q}}{n}}\right) \simeq 1 - \alpha.$$

For our particular random sample of size n, the sample proportion $\hat{p} = x/n$ is computed, and the approximate $(1 - \alpha)100\%$ confidence interval for p is given by

$$\hat{p} - z_{\alpha/2}\sqrt{\frac{\hat{p}\hat{q}}{n}} < p < \hat{p} + z_{\alpha/2}\sqrt{\frac{\hat{p}\hat{q}}{n}}.$$

CONFIDENCE INTERVAL FOR p; $n \geq 30$ $A\,(1 - \alpha)100\%$ *confidence interval for the binomial parameter p is approximately*

$$\hat{p} - z_{\alpha/2}\sqrt{\frac{\hat{p}\hat{q}}{n}} < p < \hat{p} + z_{\alpha/2}\sqrt{\frac{\hat{p}\hat{q}}{n}},$$

where $\hat{p}$ is the proportion of successes in a random sample of size n, $\hat{q} = 1 - \hat{p}$, and $z_{\alpha/2}$ is the value of the standard normal curve leaving an area of $\alpha/2$ to the right.

The methods for finding a confidence interval for the binomial parameter p is also applicable when the binomial distribution is being used to approximate the hypergeometric distribution, that is, when n is small relative to N as illustrated in Example 6.9.

Example 6.9 In a random sample of $n = 500$ families owning television sets in the city of Hamilton, Canada, it was found that $x = 160$ owned color sets. Find a 95% confidence interval for the actual proportion of families in this city with color sets.

Solution. The point estimate of p is $\hat{p} = 160/500 = 0.32$. Using Table IV we find $z_{0.025} = 1.96$. Therefore, substituting in the formula

$$\hat{p} - z_{\alpha/2}\sqrt{\frac{\hat{p}\hat{q}}{n}} < p < \hat{p} + z_{\alpha/2}\sqrt{\frac{\hat{p}\hat{q}}{n}},$$

we obtain the 95% confidence interval

$$0.32 - 1.96\sqrt{\frac{(0.32)(0.68)}{500}} < p < 0.32 + 1.96\sqrt{\frac{(0.32)(0.68)}{500}},$$

which simplifies to

$$0.28 < p < 0.36.$$

If p is the center value of a $(1 - \alpha)100\%$ confidence interval, then $\hat{p}$ estimates p without error. Most of the time, however, $\hat{p}$ will not be exactly equal to p and the point estimate is in error. The size of this error will be the difference between p and $\hat{p}$, and we can be $(1 - \alpha)100\%$ confident that this difference will be less than $z_{\alpha/2} \sqrt{\hat{p}\hat{q}/n}$. We can readily see this if we draw a diagram of the confidence interval as in Figure 6.5.

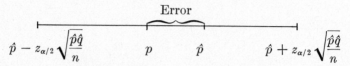

Figure 6.5 Error in estimating p by $\hat{p}$.

THEOREM 6.3 *If $\hat{p}$ is used as an estimate of p, we can be $(1 - \alpha)100\%$ confident that the error will be less than $z_{\alpha/2} \sqrt{\hat{p}\hat{q}/n}$.*

In Example 6.9 we are 95% confident that the sample proportion $\hat{p} = 0.32$ differs from the true proportion p by an amount less than 0.04.

Let us now determine how large a sample is necessary to ensure that the error in estimating p will be less than a specified amount e. By Theorem 6.3, this means we must choose n such that $z_{\alpha/2} \sqrt{\hat{p}\hat{q}/n} = e$.

THEOREM 6.4 *If $\hat{p}$ is used as an estimate of p, we can be $(1 - \alpha)100\%$ confident that the error will be less than a specified amount e when the sample size is*

$$n = \frac{z_{\alpha/2}^2 \hat{p}\hat{q}}{e^2}.$$

Theorem 6.4 is somewhat misleading in that we must use $\hat{p}$ to determine the sample size n, but $\hat{p}$ is computed from the sample. If a crude estimate of p can be made without taking a sample, we could use this value for $\hat{p}$ and then determine n. Lacking such an estimate, a preliminary sample of size $n \geq 30$ could be taken to provide an estimate of p. Then using Theorem 6.4 we could determine approximately how many observations are needed to provide the desired degree of accuracy.

An upper bound for n can be established for any degree of confidence by noting that $\hat{p}\hat{q} = \hat{p}(1 - \hat{p})$, which must be at most equal to $1/4$ since $\hat{p}$

must lie between 0 and 1. This fact may be verified by completing the square. Hence,

$$\hat{p}(1 - \hat{p}) = -(\hat{p}^2 - \hat{p}) = \tfrac{1}{4} - (\hat{p}^2 - \hat{p} + \tfrac{1}{4})$$
$$= \tfrac{1}{4} - (\hat{p} - \tfrac{1}{2})^2,$$

which is always less than 1/4 except when $\hat{p} = 1/2$ and then $\hat{p}\hat{q} = 1/4$.

THEOREM 6.5 *If $\hat{p}$ is used as an estimate of p, we can be at least $(1 - \alpha)100\%$ confident that the error will be less than a specified amount e when the sample size is*

$$n = \frac{z_{\alpha/2}^2}{4e^2}.$$

Example 6.10 How large a sample is required in Example 6.9 if we want to be (1) 95% confident that our estimate of p is off by less than 0.02 and (2) at least 95% confident?

Solution. (1) Let us treat the 500 families as a preliminary sample providing an estimate $\hat{p} = 0.32$. Then, by Theorem 6.4,

$$n = \frac{(1.96)^2(0.32)(0.68)}{(0.02)^2} = 2090.$$

Therefore, if we base our estimate of p on a random sample of size 2090, we can be 95% confident that our sample proportion will not differ from the true proportion by more than 0.02.

(2) According to Theorem 6.5, we can be at least 95% confident that our sample proportion will not differ from the true proportion by more than 0.02 if we choose a sample of size

$$n = \frac{(1.96)^2}{4(0.02)^2} = 2401.$$

Comparing the two parts of Example 6.10, we see that information concerning p, provided by a preliminary sample or perhaps from past experience, enables us to choose a smaller sample.

6.6 Estimating the Difference Between Two Proportions

Consider independent samples of size n_1 and n_2 selected at random from two binomial populations with means n_1p_1 and n_2p_2 and variances $n_1p_1q_1$ and $n_2p_2q_2$, respectively. We denote the proportion of successes in each sample by $\hat{p}_1$ and $\hat{p}_2$. A point estimator of the difference between the two proportions $p_1 - p_2$ is given by the statistic $\hat{P}_1 - \hat{P}_2$.

A confidence interval for $p_1 - p_2$ can be established by considering the sampling distribution of $\hat{P}_1 - \hat{P}_2$. From Section 6.5, we know that $\hat{P}_1$ and $\hat{P}_2$ are each approximately normally distributed, with means p_1 and p_2 and variances p_1q_1/n_1 and p_2q_2/n_2, respectively. By choosing independent samples from the two populations, the variables $\hat{P}_1$ and $\hat{P}_2$ will be independent, and then by the reproductive property of the normal distribution established in Theorem 5.11, we conclude that $\hat{P}_1 - \hat{P}_2$ is approximately normally distributed with mean

$$\mu_{\hat{P}_1-\hat{P}_2} = p_1 - p_2$$

and variance

$$\sigma^2_{\hat{P}_1-\hat{P}_2} = \frac{p_1q_1}{n_1} + \frac{p_2q_2}{n_2}.$$

Therefore, we can assert that

$$Pr(-z_{\alpha/2} < Z < z_{\alpha/2}) = 1 - \alpha,$$

where

$$Z = \frac{(\hat{P}_1 - \hat{P}_2) - (p_1 - p_2)}{\sqrt{(p_1q_1/n_1) + (p_2q_2/n_2)}}$$

and $z_{\alpha/2}$ is a value of the standard normal curve above which we find an area of $\alpha/2$. Substituting for Z, we write

$$Pr\left[-z_{\alpha/2} < \frac{(\hat{P}_1 - \hat{P}_2) - (p_1 - p_2)}{\sqrt{(p_1q_1/n_1) + (p_2q_2/n_2)}} < z_{\alpha/2}\right] = 1 - \alpha.$$

Multiplying each term of the inequality by $\sqrt{(p_1q_1/n_1) + (p_2q_2/n_2)}$, and then subtracting $\hat{P}_1 - \hat{P}_2$ and multiplying by -1, we obtain

$$Pr\left[(\hat{P}_1 - \hat{P}_2) - z_{\alpha/2}\sqrt{\frac{p_1q_1}{n_1} + \frac{p_2q_2}{n_2}} < p_1 - p_2 < (\hat{P}_1 - \hat{P}_2)\right.$$

$$\left. + z_{\alpha/2}\sqrt{\frac{p_1q_1}{n_1} + \frac{p_2q_2}{n_2}}\right] = 1 - \alpha.$$

If n_1 and n_2 are both large, we replace p_1 and p_2 under the radical sign by their estimates $\hat{p}_1 = x_1/n_1$ and $\hat{p}_2 = x_2/n_2$. Then

$$Pr\left[(\hat{P}_1 - \hat{P}_2) - z_{\alpha/2}\sqrt{\frac{\hat{p}_1\hat{q}_1}{n_1} + \frac{\hat{p}_2\hat{q}_2}{n_2}} < p_1 - p_2 < (\hat{P}_1 - \hat{P}_2)\right.$$

$$\left. + z_{\alpha/2}\sqrt{\frac{\hat{p}_1\hat{q}_1}{n_1} + \frac{\hat{p}_2\hat{q}_2}{n_2}}\right] \simeq 1 - \alpha.$$

For any two independent random samples of size n_1 and n_2, selected from two binomial populations, the difference of the sample proportions, $\hat{p}_1 - \hat{p}_2$, is computed and the $(1 - \alpha)100\%$ confidence interval is given by

$$(\hat{p}_1 - \hat{p}_2) - z_{\alpha/2}\sqrt{\frac{\hat{p}_1\hat{q}_1}{n_1} + \frac{\hat{p}_2\hat{q}_2}{n_2}} < p_1 - p_2 < (\hat{p}_1 - \hat{p}_2)$$

$$+ z_{\alpha/2}\sqrt{\frac{\hat{p}_1\hat{q}_1}{n_1} + \frac{\hat{p}_2\hat{q}_2}{n_2}}.$$

CONFIDENCE INTERVAL FOR $p_1 - p_2$; n_1 AND $n_2 \geq 30$ *A $(1 - \alpha)100\%$ confidence interval for the difference of two binomial parameters, $p_1 - p_2$, is approximately*

$$(\hat{p}_1 - \hat{p}_2) - z_{\alpha/2}\sqrt{\frac{\hat{p}_1\hat{q}_1}{n_1} + \frac{\hat{p}_2\hat{q}_2}{n_2}} < p_1 - p_2 < (\hat{p}_1 - \hat{p}_2)$$

$$+ z_{\alpha/2}\sqrt{\frac{\hat{p}_1\hat{q}_1}{n_1} + \frac{\hat{p}_2\hat{q}_2}{n_2}},$$

where $\hat{p}_1$ and $\hat{p}_2$ are the proportion of successes in random samples of size n_1 and n_2, respectively, $\hat{q}_1 = 1 - \hat{p}_1$ and $\hat{q}_2 = 1 - \hat{p}_2$, and $z_{\alpha/2}$ is the value of the standard normal curve leaving an area of $\alpha/2$ to the right.

Example 6.11 A certain change in a manufacturing procedure for component parts is being considered. Samples are taken using both the existing and the new procedure in order to determine if the new procedure results in an improvement. If 75 of 1500 items from the existing procedure were found to be defective and 80 of 2000 items from the new procedure were found to be defective, find a 90% confidence interval for the true difference in the fraction of defectives between the existing and the new process.

Solution. Let p_1 and p_2 be the true proportions of defectives for the existing and new procedures, respectively. Hence $\hat{p}_1 = 75/1500 = 0.05$ and $\hat{p}_2 = 80/2000 = 0.04$, and the point estimate of $p_1 - p_2$ is $\hat{p}_1 - \hat{p}_2 = 0.05 - 0.04 = 0.01$. Using Table IV we find $z_{0.05} = 1.645$. Therefore, substituting into this formula we obtain the 90% confidence interval

$$0.01 - 1.645 \sqrt{\frac{(0.05)(0.95)}{1500} + \frac{(0.04)(0.96)}{2000}} < p_1 - p_2$$

$$< 0.01 + 1.645 \sqrt{\frac{(0.05)(0.95)}{1500} + \frac{(0.04)(0.96)}{2000}},$$

which simplifies to

$$-0.0017 < p_1 - p_2 < 0.0217.$$

Since the interval contains the value zero, there is no reason to believe that the new procedure produced a significant decrease in the proportion of defectives over the existing method.

6.7 Estimating the Variance

An unbiased point estimate of the population variance σ^2 is provided by the sample variance s^2. Hence, the statistic S^2 is called an estimator of σ^2.

An interval estimate of σ^2 can be established by using the statistic

$$X^2 = \frac{(n-1)S^2}{\sigma^2}.$$

According to Theorem 5.16, the statistic X^2 has a chi-square distribution with $n - 1$ degrees of freedom when samples are chosen from a normal

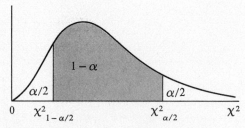

Figure 6.6 $Pr(\chi^2_{1-\alpha/2} < X^2 < \chi^2_{\alpha/2}) = 1 - \alpha.$

population. We may write (see Figure 6.6)

$$Pr(\chi^2_{1-\alpha/2} < X^2 < \chi^2_{\alpha/2}) = 1 - \alpha,$$

where $\chi^2_{1-\alpha/2}$ and $\chi^2_{\alpha/2}$ are values of the chi-square distribution with $n - 1$ degrees of freedom, leaving areas of $1 - \alpha/2$ and $\alpha/2$, respectively, to the right. Substituting for X^2, we write

$$Pr\left(\chi^2_{1-\alpha/2} < \frac{(n-1)S^2}{\sigma^2} < \chi^2_{\alpha/2}\right) = 1 - \alpha.$$

Dividing each term in the inequality by $(n - 1)S^2$, and then inverting each term (thereby changing the sense of the inequalities), we obtain

$$Pr\left[\frac{(n-1)S^2}{\chi^2_{\alpha/2}} < \sigma^2 < \frac{(n-1)S^2}{\chi^2_{1-\alpha/2}}\right] = 1 - \alpha.$$

For our particular sample of size n, the sample variance s^2 is computed, and the $(1 - \alpha)100\%$ confidence interval is given by

$$\frac{(n-1)s^2}{\chi^2_{\alpha/2}} < \sigma^2 < \frac{(n-1)s^2}{\chi^2_{1-\alpha/2}}.$$

CONFIDENCE INTERVAL FOR σ^2 *A $(1 - \alpha)100\%$ confidence interval for the variance σ^2 of a normal population is*

$$\frac{(n-1)s^2}{\chi^2_{\alpha/2}} < \sigma^2 < \frac{(n-1)s^2}{\chi^2_{1-\alpha/2}},$$

where s^2 is the variance of a random sample of size n, and $\chi^2_{\alpha/2}$ and $\chi^2_{1-\alpha/2}$ are the values of a chi-square distribution with $\nu = n - 1$ degrees of freedom leaving areas of $\alpha/2$ and $1 - \alpha/2$, respectively, to the right.

Example 6.12 The following are the weights, in ounces, of 10 packages of grass seed distributed by a certain company: 16.4, 16.1, 15.8, 17.0, 16.1, 15.9, 15.8, 16.9, 15.2, and 16.0. Find a 95% confidence interval for the variance of all such packages of grass seed distributed by this company.

Solution. First we find

$$s^2 = \frac{n\sum_{i=1}^{n} x_i^2 - \left(\sum_{i=1}^{n} x_i\right)^2}{n(n-1)} = \frac{(10)(2.72) - (1.2)^2}{(10)(9)}$$

$$= 0.286.$$

To obtain a 95% confidence interval we choose $\alpha = 0.05$. Then using Table VI with $\nu = 9$ degrees of freedom, we find $\chi^2_{0.025} = 19.023$ and $\chi^2_{0.975} = 2.700$. Substituting in the formula

$$\frac{(n-1)s^2}{\chi^2_{\alpha/2}} < \sigma^2 < \frac{(n-1)s^2}{\chi^2_{1-\alpha/2}}$$

we obtain the 95% confidence interval

$$\frac{(9)(0.286)}{19.023} < \sigma^2 < \frac{(9)(0.286)}{2.700},$$

or simply

$$0.135 < \sigma^2 < 0.953.$$

6.8 Estimating the Ratio of Two Variances

A point estimate of the ratio of two population variances σ_1^2/σ_2^2 is given by the ratio s_1^2/s_2^2 of the sample variances. Hence, the statistic S_1^2/S_2^2 is called an estimator of σ_1^2/σ_2^2.

If σ_1^2 and σ_2^2 are the variances of normal populations, we can establish an interval estimate of σ_1^2/σ_2^2 by using the statistic

$$F = \frac{\sigma_2^2 S_1^2}{\sigma_1^2 S_2^2}.$$

According to Theorem 5.20, the random variable F has an F distribution with $\nu_1 = n_1 - 1$ and $\nu_2 = n_2 - 1$ degrees of freedom. Therefore we may write (see Figure 6.7)

$$Pr[f_{1-\alpha/2}(\nu_1,\nu_2) < F < f_{\alpha/2}(\nu_1,\nu_2)] = 1 - \alpha,$$

where $f_{1-\alpha/2}(\nu_1,\nu_2)$ and $f_{\alpha/2}(\nu_1,\nu_2)$ are the values of the F distribution with

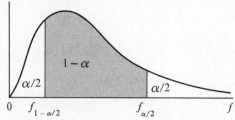

Figure 6.7 $Pr[f_{1-\alpha/2}(\nu_1,\nu_2) < F < f_{\alpha/2}(\nu_1,\nu_2)] = 1 - \alpha.$

ν_1 and ν_2 degrees of freedom, leaving areas of $1 - \alpha/2$ and $\alpha/2$, respectively, to the right. Substituting for F, we write

$$Pr\left[f_{1-\alpha/2}(\nu_1,\nu_2) < \frac{\sigma_2^2 S_1^2}{\sigma_1^2 S_2^2} < f_{\alpha/2}(\nu_1,\nu_2)\right] = 1 - \alpha.$$

Multiplying each term in the inequality by S_2^2/S_1^2, and then inverting each term (again changing the sense of the inequalities), we obtain

$$Pr\left[\frac{S_1^2}{S_2^2}\frac{1}{f_{\alpha/2}(\nu_1,\nu_2)} < \frac{\sigma_1^2}{\sigma_2^2} < \frac{S_1^2}{S_2^2}\frac{1}{f_{1-\alpha/2}(\nu_1,\nu_2)}\right] = 1 - \alpha.$$

The results of Theorem 5.19 enable us to replace $f_{1-\alpha/2}(\nu_1,\nu_2)$ by $1/f_{\alpha/2}(\nu_2,\nu_1)$. Therefore

$$Pr\left[\frac{S_1^2}{S_2^2}\frac{1}{f_{\alpha/2}(\nu_1,\nu_2)} < \frac{\sigma_1^2}{\sigma_2^2} < \frac{S_1^2}{S_2^2}f_{\alpha/2}(\nu_2,\nu_1)\right] = 1 - \alpha.$$

For any two independent random samples of size n_1 and n_2 selected from two normal populations, the ratio of the sample variances, s_1^2/s_2^2, is computed and the $(1 - \alpha)100\%$ confidence interval for σ_1^2/σ_2^2 is

$$\frac{s_1^2}{s_2^2}\frac{1}{f_{\alpha/2}(\nu_1,\nu_2)} < \frac{\sigma_1^2}{\sigma_2^2} < \frac{s_1^2}{s_2^2}f_{\alpha/2}(\nu_2,\nu_1).$$

CONFIDENCE INTERVAL FOR σ_1^2/σ_2^2 *A* $(1 - \alpha)100\%$ *confidence interval for the ratio* σ_1^2/σ_2^2 *is*

$$\frac{s_1^2}{s_2^2}\frac{1}{f_{\alpha/2}(\nu_1,\nu_2)} < \frac{\sigma_1^2}{\sigma_2^2} < \frac{s_1^2}{s_2^2}f_{\alpha/2}(\nu_2,\nu_1),$$

where s_1^2 and s_2^2 are the variances of independent samples of size n_1 and n_2, respectively, from normal populations, $f_{\alpha/2}(\nu_1,\nu_2)$ is a value of the F distribution with $\nu_1 = n_1 - 1$ and $\nu_2 = n_2 - 1$ degrees of freedom leaving an area of $\alpha/2$ to the right, and $f_{\alpha/2}(\nu_2,\nu_1)$ is a similar f value with $\nu_2 = n_2 - 1$ and $\nu_1 = n_1 - 1$ degrees of freedom.

Example 6.13 A standardized placement test in mathematics was given to 25 boys and 16 girls. The boys made an average grade of 82 with a standard deviation of 8, while the girls made an average grade of 78 with a standard deviation of 7. Find a 98% confidence interval for

σ_1^2/σ_2^2 and σ_1/σ_2, where σ_1^2 and σ_2^2 are the variances of the populations of grades for all boys and girls, respectively, who at some time have taken or will take this test.

Solution. We have $n_1 = 25$, $n_2 = 16$, $s_1 = 8$, and $s_2 = 7$. For a 98% confidence interval, $\alpha = 0.02$. Using Table VII, $f_{0.01}(24,15) = 3.29$, and $f_{0.01}(15,24) = 2.89$. Substituting in the formula

$$\frac{s_1^2}{s_2^2}\frac{1}{f_{\alpha/2}(\nu_1,\nu_2)} < \frac{\sigma_1^2}{\sigma_2^2} < \frac{s_1^2}{s_2^2}f_{\alpha/2}(\nu_2,\nu_1),$$

we obtain the 98% confidence interval

$$\frac{64}{49}\left(\frac{1}{3.29}\right) < \frac{\sigma_1^2}{\sigma_2^2} < \frac{64}{49}(2.89),$$

which simplifies to

$$0.397 < \frac{\sigma_1^2}{\sigma_2^2} < 3.775.$$

Taking square roots of the confidence limits, a 98% confidence interval for σ_1/σ_2 is

$$0.630 < \frac{\sigma_1}{\sigma_2} < 1.943.$$

6.9 Bayesian Methods of Estimation

The classical methods of estimation that we have studied so far are based solely on information provided by the random sample. These methods essentially interpret probabilities as relative frequencies. For example, in arriving at a 95% confidence interval for μ, we interpret the statement $Pr(-1.96 < Z < 1.96) = 0.95$ to mean that 95% of the time in repeated experiments Z will fall between -1.96 and 1.96. Probabilities of this type that can be interpreted in the frequency sense will be referred to as *objective probabilities*. The Bayesian approach to statistical methods of estimation combines sample information with other available prior information that may appear to be pertinent. The probabilities associated with this prior information are called *subjective probabilities* in that they measure a person's *degree of belief* in a proposition. The person uses his own experience and knowledge as the basis for arriving at a subjective probability.

Consider the problem of finding a point estimate of the parameter θ for the population $f(x;\theta)$. The classical approach would be to take a random sample of size n and substitute the information provided by the sample into the appropriate estimator or decision function. Thus in the case of a binomial population $b(x;n,p)$ our estimate of the proportion of successes would be $\hat{p} = x/n$. Suppose, however, that additional information is given about θ, namely, that it is known to vary according to the probability distribution $f(\theta)$. That is, we are now assuming θ to be a value of a random variable Θ with probability distribution $f(\theta)$ and we wish to estimate the particular value θ for the population from which we selected our random sample. We define $f(\theta)$ to be the *prior distribution* for the unknown variable parameter Θ. Note that $f(\theta)$ expresses our degree of belief of the location of Θ prior to sampling. Bayesian techniques use the prior distribution $f(\theta)$ along with the joint distribution of the sample, $f(x_1,x_2,\ldots,x_n;\theta)$ to compute the *posterior distribution* $f(\theta \mid x_1,x_2,\ldots,x_n)$. The posterior distribution consists of information from both the subjective prior distribution and the objective sample distribution and expresses our degree of belief of the location of Θ after we have observed the sample.

Let us write $f(x_1,x_2,\ldots,x_n \mid \theta)$ instead of $f(x_1,x_2,\ldots,x_n;\theta)$ for the joint probability distribution of our sample whenever we wish to indicate that the parameter is also a random variable. The joint distribution of the sample $X_1, X_2, \ldots, X_n$ and the parameter Θ is then

$$f(x_1,x_2,\ldots,x_n,\theta) = f(x_1,x_2,\ldots,x_n \mid \theta)f(\theta)$$

from which we readily obtain the marginal distribution

$$g(x_1,x_2,\ldots,x_n) = \sum_\theta f(x_1,x_2,\ldots,x_n,\theta) \qquad \text{(discrete case)}$$

$$= \int_{-\infty}^{\infty} f(x_1,x_2,\ldots,x_n,\theta)\, d\theta \qquad \text{(continuous case)}.$$

Hence, the posterior distribution may be written

$$f(\theta \mid x_1,x_2,\ldots,x_n) = \frac{f(x_1,x_2,\ldots,x_n,\theta)}{g(x_1,x_2,\ldots,x_n)}.$$

DEFINITION 6.4 *The mean of the posterior distribution* $f(\theta \mid x_1,x_2,\ldots,x_n)$, *denoted by* θ^*, *is called the* Bayes estimate *for* θ.

Example 6.14 Using a random sample of size 2, estimate the proportion p of defectives produced by a machine when we assume our prior dis-

214

Estimation Theory [Ch. 6

tribution to be

p	0.1	0.2
$f(p)$	0.6	0.4

Solution. Let X be the number of defectives in our sample. Then the probability distribution for our sample is

$$f(x \mid p) = b(x;n,p) = \binom{2}{x} p^x q^{2-x}, \qquad x = 0, 1, 2.$$

From the fact that $f(x,p) = f(x \mid p)f(p)$, we can set up the following table:

$f(x,p)$	x		
	0	*1*	*2*
p			
0.1	0.486	0.108	0.006
0.2	0.256	0.128	0.016

The marginal distribution for X is then

x	0	1	2
$g(x)$	0.742	0.236	0.022

We obtain the posterior distribution from the formula $f(p \mid x) = f(x,p)/g(x)$. Hence, we have

p	0.1	0.2
$f(p \mid x = 0)$	0.655	0.345

p	0.1	0.2
$f(p \mid x = 1)$	0.458	0.542

p	0.1	0.2
$f(p \mid x = 2)$	0.273	0.727

from which we get

$$
\begin{aligned}
p^* &= (0.1)(0.655) + (0.2)(0.345) = 0.1345, &&\text{if } x = 0 \\
&= (0.1)(0.458) + (0.2)(0.542) = 0.1542, &&\text{if } x = 1 \\
&= (0.1)(0.273) + (0.2)(0.727) = 0.1727, &&\text{if } x = 2.
\end{aligned}
$$

Example 6.15 Repeat Example 6.14 using the uniform prior distribution $f(p) = 1, 0 < p < 1$.

Solution. As before, we find

$$f(x \mid p) = \binom{2}{x} p^x q^{2-x}, \qquad x = 0, 1, 2.$$

Now

$$
\begin{aligned}
f(x,p) &= f(x \mid p)f(p) \\
&= \binom{2}{x} p^x q^{2-x} \\
&= (1 - p)^2, && \text{for } x = 0, 0 < p < 1 \\
&= 2p(1 - p), && \text{for } x = 1, 0 < p < 1 \\
&= p^2, && \text{for } x = 2, 0 < p < 1
\end{aligned}
$$

and the marginal distribution for X is obtained by evaluating the integral

$$
\begin{aligned}
g(x) &= \int_0^1 (1 - p)^2 \, dp = \frac{1}{3}, && \text{for } x = 0 \\
&= \int_0^1 2p(1 - p) \, dp = \frac{1}{3}, && \text{for } x = 1 \\
&= \int_0^1 p^2 \, dp = \frac{1}{3}, && \text{for } x = 2.
\end{aligned}
$$

The posterior distribution is then

$$
\begin{aligned}
f(p \mid x) &= \frac{f(x,p)}{g(x)} \\
&= 3 \binom{2}{x} p^x q^{2-x} \\
&= 3(1 - p)^2, && x = 0, 0 < p < 1 \\
&= 6p(1 - p), && x = 1, 0 < p < 1 \\
&= 3p^2, && x = 2, 0 < p < 1
\end{aligned}
$$

from which we evaluate the point estimate of our parameter to be

$$
\begin{aligned}
p^* &= 3 \int_0^1 p(1 - p)^2 \, dp = \frac{1}{4}, && \text{if } x = 0 \\
&= 6 \int_0^1 p^2(1 - p) \, dp = \frac{1}{2}, && \text{if } x = 1 \\
&= 3 \int_0^1 p^2 \, dp = 1, && \text{if } x = 2.
\end{aligned}
$$

Comparing these estimates with the values obtained by classical procedures, we see that p^* and $\hat{p}$ are equivalent if $x = 1$ or $x = 2$ but that
$\hat{p} = 0$ for $x = 0$.

A $(1 - \alpha)100\%$ Bayesian interval for the parameter θ can be constructed by finding an interval centered at the posterior mean that contains $(1 - \alpha)100\%$ of the posterior probability.

DEFINITION 6.5 *The interval $a < \theta < b$ will be called a $(1 - \alpha)100\%$
Bayes interval for θ if*

$$\int_{\theta*}^{b} f(\theta \mid x_1, x_2, \ldots, x_n)\, d\theta = \int_{a}^{\theta*} f(\theta \mid x_1, x_2, \ldots, x_n)\, d\theta = \frac{1 - \alpha}{2}.$$

Example 6.16 An electrical firm manufactures light bulbs that have a
length of life that is approximately normally distributed with a standard
deviation of 100 hours. Prior experience leads us to assume μ to be a
value of a normal random variable M with a mean equal to 800 hours and
a standard deviation of 10 hours. If a random sample of 25 bulbs have an
average life of 780 hours, find a 95% Bayes interval for μ.

Solution. Multiplying the density of our sample

$$f(x_1, x_2, \ldots, x_{25} \mid \mu) = \frac{1}{(2\pi)^{25/2} \cdot 100^{25}} e^{-(1/2) \sum_{i=1}^{25} [(x_i - \mu)/100]^2},$$

$$-\infty < x_i < \infty,\ i = 1, 2, \ldots, 25$$

by our prior

$$f(\mu) = \frac{1}{\sqrt{2\pi} \cdot 10} e^{-(1/2)[(\mu - 800)/10]^2}, \qquad -\infty < \mu < \infty,$$

we obtain the joint density of the random sample and M. That is,

$$f(x_1, x_2, \ldots, x_{25}, \mu) = \frac{1}{(2\pi)^{13} \cdot 10^{51}} e^{-(1/2)\left\{ \sum_{i=1}^{25} [(x_i - \mu)/100]^2 + [(\mu - 800)/10]^2 \right\}}.$$

In Chapter 5, Section 5.6, we established the identity

$$\sum_{i=1}^{n} (x_i - \mu)^2 = \sum_{i=1}^{n} (x_i - \bar{x})^2 + n(\bar{x} - \mu)^2,$$

which enables us to write

$f(x_1,x_2,\ldots,x_{25},\mu)$

$$= \frac{1}{(2\pi)^{13} \cdot 10^{51}} e^{-(1/2) \sum\limits_{i=1}^{25} [(x_i-780)/100]^2} e^{-(1/2)\{25[(780-\mu)/100]^2+[(\mu-800)/10]^2\}}.$$

Completing the square in the second exponent we have

$$25\left(\frac{780-\mu}{100}\right)^2 + \left(\frac{\mu-800}{10}\right)^2 = \frac{\mu^2 - 1592\mu + 635{,}280}{80}$$

$$= \frac{(\mu-796)^2 + 1664}{80}.$$

The joint density of the sample and M can now be written

$$f(x_1,x_2,\ldots,x_{25},\mu) = Ke^{-(1/2)[(\mu-796)/\sqrt{80}]^2},$$

where K is a function of the sample values. The marginal distribution of the sample is then

$$g(x_1,x_2,\ldots,x_{25}) = K\sqrt{2\pi}\,\sqrt{80}\int_{-\infty}^{\infty} \frac{1}{\sqrt{2\pi}\,\sqrt{80}} e^{-(1/2)[(\mu-796)/\sqrt{80}]^2}\,d\mu$$

$$= K\sqrt{2\pi}\,\sqrt{80},$$

and the posterior distribution is

$$f(\mu \mid x_1,x_2,\ldots,x_{25}) = \frac{f(x_1,x_2,\ldots,x_{25},\mu)}{g(x_1,x_2,\ldots,x_{25})}$$

$$= \frac{1}{\sqrt{2\pi}\,\sqrt{80}} e^{-(1/2)[(\mu-796)/\sqrt{80}]^2}, \qquad -\infty < \mu < \infty,$$

which is normal with mean $\mu^* = 796$ and standard deviation $\sigma^* = \sqrt{80}$. The 95% Bayes interval for μ is then given by

$$\mu^* - 1.96\sigma^* < \mu < \mu^* + 1.96\sigma^*.$$

That is,

$$796 - 1.96\sqrt{80} < \mu < 796 + 1.96\sqrt{80}$$

or

$$778.5 < \mu < 813.5.$$

Ignoring the prior information about μ and comparing this result with that given by the classical 95% confidence interval

$$780 - (1.96)\,\tfrac{100}{5} < \mu < 780 + (1.96)\,\tfrac{100}{5}$$

or

$$740.8 < \mu < 819.2,$$

we notice that the Bayes interval is shorter than the classical confidence interval.

6.10 Decision Theory

In our discussion of the classical approach to point estimation we adopted the criterion that selects the decision function that is most efficient. That is, we choose from all possible unbiased estimators the one with the smallest variance as our "best" estimator. In *decision theory* we also take into account the rewards for making correct decisions and the penalties for making incorrect decisions. This leads to a new criterion that chooses the decision function $\hat{\theta}$ that penalizes us the least when the action taken is incorrect. It is convenient now to introduce a *loss function* whose values depend on the true value of the parameter θ and the action $\hat{\theta}$. This is usually written in functional notation as $L(\hat{\theta};\theta)$. In many decision-making problems it is desirable to use a loss function of the form

$$L(\hat{\theta};\theta) = |\hat{\theta} - \theta|$$

or perhaps

$$L(\hat{\theta};\theta) = (\hat{\theta} - \theta)^2$$

in arriving at a choice between two or more decision functions.

Since θ is unknown, it must be assumed that it can equal any of several possible values. The set of all possible values that θ can assume is called the *parameter space*. For each possible value of θ in the parameter space, the loss function will vary from sample to sample. We define the *risk function* for the decision function $\hat{\theta}$ to be the expected value of the loss function when the value of the parameter is θ and denote this function by $R(\hat{\theta};\theta)$. Hence we have

$$R(\hat{\theta};\theta) = E[L(\hat{\theta};\theta)].$$

One method of arriving at a choice between $\hat{\theta}_1$ and $\hat{\theta}_2$ as an estimator for θ would be to apply the *minimax criterion*. Essentially we determine the maximum value of $R(\hat{\theta}_1;\theta)$ and the maximum value of $R(\hat{\theta}_2;\theta)$ in the parameter space and then choose the decision function that provided the minimum of these two maximum risks.

Example 6.17 According to the minimax criterion, is $\bar{X}$ or $\tilde{X}$ a better estimator of the mean μ of a normal population with known variance σ^2, based on a random sample of size n when the loss function is of the form $L(\hat{\theta};\theta) = (\hat{\theta} - \theta)^2$?

Solution. The loss function corresponding to $\bar{X}$ is given by

$$L(\bar{X};\mu) = (\bar{X} - \mu)^2.$$

Hence the risk function is

$$R(\bar{X};\mu) = E[(\bar{X} - \mu)^2] = \frac{\sigma^2}{n}$$

for every μ in the parameter space. Similarly, one can show that the risk function corresponding to $\tilde{X}$ is given by

$$R(\tilde{X};\mu) = E[(\tilde{X} - \mu)^2] \simeq \frac{\pi\sigma^2}{2n}$$

for every μ in the parameter space. In view of the fact that $\sigma^2/n < \pi\sigma^2/2n$, the minimax criterion selects $\bar{X}$ rather than $\tilde{X}$ as the better estimator for μ.

In some practical situations we may have additional information concerning the unknown parameter θ. For example, suppose we wish to estimate the binomial parameter p, the proportion of defectives produced by a machine during a certain day when we know that p varies from day to day. If we can write down the prior distribution $f(p)$, then it is possible to determine the expected value of the risk function for each decision function. The expected risk corresponding to the estimator $\hat{P}$, often referred to as the *Bayes risk*, is written $B(\hat{P}) = E[R(\hat{P};P)]$, where we are now treating the true proportion of defectives as a random variable. In general, when the unknown parameter is treated as a random variable with a prior distribution given by $f(\theta)$, the Bayes risk in estimating θ by means of the estimator $\hat{\theta}$ is given by

$$B(\hat{\theta}) = E[R(\hat{\theta};\theta)] = \sum_i R(\hat{\theta};\theta_i)f(\theta_i) \qquad \text{(discrete case)}$$

$$= \int_{-\infty}^{\infty} R(\hat{\theta};\theta)f(\theta) \, d\theta \qquad \text{(continuous case).}$$

The decision function $\hat{\theta}$ that minimizes $B(\hat{\theta})$ is the *optimal estimator* of θ. We shall make no attempt in this text to derive an optimal estimator, but instead we shall employ the Bayes risk to establish a criterion for choosing between two estimators.

> BAYES' CRITERION *Let $\hat{\Theta}_1$ and $\hat{\Theta}_2$ be two estimators of the unknown parameter θ, which may be looked upon as a value of the random variable Θ with probability distribution $f(\theta)$. If $B(\hat{\Theta}_1) < B(\hat{\Theta}_2)$, then $\hat{\Theta}_1$ is selected as the better estimator for θ.*

The foregoing discussion on decision theory might better be understood if one considers the following two examples.

Example 6.18 Suppose a friend has three similar coins except for the fact that the first one has two heads, the second one has two tails, and the third one is honest. We wish to estimate which coin our friend is flipping on the basis of two flips of the coin. Let θ be the number of heads on the coin. Consider two decision functions $\hat{\Theta}_1$ and $\hat{\Theta}_2$, where $\hat{\Theta}_1$ is the estimator that assigns to θ the number of heads that occur when the coin is flipped twice and $\hat{\Theta}_2$ is the estimator that assigns the value of 1 to θ no matter what the experiment yields. If the loss function is of the form $L(\hat{\Theta};\theta) = (\hat{\Theta} - \theta)^2$, which estimator is better according to the minimax procedure?

Solution. For the estimator $\hat{\Theta}_1$, the loss function assumes the values $L(\hat{\theta}_1;\theta) = (\hat{\theta}_1 - \theta)^2$, where $\hat{\theta}_1$ may be 0, 1, or 2, depending on the true value of θ. Clearly, if $\theta = 0$ or 2, both flips will yield all tails or all heads and our decision will be a correct one. Hence $L(0;0) = 0$ and $L(2;2) = 0$, from which one may easily conclude that $R(\hat{\theta}_1;0) = 0$ and $R(\hat{\theta}_1;2) = 0$. However, when $\theta = 1$ we could obtain 0, 1, or 2 heads in the two flips with probabilities 1/4, 1/2, and 1/4, respectively. In this case we have $L(0;1) = 1$, $L(1;1) = 0$, and $L(2;1) = 1$, from which we find

$$R(\hat{\theta}_1;1) = 1 \times \tfrac{1}{4} + 0 \times \tfrac{1}{2} + 1 \times \tfrac{1}{4} = \tfrac{1}{2}.$$

For the estimator $\hat{\Theta}_2$, the loss function assumes values given by $L(\hat{\theta}_2;\theta) = (\hat{\theta}_2 - \theta)^2 = (1 - \theta)^2$. Hence $L(1;0) = 1$, $L(1;1) = 0$, and $L(1;2) = 1$, and the corresponding risks are $R(\hat{\Theta}_2;0) = 1$, $R(\hat{\Theta}_2;1) = 0$, and $R(\hat{\Theta}_2;2) = 1$. Since the maximum risk is $1/2$ for the estimator $\hat{\Theta}_1$ compared to a maximum risk of 1 for $\hat{\Theta}_2$, the minimax criterion selects $\hat{\Theta}_1$ as the better of the two estimators.

Example 6.19 Referring to Example 6.18, let us suppose that our friend flips the honest coin 80% of the time and the other two coins each about 10% of the time. Does the Bayes criterion select $\hat{\Theta}_1$ or $\hat{\Theta}_2$ as the better estimator?

Solution. The parameter Θ may now be treated as a random variable with the following probability distribution:

θ	0	1	2
$f(\theta)$	0.1	0.8	0.1

For the estimator $\hat{\Theta}_1$, the Bayes risk is

$$B(\hat{\Theta}_1) = R(\hat{\Theta}_1;0)f(0) + R(\hat{\Theta}_1;1)f(1) + R(\hat{\Theta}_1;2)f(2)$$
$$= (0)(0.1) + \tfrac{1}{2}(0.8) + (0)(0.1) = 0.4.$$

Similarly, for the estimator $\hat{\Theta}_2$, we have

$$B(\hat{\Theta}_2) = R(\hat{\Theta}_2;0)f(0) + R(\hat{\Theta}_2;1)f(1) + R(\hat{\Theta}_2;2)f(2)$$
$$= (1)(0.1) + (0)(0.8) + (1)(0.1) = 0.2.$$

Since $B(\hat{\Theta}_2) < B(\hat{\Theta}_1)$, the Bayes criterion selects $\hat{\Theta}_2$ as the better estimator for the parameter θ.

EXERCISES

1. Let $S'^2 = \sum_{i=1}^{n} (X_i - \bar{X})^2/n$. Show that $E(S'^2) = [(n-1)/n]\sigma^2$, and hence S'^2 is a biased estimator for σ^2.

2. An electrical firm manufactures light bulbs that have a length of life that is approximately normally distributed with a standard deviation of 40 hours. If a random sample of 30 bulbs has an average life of 780 hours, find a 96% confidence interval for the population mean of all bulbs produced by this firm.

3. A soft-drink machine is regulated so that the amount of drink dispensed is approximately normally distributed with a standard deviation equal to 0.5 ounce. Find a 95% confidence interval for the mean of all drinks dispensed by this machine if a random sample of 36 drinks had an average content of 7.4 ounces.

4. The heights of a random sample of 50 college students showed a mean of 68.5 inches and a standard deviation of 2.7 inches.
   (a) Construct a 98% confidence interval for the mean height of all college students.
   (b) What can we assert with 98% confidence about the possible size of our error if we estimate the mean height of all college students to be 68.5 inches?

5. A random sample of 100 automobile owners shows that an automobile is driven on the average 14,500 miles per year, in the state of Virginia, with a standard deviation of 2400 miles.

(a) Construct a 99% confidence interval for the average number of miles an automobile is driven annually in Virginia.

(b) What can we assert with 99% confidence about the possible size of our error if we estimate the average number of miles driven by car owners in Virginia as 14,500 miles per year?

6. How large a sample is needed in Exercise 2 if we wish to be 96% confident that our sample mean will be within 10 hours of the true mean?

7. How large a sample is needed in Exercise 3 if we wish to be 95% confident that our sample mean will be within 0.3 ounce of the true mean?

8. An efficiency expert wishes to determine the average time that it takes to drill three holes in a certain metal clamp. How large a sample will he need to be 95% confident that his sample mean will be within 15 seconds of the true mean? Assume that it is known from previous studies that $\sigma = 40$ seconds.

9. A machine is producing metal pieces that are cylindrical in shape. A sample of pieces is taken and the diameters are 1.01, 0.97, 1.03, 1.04, 0.99, 0.98, 0.99, 1.01, and 1.03 inches. Find a 99% confidence interval for the mean diameter of pieces from this machine, assuming an approximate normal distribution.

10. A random sample of size 20 from a normal distribution has a mean $\bar{x} = 32.8$ and a standard deviation $s = 4.51$. Construct a 95% confidence interval for μ.

11. A random sample of eight cigarettes of a certain brand has an average nicotine content of 18.6 milligrams and a standard deviation of 2.4 milligrams. Construct a 99% confidence interval for the true average nicotine content of this particular brand of cigarettes, assuming an approximate normal distribution.

12. A random sample of 12 shearing pins are taken in a study of the Rockwell hardness of the head on the pin. Measurements on the Rockwell hardness were made for each of the 12, yielding an average value of 48.50 with a sample standard deviation of 1.5. Assuming the measurements to be normally distributed, construct a 90% confidence interval for the mean Rockwell hardness.

13. A random sample of size $n_1 = 25$ taken from a normal population with a standard deviation $\sigma_1 = 5$ has a mean $\bar{x}_1 = 80$. A second random sample of size $n_2 = 36$, taken from a different normal population with a standard deviation $\sigma_2 = 3$, has a mean $\bar{x}_2 = 75$. Find a 94% confidence interval for $\mu_1 - \mu_2$.

14. Two varieties of corn are being compared for yield. Fifty acres of each variety are planted and grown under similar conditions. Variety A yielded, on the average, 78.3 bushels per acre with a standard deviation of 5.6 bushels per acre, while variety B yielded, on the average, 87.2 bushels per arce with a standard deviation of 6.3 bushels per acre. Construct a 95% confidence interval for the difference of the population means.

15. A study was made to determine if a certain metal treatment has any effect on the amount of metal removed in a pickling operation. A random sample of 100 pieces was immersed in a bath for 24 hours without the treatment,

yielding an average of 0.0048 inch of metal removed and a sample standard deviation of 0.00041 inch. A second sample of 200 pieces was exposed to the treatment followed by the 24-hour immersion in the bath, resulting in an average removal of 0.0036 inch of metal with a sample standard deviation of 0.00034 inch. Compute a 98% confidence interval estimate for the difference between the population means. Does the treatment appear to reduce the mean amount of metal removed?

16. Given two random samples of size $n_1 = 9$ and $n_2 = 16$, from two independent normal populations, with $\bar{x}_1 = 64$, $\bar{x}_2 = 59$, $s_1 = 6$, and $s_2 = 5$, find a 95% confidence interval for $\mu_1 - \mu_2$, assuming that $\sigma_1 = \sigma_2$.

17. Students may choose between a 3-semester-hour course in physics without labs and a 4-semester-hour course with labs. The final written examination is the same for each section. If 12 students in the section with labs made an average examination grade of 84 with a standard deviation of 4 and 18 students in the section without labs made an average grade of 77 with a standard deviation of 6, find a 99% confidence interval for the difference between the average grades for the two courses. Assume the populations to be approximately normally distributed with equal variances.

18. A taxi company is trying to decide whether to purchase brand A or brand B tires for its fleet of taxis. To estimate the difference in the two brands, an experiment is conducted using 12 of each brand. The tires are run until they wear out. The results are brand A: $\bar{x}_1 = 22{,}500$ miles, $s_1 = 3100$ miles; brand B: $\bar{x}_2 = 23{,}600$ miles, $s_2 = 3800$ miles. Compute a 95% confidence interval for $\mu_1 - \mu_2$, assuming the populations to be approximately normally distributed.

19. The following data represent the running times of films produced by two different motion-picture companies.

	Time (minutes)						
Company I	103	94	110	87	98		
Company II	97	82	123	92	175	88	118

Compute a 90% confidence interval for the difference between the average running times of films produced by the two companies. Assume that the running times are approximately normally distributed.

20. The government awarded grants to the agricultural departments of nine different universities to test the yield capabilities of two new varieties of wheat. Five acres of each variety are planted at each university and the yields, in bushels per acre, recorded as follows:

	University								
	1	2	3	4	5	6	7	8	9
Variety 1	38	23	35	41	44	29	37	31	38
Variety 2	45	25	31	38	50	33	36	40	43

Find a 95% confidence interval for the mean difference between the yields
of the two varieties assuming the distributions of yields to be approximately
normal. Explain why pairing is necessary in this problem.

21. Referring to Exercise 18, find a 99% confidence interval for $\mu_1 - \mu_2$ if a
tire from each company is assigned at random to the rear wheels of eight
taxis and the following results recorded:

Taxi	Brand A	Brand B
1	21,400	22,800
2	28,300	29,100
3	22,800	23,400
4	19,900	19,300
5	30,100	29,700
6	20,400	22,600
7	23,700	24,200
8	18,700	19,600

22. It is claimed that a new diet will reduce a person's weight by 10 pounds
on the average in a period of 2 weeks. The weights of seven women who
followed this diet were recorded before and after a 2-week period.

	Woman						
	1	2	3	4	5	6	7
Weight before	129	133	136	152	141	138	125
Weight after	130	121	128	137	129	132	120

Test a manufacturer's claim by computing a 95% confidence interval for
the mean difference in the weight. Assume the distributions of weights to
be approximately normal.

23. (a) A random sample of 200 voters is selected and 114 are found to support
an annexation suit. Find the 96% confidence interval for the fraction
of the voting population favoring the suit.

   (b) What can we assert with 96% confidence about the possible size of
our error if we estimate the fraction of voters favoring the annexation
suit to be 0.57?

24. (a) A random sample of 500 cigarette smokers is selected and 86 are found
to have a preference for brand X. Find the 90% confidence interval for
the fraction of the population of cigarette smokers who prefer brand X.

   (b) What can we assert with 90% confidence about the possible size of
our error if we estimate the fraction of cigarette smokers who prefer
brand X to be 0.172?

25. Compute a 98% confidence interval for the proportion of defective items
in a process when it is found that a sample of size 100 yields 8 defectives.

26. A certain new rocket-launching system is being considered for deployment of small short-range launches. The existing system has $p = 0.8$ as the probability of a successful launch. A sample of 40 experimental launches is made with the new system and 34 are successful.
    (a) Give a point estimate of the probability of a successful launch using the new system.
    (b) Construct a 95% confidence interval for this probability.
    (c) Does the evidence strongly indicate that the new system is better? Explain.

27. How large a sample is needed in Exercise 23 if we wish to be 96% confident that our sample proportion will be within 0.02 of the true fraction of the voting population?

28. How large a sample is needed in Exercise 25 if we wish to be 98% confident that our sample proportion will be within 0.05 of the true proportion defective?

29. A study is to be made to estimate the percentage of citizens in a town who favor having their water fluoridated. How large a sample is needed if one wishes to be at least 95% confident that our estimate is within 1% of the true percentage?

30. A study is to be made to estimate the proportion of housewives who own an automatic dryer. How large a sample is needed if one wishes to be at least 99% confident that our estimate differs from the true proportion by an amount less than 0.01?

31. A certain geneticist is interested in the proportion of males and females in the population that have a certain minor blood disorder. In a random sample of 1000 males 250 are found to be afflicted, whereas 275 of 1000 females tested appear to have the disorder. Compute a 95% confidence interval for the difference between the proportion of males and females that have the blood disorder.

32. A cigarette-manufacturing firm claims that its brand A line of cigarettes outsells its brand B line by 8%. If it is found that 42 of 200 smokers prefer brand A and 18 of 150 smokers prefer brand B, compute a 94% confidence interval for the difference between the proportions of sales of the two brands and decide if the 8% difference is a valid claim.

33. A clinical trial is conducted to determine if a certain type of inoculation has an effect on the incidence of a certain disease. A sample of 1000 rats was kept in a controlled environment for a period of 1 year and 500 of the rats were given the inoculation. Of the group not given the drug, there were 120 incidences of the disease, while 98 of the inoculated group contracted it. If we call p_1 the probability of incidence of the disease in uninoculated rats and p_2 the probability of incidence after receiving the drug, compute a 90% confidence interval for $p_1 - p_2$.

34. A study is made to determine if a cold climate results in more students being absent from school during a semester than a warmer climate. Two groups of students are selected at random, one group from Vermont and the other group from Georgia. Of the 300 students from Vermont, 64 were

absent at least 1 day during the semester, and of the 400 students from Georgia, 51 were absent 1 or more days. Find a 95% confidence interval for the difference between the fractions of the students who are absent in the two states.

35. Construct a 99% confidence interval for σ^2 in Exercise 9.

36. Construct a 95% confidence interval for σ in Exercise 10.

37. Construct a 99% confidence interval for σ in Exercise 11.

38. Construct a 90% confidence interval for σ^2 in Exercise 12.

39. Construct a 98% confidence interval for σ_1/σ_2 in Exercise 16. Were we justified in assuming $\sigma_1 = \sigma_2$?

40. Construct a 90% confidence interval for σ_1^2/σ_2^2 in Exercise 18.

41. Construct a 90% confidence interval for σ_1^2/σ_2^2 in Exercise 19.

42. Using a random sample of size 2, estimate the proportion p of defectives produced by a machine when we assume the prior distribution to be

p	0.05	0.10	0.15
$f(p)$	0.3	0.5	0.2

43. Repeat Exercise 42 using the uniform prior distribution $f(p) = 10$, $0.05 < p < 0.15$.

44. The time of burn for the first stage of a rocket is a normal random variable with a standard deviation of 0.8 minute. Assume a normal prior distribution for μ with a mean of 8 minutes and a standard deviation of 0.2 minute. If 10 of these rockets are fired and the first stage has an average burning time of 9 minutes, find a 95% Bayes interval for μ.

45. Suppose that in Example 6.16 the electrical firm does not have enough prior information regarding the population mean length of life to be able to assume a normal distribution for μ. The firm believes, however, that μ is surely between 770 and 830 hours and it is felt that a more realistic Bayesian approach would be to assume the prior distribution

$$f(\mu) = \tfrac{1}{60}, \qquad 770 < \mu < 830.$$

If a random sample of 25 bulbs gives an average life of 780 hours, find the posterior distribution $f(\mu \mid x_1, x_2, \ldots, x_{25})$.

46. Suppose the time to failure T of a certain hinge is an exponential random variable with probability density

$$f(t) = \theta e^{-\theta t}, \qquad t > 0.$$

From prior experience we are led to believe that θ is a value of an exponential random variable with probability density

$$f(\theta) = 2e^{-2\theta}, \qquad \theta > 0.$$

If we have a sample of n observations on T, show that the posterior distribution of Θ is a gamma distribution with parameters $\alpha = n + 1$ and
$$\beta = 1 \Big/ \Big(\sum_{i=1}^{n} t_i + 2 \Big).$$

47. We wish to estimate the binomial parameter p by the decision function $\hat{P}$, the proportion of successes in a binomial experiment consisting of n trials. Find $R(\hat{P};p)$ when the loss function is of the form $L(\hat{P};p) = (\hat{P} - p)^2$.

48. Suppose an urn contains three balls, of which θ are red and the remainder black, where θ can vary from zero to 3. We wish to estimate θ by selecting two balls in succession without replacement. Let $\hat{\Theta}_1$ be the decision function that assigns to θ the value zero if neither ball is red, the value 1 if the first ball only is red, the value 2 if the second ball only is red, and the value 3 if both balls are red. Using a loss function of the form $L(\hat{\Theta}_1;\theta) = |\hat{\Theta}_1 - \theta|$, find $R(\hat{\Theta}_1;\theta)$.

49. In Exercise 48, consider the estimator $\hat{\Theta}_2 = X(X + 1)/2$, where X is the number of red balls in our sample. Find $R(\hat{\Theta}_2;\theta)$.

50. Use the minimax criterion to determine whether the estimator $\hat{\Theta}_1$ of Exercise 48 or the estimator $\hat{\Theta}_2$ of Exercise 49 is the better estimator.

51. Use the Bayes criterion to determine whether the estimator $\hat{\Theta}_1$ of Exercise 48 or the estimator $\hat{\Theta}_2$ of Exercise 49 is the better estimator, given the following additional information:

θ	0	1	2	3
$f(\theta)$	0.1	0.5	0.1	0.3

7 TESTS OF HYPOTHESES

7.1 Statistical Hypotheses

The testing of statistical hypotheses is perhaps the most important area of decision theory. First, let us define precisely what we mean by a statistical hypothesis.

> DEFINITION 7.1 *A statistical hypothesis* is an assumption or statement, which may or may not be true, concerning one or more populations.

The truth or falsity of a statistical hypothesis is never known with certainty unless we examine the entire population. This, of course, would be impractical in most situations. Instead, we take a random sample from the population of interest and use the information contained in this sample to decide whether the hypothesis is likely to be true or false. Evidence from the sample that is inconsistent with the stated hypothesis leads to a rejection of the hypothesis, whereas evidence supporting the hypothesis leads to its acceptance. We should make it clear at this point that the acceptance of a statistical hypothesis is a result of insufficient evidence to reject it and does not necessarily imply that it is true. For example, in tossing a coin 100 times we might test the hypothesis that the coin is balanced. In terms of population parameters, we are testing the hypothesis that the proportion of heads is $p = 0.5$ if the coin were tossed indefinitely. An outcome of 48 heads would not be surprising if the coin is balanced. Such a result would surely support the hypothesis $p = 0.5$. One might argue that such an occurrence is also consistent with the hypothesis that $p = 0.45$. Thus, in accepting the hypothesis, the only thing we can be reasonably certain about is that the true proportion of heads is not a great deal different from one half. If the 100 trials had resulted in only 35 heads, we would then have evidence to support the rejection of our hypothesis. In view of the fact that the probability of obtaining 35 or fewer heads in 100 tosses of a balanced coin is approximately 0.002, either a very rare event has occurred or we are right in concluding that $p \neq 0.5$.

Although we shall use the terms *accept* and *reject* frequently throughout this chapter, it is important to understand that the rejection of a hypothesis is to conclude that it is false, while the acceptance of a hypothesis merely implies that we have no evidence to believe otherwise. Because of this terminology, the statistician or experimenter should always state as his hypothesis that which he hopes to reject. If he is interested in a new cold vaccine, he should assume that it is no better than the vaccine now on the market and then set out to reject this contention. Similarly, to prove that one teaching technique is superior to another, we test the hypothesis that there is no difference in the two techniques.

Hypotheses that we formulate with the hope of rejecting are called *null hypotheses* and are denoted by H_0. The rejection of H_0 leads to the acceptance of an *alternative hypothesis* denoted by H_1. Hence, if H_0 is the null hypothesis $p = 0.5$ for a binomial population, the alternative hypothesis H_1 might be $p = 0.75$, $p > 0.5$, $p < 0.5$, or $p \neq 0.5$.

7.2 Type I and Type II Errors

To illustrate the concepts used in testing a statistical hypothesis about a population, consider the following example. A certain type of cold vaccine is known to be only 25% effective after a period of 2 years. To determine if a new and somewhat more expensive vaccine is superior in providing protection against the same virus for a longer period of time, 20 people are chosen at random and inoculated. If fewer than 12 of those receiving the new vaccine contract the virus within a 2-year period, the new vaccine will be considered superior to the one presently in use. We shall test the null hypothesis that the new vaccine is equally effective after a period of 2 years as the one now commonly used against the alternative hypothesis that the new vaccine is in fact superior. This is equivalent to testing the hypothesis that the binomial parameter for the probability of a success on a given trial is $p = 1/4$ against the alternative that $p > 1/4$. This is usually written as follows:

$$H_0: \quad p = 1/4$$

$$H_1: \quad p > 1/4.$$

This decision procedure could lead to either of two wrong conclusions. For instance, the new vaccine may be no better than the one now in use and, for this particular randomly selected group of individuals, 9 or more people may surpass the 2-year period without contracting the virus. We would be committing an error by rejecting H_0 in favor of H_1 when, in fact, H_0 is true. Such an error is called a *type I error*.

DEFINITION 7.2 *A type I error* has been committed if we reject the null hypothesis when it is true.

A second kind of error is committed if fewer than 9 of the group surpass the 2-year period successfully and we conclude that the new vaccine is no better when it actually is. In this case we would accept H_0 when it is false. This is called a *type II error*.

DEFINITION 7.3 *A type II error* has been committed if we accept the null hypothesis when it is false.

The statistic on which we base our decision is the number of individuals who receive protection from the new vaccine for a period of at least 2 years. The possible values, from zero to 20, are divided into two groups, namely, those numbers less than 9 and those greater than or equal to 9. All possible scores above 8.5 constitute the *critical region* and all possible scores below 8.5 determine the *acceptance region*. The number 8.5 separating these two regions is called the *critical value*. If the statistic on which we base our decision falls in the critical region, we reject H_0 in favor of the alternative hypothesis H_1. If it falls in the acceptance region, we accept H_0.

DEFINITION 7.4 *The probability of committing a type I error is called the* level of significance *of the test and is denoted by* α.

In our example, a type I error will occur when 9 or more individuals surpass the 2-year period without contracting the virus using a new vaccine that is actually equivalent to the one in use. Hence, if X is the number of individuals who remain healthy for at least 2 years,

$$\alpha = Pr(\text{type I error})$$
$$= Pr\left(X \geq 9 \mid p = \frac{1}{4}\right)$$
$$= \sum_{x=9}^{20} b\left(x;20,\frac{1}{4}\right)$$
$$= 1 - \sum_{x=0}^{8} b\left(x;20,\frac{1}{4}\right)$$
$$= 1 - 0.9591$$
$$= 0.0409.$$

We say that the null hypothesis, $p = 1/4$, is being tested at the $\alpha = 0.0409$ level of significance. Sometimes the level of significance is called the *size* of the critical region. A critical region of size 0.0409 is very small and therefore it is unlikely that a type I error will be committed. Consequently, it would be most unusual for 9 or more individuals to remain immune to a virus for a 2-year period using a new vaccine that is essentially equivalent to the one now on the market.

The probability of committing a type II error, denoted by β, is impossible to compute unless we have a specific alternative hypothesis. If we test the null hypothesis that $p = 1/4$ against the alternative hypothesis that $p = 1/2$, then we are able to compute the probability of accepting H_0 when it is false. We simply find the probability of obtaining fewer than 9 in the group that surpass the 2-year period when $p = 1/2$. In this case

$$\beta = Pr(\text{type II error})$$
$$= Pr\left(X < 9 \mid p = \frac{1}{2}\right)$$
$$= \sum_{x=0}^{8} b\left(x;20,\frac{1}{2}\right)$$
$$= 0.2517.$$

This is a rather high probability indicating a poor test procedure. It is quite likely that we will reject the new vaccine when, in fact, it is superior to that now in use. Ideally we like to use a test procedure for which both the type I and type II errors are small.

It is possible that the director of the testing program is willing to make a type II error if the more expensive vaccine is not significantly superior. The only time he wishes to guard against the type II error is when the true value of p is at least 0.7. Letting $p = 0.7$, this test procedure gives

$$\beta = Pr(\text{type II error})$$
$$= Pr(X < 9 \mid p = 0.7)$$
$$= \sum_{x=0}^{8} b(x;20,0.7)$$
$$= 0.0051.$$

With such a small probability of committing a type II error it is extremely unlikely that the new vaccine would be rejected when it is 70% effective after a period of 2 years. As the alternative hypothesis approaches unity, the value of β diminishes to zero.

Let us assume that the director of the testing program is unwilling to commit a type II error when the alternative hypothesis $p = 1/2$ is true even though we have found the probability of such an error to be $\beta = 0.2517$. A reduction in β is always possible by increasing the size of the critical region. For example, consider what happens to the values of α and β when we change our critical value to 7.5 so that all scores of 8 or more fall in the critical region and those below 8 fall in the acceptance region. Now, in testing $p = 1/4$ against the alternative hypothesis that $p = 1/2$, we find

$$\alpha = \sum_{x=8}^{20} b\left(x;20,\frac{1}{4}\right)$$
$$= 1 - \sum_{x=0}^{7} b\left(x;20,\frac{1}{4}\right)$$
$$= 1 - 0.8982$$
$$= 0.1018$$

and

$$\beta = \sum_{x=0}^{7} b\left(x;20,\frac{1}{2}\right)$$
$$= 0.1316.$$

By adopting a new decision procedure we have reduced the probability of committing a type II error at the expense of increasing the probability of committing a type I error. For a fixed sample size, a decrease in the probability of one error will usually result in an increase in the probability of the other error. Fortunately the probability of committing both types of error can be reduced by increasing the sample size. Consider the same problem using a random sample of 100 individuals. If 37 or more of the group surpass the 2-year period, we reject the null hypothesis that $p = 1/4$ and accept the alternative hypothesis $p > 1/4$. The critical value is now 36.5. All possible scores above 36.5 constitute the critical region and all possible scores below 36.5 fall in the acceptance region.

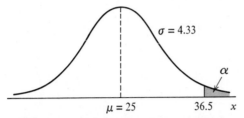

Figure 7.1 Probability of a type I error.

To determine the probability of committing a type I error we shall use the normal curve approximation with

$$\mu = np = (100)(\tfrac{1}{4}) = 25$$

and

$$\sigma = \sqrt{npq} = \sqrt{(100)(\tfrac{1}{4})(\tfrac{3}{4})} = 4.33.$$

Referring to Figure 7.1,

$$\alpha = Pr(\text{type I error})$$
$$= Pr(X > 36.5 \mid H_0 \text{ is true}).$$

The z value corresponding to $x = 36.5$ is

$$z = \frac{36.5 - 25}{4.33} = 2.656.$$

Therefore

$$\alpha = Pr(Z > 2.656)$$
$$= 1 - Pr(Z < 2.656)$$
$$= 1 - 0.9961$$
$$= 0.0039.$$

If H_0 is false and the true value of H_1 is $p = 1/2$, we can determine the probability of a type II error using the normal curve approximation with

$$\mu = np = (100)(\tfrac{1}{2}) = 50$$

and

$$\sigma = \sqrt{npq} = \sqrt{(100)(\tfrac{1}{2})(\tfrac{1}{2})} = 5.$$

The probability of falling in the acceptance region when H_1 is true is

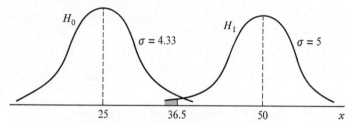

Figure 7.2 Probability of a type II error.

given by the area of the shaded region in Figure 7.2. Hence,

$$\beta = Pr(\text{type II error})$$
$$= Pr(X < 36.5 \mid H_1 \text{ is true}).$$

The z value corresponding to $x = 36.5$ is

$$z = \frac{36.5 - 50}{5} = -2.7.$$

Therefore

$$\beta = Pr(Z < -2.7)$$
$$= 0.0035.$$

Obviously, the type I and type II errors will rarely occur if the experiment consists of 100 individuals.

The concepts discussed above can easily be seen graphically when the population is continuous. Consider the null hypothesis that the average height of students in a certain college is 68 inches against the alternative hypothesis that it is unequal to 68. That is, we wish to test

$$H_0: \quad \mu = 68$$
$$H_1: \quad \mu \neq 68.$$

The alternative hypothesis allows for the possibility that $\mu < 68$ or $\mu > 68$.

Assume the standard deviation of the population of heights to be $\sigma = 3.6$. Our decision statistic, based on a sample of size $n = 36$, will be $\bar{X}$, the most efficient estimator of μ. From Chapter 5 we know that the sampling distribution of $\bar{X}$ is approximately normally distributed with standard deviation $\sigma_{\bar{X}} = \sigma/\sqrt{n} = 3.6/6 = 0.6$.

A sample mean that falls close to the hypothesized value of 68 would be considered evidence in favor of H_0. On the other hand, a sample mean that is considerably less than or more than 68 would be evidence inconsistent with H_0 and therefore favoring H_1. A critical region, indicated by the shaded area in Figure 7.3, is arbitrarily chosen to be $\bar{X} < 67$ and $\bar{X} > 69$. The acceptance region will therefore be $67 < \bar{X} < 69$. Hence, if our sample mean $\bar{x}$ falls inside the critical region, H_0 is rejected; otherwise we accept H_0.

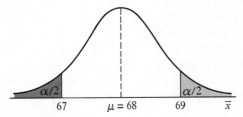

Figure 7.3 Critical region for testing $\mu = 68$ versus $\mu \neq 68$.

The probability of committing a type I error, or the level of significance of our test, is equal to the sum of the areas that have been shaded in each tail of the distribution in Figure 7.3. Therefore,

$$\alpha = Pr(\bar{X} < 67 \mid H_0 \text{ is true}) + Pr(\bar{X} > 69 \mid H_0 \text{ is true}).$$

The z values corresponding to $\bar{x}_1 = 67$ and $\bar{x}_2 = 69$ when H_0 is true are

$$z_1 = \frac{67 - 68}{0.6} = -1.67$$

$$z_2 = \frac{69 - 68}{0.6} = 1.67.$$

Therefore

$$\alpha = Pr(Z < -1.67) + Pr(Z > 1.67)$$
$$= 2Pr(Z < -1.67)$$
$$= 0.0950.$$

Thus 9.5% of all samples of size 36 would lead us to reject $\mu = 68$ inches when it is true. To reduce α we have a choice of increasing the sample size or widening the acceptance region. Suppose we increase the sample size to $n = 64$. Then $\sigma_{\bar{x}} = 3.6/8 = 0.45$. Now

$$z_1 = \frac{67 - 68}{0.45} = -2.22$$

$$z_2 = \frac{69 - 68}{0.45} = 2.22.$$

Hence

$$\alpha = Pr(Z < -2.22) + Pr(Z > 2.22)$$

$$= 2Pr(Z < -2.22)$$

$$= 0.0264.$$

The reduction in α is not sufficient by itself to guarantee a good testing procedure. We must evaluate β for various alternative hypotheses that we feel should be accepted if true. Therefore, if it is important to reject H_0 when the true mean is some value $\mu \geq 70$ or $\mu \leq 66$, then the probability of committing a type II error should be computed and examined for the alternatives $\mu = 66$ and $\mu = 70$. Owing to symmetry, it is only necessary to consider the probability of accepting the null hypothesis

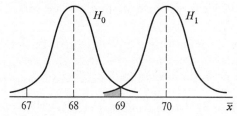

Figure 7.4 Type II error for testing $\mu = 68$ versus $\mu = 70$.

that $\mu = 68$ when the alternative $\mu = 70$ is true. A type II error will result when the sample mean $\bar{x}$ falls between 67 and 69 when H_1 is true. Therefore, referring to Figure 7.4,

$$\beta = Pr(67 < \bar{X} < 69 \mid H_1 \text{ is true}).$$

The z values corresponding to $\bar{x}_1 = 67$ and $\bar{x}_2 = 69$ when H_1 is true are

$$z_1 = \frac{67 - 70}{0.45} = -6.67$$

$$z_2 = \frac{69 - 70}{0.45} = -2.22.$$

Therefore

$$\beta = Pr(-6.67 < Z < -2.22)$$
$$= Pr(Z < -2.22) - Pr(Z < -6.67)$$
$$= 0.0132 - 0.0000$$
$$= 0.0132.$$

If the true value of μ is the alternative $\mu = 66$, the value of β will again be 0.0132. For all possible values of $\mu < 66$ or $\mu > 70$, the value of β will be even smaller when $n = 64$, and consequently there would be little chance of accepting H_0 when it is false.

The probability of committing a type II error increases rapidly when the true value of μ approaches, but is not equal to, the hypothesized value. Of course, this is usually the situation where we do not mind making a type II error. For example, if the alternative hypothesis $\mu = 68.5$ is true, we do not mind committing a type II error by concluding

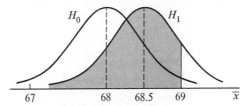

$$H_0 \qquad H_1$$
$$67 \qquad 68 \quad 68.5 \quad 69 \qquad \bar{x}$$

Figure 7.5 Type II error for testing $\mu = 68$ versus $\mu = 68.5$.

that the true answer is $\mu = 68$. The probability of making such an error will be high when $n = 64$. Referring to Figure 7.5, we have

$$\beta = Pr(67 < \bar{X} < 69 \mid H_1 \text{ is true}).$$

The z values corresponding to $\bar{x}_1 = 67$ and $\bar{x}_2 = 69$ when $\mu = 68.5$ are

$$z_1 = \frac{67 - 68.5}{0.45} = -3.33$$

$$z_2 = \frac{69 - 68.5}{0.45} = 1.11.$$

Therefore

$$\beta = Pr(-3.33 < Z < 1.11)$$
$$= Pr(Z < 1.11) - Pr(Z < -3.33)$$
$$= 0.8665 - 0.0004$$
$$= 0.8661.$$

The preceding examples illustrate the following important properties:

1. The type I error and type II error are related. A decrease in the probability of one generally results in an increase in the probability of the other.
2. The size of the critical region, and therefore the probability of committing a type I error, can always be reduced by adjusting the critical value(s).
3. An increase in the sample size n will reduce α and β simultaneously.
4. If the null hypothesis is false, β is a maximum when the true value of a parameter is close to the hypothesized value. The greater the distance between the true value and the hypothesized value, the smaller β will be.

7.3 One-Tailed and Two-Tailed Tests

A test of any statistical hypothesis where the alternative is *one-sided* such as

$$H_0: \quad \theta = \theta_0$$
$$H_1: \quad \theta > \theta_0$$

or perhaps

$$H_0: \quad \theta = \theta_0$$
$$H_1: \quad \theta < \theta_0$$

is called a *one-tailed test*. The critical region for the alternative hypothesis $\theta > \theta_0$ lies entirely in the right tail of the distribution, while the critical region for the alternative hypothesis $\theta < \theta_0$ lies entirely in the left tail. A one-tailed test was used by the director of the testing program to test the hypothesis $p = 1/4$ against the one-sided alternative $p > 1/4$ for the binomial distribution.

A test of any statistical hypothesis where the alternative is *two-sided* such as

$$H_0: \quad \theta = \theta_0$$
$$H_1: \quad \theta \neq \theta_0$$

is called a *two-tailed test*. The alternative hypothesis states that either $\theta < \theta_0$ or $\theta > \theta_0$. Values in both tails of the distribution constitute the critical region. A two-tailed test was used to test the null hypothesis that $\mu = 68$ inches against the two-tailed alternative $\mu \neq 68$ for the continuous population of students' heights.

Whether one sets up a one-sided or a two-sided alternative hypothesis will depend on the conclusion to be drawn if H_0 is rejected. The location of the critical region can be determined only after H_1 has been stated. For example, in testing a new drug, one sets up the hypothesis that it is no better than similar drugs now on the market and tests this against the alternative hypothesis that the new drug is superior. Such an alternative hypothesis will result in a one-tailed test with the critical region in the right tail. However, if we wish to determine whether two teaching procedures are equally effective, the alternative hypothesis should allow for either procedure to be superior. Hence the test is two-tailed with the critical region divided so as to fall in the extreme left and right tails of the distribution.

In testing hypotheses about discrete populations the critical region is chosen arbitrarily and its size determined. If the size α is too large, it can be reduced by making an adjustment in the critical value. It may be necessary to increase the sample size to offset the increase that automatically occurs in β. In testing hypotheses about continuous populations it is customary to choose the value of α to be 0.05 or 0.01 and then find the critical value(s). For example, in a two-tailed test at the 0.05 level of significance the critical values for a statistic having a standard normal distribution will be $-z_{0.025} = -1.96$ and $z_{0.025} = 1.96$. In terms of z values, the critical region of size 0.05 will be $Z < -1.96$ and $Z > 1.96$ and the acceptance region is $-1.96 < Z < 1.96$. A test is said to be *significant* if the null hypothesis is rejected at the 0.05 level of significance, and is considered *highly significant* if the null hypothesis is rejected at the 0.01 level of significance.

In the remaining sections of this chapter we shall consider several special tests of hypotheses that are frequently used by statisticians.

7.4 Tests Concerning Means and Variances

Consider the problem of testing the hypothesis that the mean μ of a population, with known variance σ^2, equals a specified value μ_0 against the two-sided alternative that the mean is not equal to μ_0; that is, we shall test

$$H_0: \quad \mu = \mu_0$$
$$H_1: \quad \mu \neq \mu_0.$$

An appropriate statistic on which we base our decision criterion is the random variable $\bar{X}$. From Chapter 5 we already know that the sampling distribution of $\bar{X}$ is approximately normally distributed with mean $\mu_{\bar{X}} = \mu$ and variance $\sigma_{\bar{X}}^2 = \sigma^2/n$, where μ and σ^2 are the mean and variance of the population from which we select random samples of size n. Using a significance level of α, it is possible to find two critical values $\bar{x}_1$ and $\bar{x}_2$

such that the interval $\bar{x}_1 < \bar{X} < \bar{x}_2$ defines the acceptance region and the two tails of the distribution, $\bar{X} < \bar{x}_1$ and $\bar{X} > \bar{x}_2$, constitute the critical region.

The critical region can be given in terms of z values by means of the transformation

$$z = \frac{\bar{x} - \mu_0}{\sigma/\sqrt{n}}.$$

Hence, for an α level of significance, the critical values of the random

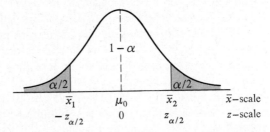

Figure 7.6 Critical region for alternative hypothesis $\mu \neq \mu_0$.

variable Z, corresponding to $\bar{x}_1$ and $\bar{x}_2$, are shown in Figure 7.6 to be

$$-z_{\alpha/2} = \frac{\bar{x}_1 - \mu_0}{\sigma/\sqrt{n}}$$

$$z_{\alpha/2} = \frac{\bar{x}_2 - \mu_0}{\sigma/\sqrt{n}}.$$

From the population we select a random sample of size n and compute the sample mean $\bar{x}$. If $\bar{x}$ falls in the acceptance region, $\bar{x}_1 < \bar{X} < \bar{x}_2$, then

$$z = \frac{\bar{x} - \mu_0}{\sigma/\sqrt{n}}$$

will fall in the region $-z_{\alpha/2} < Z < z_{\alpha/2}$ and we conclude that $\mu = \mu_0$; otherwise we reject H_0 and accept the alternative hypothesis that $\mu \neq \mu_0$. The critical region is usually stated in terms of Z rather than $\bar{X}$.

The test procedure just described is equivalent to finding a $(1-\alpha)100\%$ confidence interval for μ and accepting H_0 if μ_0 lies in the interval. If μ_0 lies outside the interval, we reject H_0 in favor of the alternative hypothesis H_1. Consequently, when one makes inferences about the mean μ from a population with known variance σ^2, whether it be by the construction of a confidence interval or through the testing of a statistical hypothesis, the same statistic $Z = (\bar{X} - \mu)/(\sigma/\sqrt{n})$ is employed.

In general, if one uses a statistic to construct a confidence interval for a parameter θ, whether it be the statistic Z, T, X^2, or F, we can use that same statistic to test the hypothesis that the parameter equals some specified value θ_0 against an appropriate alternative. Of course all the underlying assumptions made in Chapter 6 relative to the use of a given statistic will apply to the tests described here. This essentially means that all our samples are selected from approximately normal populations. However, a Z statistic may be used to test hypotheses about means from nonnormal populations when $n \geq 30$.

In Table 7.1 we list the statistic used to test a specified hypothesis H_0 and give the appropriate critical regions for one- and two-sided alternative hypotheses H_1. The steps for testing a hypothesis concerning a population parameter θ against some alternative hypothesis may be summarized as follows:

1. H_0: $\theta = \theta_0$.
2. H_1: Alternatives are $\theta < \theta_0$, $\theta > \theta_0$, or $\theta \neq \theta_0$.
3. Choose a level of significance equal to α.
4. Select the appropriate test statistic and establish the critical region.
5. Compute the value of the statistic from a random sample of size n.
6. Conclusion: Reject H_0 if the statistic has a value in the critical region; otherwise accept H_0.

Example 7.1 A manufacturer of sports equipment has developed a new synthetic fishing line that he claims has a mean breaking strength of 15 pounds with a standard deviation of 0.5 pound. Test the hypothesis that $\mu = 15$ pounds against the alternative that $\mu \neq 15$ pounds if a random sample of 50 lines is tested and found to have a mean breaking strength of 14.8 pounds. Use a 0.01 level of significance.

Solution.

1. H_0: $\mu = 15$ pounds.
2. H_1: $\mu \neq 15$ pounds.
3. $\alpha = 0.01$.
4. Critical region: $Z < -2.58$ and $Z > 2.58$, where

$$Z = \frac{\bar{X} - \mu_0}{\sigma/\sqrt{n}}.$$

5. Computations:

$$\bar{x} = 14.8 \text{ pounds}, \qquad n = 50$$

$$z = \frac{14.8 - 15}{0.5/\sqrt{50}} = -2.828.$$

6. Conclusion: Reject H_0 and conclude that the average breaking strength is not equal to 15 but is in fact less than 15 pounds.

Table 7.1 Tests Concerning Means and Variances

H_0	Test statistic	H_1	Critical region
$\mu = \mu_0$	$Z = \dfrac{\bar{X} - \mu_0}{\sigma/\sqrt{n}}$; σ known	$\mu < \mu_0$ $\mu > \mu_0$ $\mu \neq \mu_0$	$Z < -z_\alpha$ $Z > z_\alpha$ $Z < -z_{\alpha/2}$ and $Z > z_{\alpha/2}$
$\mu = \mu_0$	$T = \dfrac{\bar{X} - \mu_0}{S/\sqrt{n}}$; $\nu = n-1$, σ unknown	$\mu < \mu_0$ $\mu > \mu_0$ $\mu \neq \mu_0$	$T < -t_\alpha$ $T > t_\alpha$ $T < -t_{\alpha/2}$ and $T > t_{\alpha/2}$
$\mu_1 - \mu_2 = d_0$	$Z = \dfrac{(\bar{X}_1 - \bar{X}_2) - d_0}{\sqrt{(\sigma_1^2/n_1) + (\sigma_2^2/n_2)}}$; σ_1 and σ_2 known	$\mu_1 - \mu_2 < d_0$ $\mu_1 - \mu_2 > d_0$ $\mu_1 - \mu_2 \neq d_0$	$Z < -z_\alpha$ $Z > z_\alpha$ $Z < -z_{\alpha/2}$ and $Z > z_{\alpha/2}$
$\mu_1 - \mu_2 = d_0$	$T = \dfrac{(\bar{X}_1 - \bar{X}_2) - d_0}{S_p \sqrt{(1/n_1) + (1/n_2)}}$; $\nu = n_1 + n_2 - 2$, $\sigma_1 = \sigma_2$ but unknown, $S_p^2 = \dfrac{(n_1-1)S_1^2 + (n_2-1)S_2^2}{n_1 + n_2 - 2}$	$\mu_1 - \mu_2 < d_0$ $\mu_1 - \mu_2 > d_0$ $\mu_1 - \mu_2 \neq d_0$	$T < -t_\alpha$ $T > t_\alpha$ $T < -t_{\alpha/2}$ and $T > t_{\alpha/2}$
$\mu_1 - \mu_2 = d_0$	$T' = \dfrac{(\bar{X}_1 - \bar{X}_2) - d_0}{\sqrt{(S_1^2/n_1) + (S_2^2/n_2)}}$; $\nu = \dfrac{(s_1^2/n_1 + s_2^2/n_2)^2}{\dfrac{(s_1^2/n_1)^2}{(n_1-1)} + \dfrac{(s_2^2/n_2)^2}{(n_2-1)}}$; $\sigma_1 \neq \sigma_2$ and unknown	$\mu_1 - \mu_2 < d_0$ $\mu_1 - \mu_2 > d_0$ $\mu_1 - \mu_2 \neq d_0$	$T' < -t_\alpha$ $T' > t_\alpha$ $T' < -t_{\alpha/2}$ and $T' > t_{\alpha/2}$
$\mu_D = d_0$	$T = \dfrac{\bar{D} - d_0}{S_d/\sqrt{n}}$; $\nu = n-1$, paired observations	$\mu_D < d_0$ $\mu_D > d_0$ $\mu_D \neq d_0$	$T < -t_\alpha$ $T > t_\alpha$ $T < -t_{\alpha/2}$ and $T > t_{\alpha/2}$
$\sigma^2 = \sigma_0^2$	$X^2 = \dfrac{(n-1)S^2}{\sigma_0^2}$; $\nu = n-1$	$\sigma^2 < \sigma_0^2$ $\sigma^2 > \sigma_0^2$ $\sigma^2 \neq \sigma_0^2$	$X^2 < \chi_{1-\alpha}^2$ $X^2 > \chi_\alpha^2$ $X^2 < \chi_{1-\alpha/2}^2$ and $X^2 > \chi_{\alpha/2}^2$
$\sigma_1^2 = \sigma_2^2$	$F = S_1^2/S_2^2$; $\nu_1 = n_1 - 1$ and $\nu_2 = n_2 - 1$	$\sigma_1^2 < \sigma_2^2$ $\sigma_1^2 > \sigma_2^2$ $\sigma_1^2 \neq \sigma_2^2$	$F < f_{1-\alpha}(\nu_1,\nu_2)$ $F > f_\alpha(\nu_1,\nu_2)$ $F < f_{1-\alpha/2}(\nu_1,\nu_2)$ and $F > f_{\alpha/2}(\nu_1,\nu_2)$

Example 7.2 The average length of time for students to register for fall classes at a certain college has been 50 minutes with a standard deviation of 10 minutes. A new registration procedure using modern computing machines is being tried. If a random sample of 12 students had an average registration time of 42 minutes with a standard deviation

of 11.9 minutes under the new system, test the hypothesis that the population mean is now less than 50, using a level of significance of (1) 0.05 and (2) 0.01. Assume the population of times to be normal.

Solution.

1. H_0: $\mu = 50$ minutes.
2. H_1: $\mu < 50$ minutes.
3. (1) $\alpha = 0.05$, (2) $\alpha = 0.01$.
4. Critical region: (1) $T < -1.796$, (2) $T < -2.718$, where $T = (\bar{X} - \mu_0)/(S/\sqrt{n})$ with $\nu = 11$ degrees of freedom.
5. Computations: $\bar{x} = 42$ minutes, $s = 11.9$ minutes, and $n = 12$. Hence

$$t = \frac{42 - 50}{11.9/\sqrt{12}} = -2.33.$$

6. Conclusion: Reject H_0 at the 0.05 level of significance but not at the 0.01 level. This essentially means that the true mean is likely to be less than 50 minutes but does not differ sufficiently to warrant the high cost that would be required to operate a computer.

Example 7.3 An experiment was performed to compare the abrasive wear of two different laminated materials. Twelve pieces of material 1 were tested, by exposing each piece to a machine measuring wear. Ten pieces of material 2 were similarly tested. In each case, the depth of wear was observed. The samples of material 1 gave an average (coded) wear of 85 units with a standard deviation of 4, while the samples of material 2 gave an average of 81 and a standard deviation of 5. Test the hypothesis that the two types of material exhibit the same mean abrasive wear at the 0.10 level of significance. Assume the populations to be approximately normal with equal variances.

Solution. Let μ_1 and μ_2 represent the population means of material 1 and material 2, respectively. Using the six-step procedure, we have

1. H_0: $\mu_1 = \mu_2$ or $\mu_1 - \mu_2 = 0$.
2. H_1: $\mu_1 \neq \mu_2$ or $\mu_1 - \mu_2 \neq 0$.
3. $\alpha = 0.10$.
4. Critical region: $T < -1.725$ and $T > 1.725$, where

$$T = \frac{(\bar{X}_1 - \bar{X}_2) - d_0}{S_p \sqrt{(1/n_1) + (1/n_2)}}$$

with $\nu = 20$ degrees of freedom.

5. Computations: $\bar{x}_1 = 85$, $s_1 = 4$, $n_1 = 12$, and $\bar{x}_2 = 81$, $s_2 = 5$, $n_2 = 10$. Hence

$$s_p = \sqrt{\frac{(11)(16) + (9)(25)}{12 + 10 - 2}} = 4.478$$

$$t = \frac{(85 - 81) - 0}{4.478\sqrt{(1/12) + (1/10)}} = 2.07.$$

6. Conclusion: Reject H_0 and conclude that the two materials do not exhibit the same abrasive wear.

Example 7.4 Five samples of a ferrous-type substance are to be used to determine if there is a difference between a laboratory chemical analysis and an X-ray fluorescence analysis of the iron content. Each sample was split into two subsamples and the two types of analysis were applied. The following is the coded data showing the iron content analysis:

| | *Sample* | | | | |
Analysis	*1*	*2*	*3*	*4*	*5*
X-ray	2.0	2.0	2.3	2.1	2.4
Chemical	2.2	1.9	2.5	2.3	2.4

Assuming the populations normal, test at the 0.05 level of significance whether the two methods of analysis give, on the average, the same result.

Solution. Let μ_1 and μ_2 be the average iron content determined by the chemical and X-ray analyses, respectively. We proceed as follows:

1. H_0: $\mu_1 = \mu_2$ or $\mu_D = 0$.
2. H_1: $\mu_1 \neq \mu_2$ or $\mu_D \neq 0$.
3. $\alpha = 0.05$.
4. Critical region: $T < -2.776$ and $T > 2.776$, where $T = (\bar{D} - d_0)/(S_d/\sqrt{n})$ with $\nu = 4$ degrees of freedom.
5. Computations:

X-ray	*Chemical*	d_i	d_i^2
2.0	2.2	−0.2	0.04
2.0	1.9	0.1	0.01
2.3	2.5	−0.2	0.04
2.1	2.3	−0.2	0.04
2.4	2.4	0.0	0.00
		−0.5	0.13

We find $\bar{d} = -0.5/5 = -0.1$ and $s_d^2 = [5(0.13) - (-0.5)^2]/(5)(4) = 0.02$. Taking the square root, we have $s_d = 0.14142$. Hence

$$t = \frac{-0.1 - 0}{0.14142/\sqrt{5}} = -1.6.$$

6. Conclusion: Accept H_0 and conclude that the two methods of analysis are not significantly different.

Example 7.5 A manufacturer of car batteries claims that the life of his batteries are approximately normally distributed with a standard deviation equal to 0.9 year. If a random sample of 10 of these batteries have a standard deviation of 1.2 years, do you think that $\sigma > 0.9$ year? Use a 0.05 level of significance.

Solution.

1. H_0: $\sigma^2 = 0.81$.
2. H_1: $\sigma^2 > 0.81$.
3. $\alpha = 0.05$.
4. Critical region: $X^2 > 16.919$, where $X^2 = (n - 1)S^2/\sigma_0^2$ with $\nu = 9$ degrees of freedom.
5. Computations: $s^2 = 1.44$, $n = 10$, and

$$\chi^2 = \frac{(9)(1.44)}{0.81} = 16.0.$$

6. Conclusion: Accept H_0 and conclude that there is no reason to doubt that the standard deviation is 0.9 year.

7.5 Choice of Sample Size for Testing Means

The significance level for testing a statistical hypothesis is normally controlled by the experimenter, while β, or the *power* of the test defined by $1 - \beta$, is controlled by using the proper sample size. In this section we shall discuss the choice of sample size for tests involving one and two population means. For situations in which the normal distribution is used and the population variance or variances are known, it is a simple matter to determine the sample size necessary to attain the desired power.

Suppose we wish to test the hypothesis

$$H_0: \quad \mu = \mu_0$$
$$H_1: \quad \mu > \mu_0$$

with a significance level α when the variance σ^2 is known. For a specific alternative, say $\mu = \mu_0 + \delta$, the power of our test is given by

$$1 - \beta = Pr\left[\frac{\bar{X} - \mu_0}{\sigma/\sqrt{n}} > z_\alpha \mid H_1 \text{ is true}\right]$$

$$= Pr\left[\frac{\bar{X} - (\mu_0 + \delta)}{\sigma/\sqrt{n}} > z_\alpha - \frac{\delta}{\sigma/\sqrt{n}} \mid \mu = \mu_0 + \delta\right].$$

Under the alternative hypothesis $\mu = \mu_0 + \delta$, the statistic

$$\frac{\bar{X} - (\mu_0 + \delta)}{\sigma/\sqrt{n}}$$

is a standard normal variable. Therefore

$$1 - \beta = Pr\left(Z > z_\alpha - \frac{\delta\sqrt{n}}{\sigma}\right),$$

from which we conclude that

$$-z_\beta = z_\alpha - \frac{\delta\sqrt{n}}{\sigma}$$

and hence

$$n = \frac{(z_\alpha + z_\beta)^2\sigma^2}{\delta^2},$$

a result that is also true when the alternative hypothesis is $\mu < \mu_0$.

In the case of a two-tailed test we obtain the power $1 - \beta$ for a specified alternative when

$$n \simeq \frac{(z_{\alpha/2} + z_\beta)^2\sigma^2}{\delta^2}.$$

Example 7.6 Suppose we wish to test the hypothesis

$$H_0: \quad \mu = 68$$

$$H_1: \quad \mu > 68$$

with a significance level $\alpha = 0.05$ when $\sigma = 5$. Find the sample size required if the power of our test is to be 0.95 when the true mean is 69.

Solution. Since $\alpha = \beta = 0.05$, we have $z_\alpha = z_\beta = 1.645$. For the alternative $\mu = 69$, we take $\delta = 1$ and then

$$n = \frac{(1.645 + 1.645)^2(25)}{1} = 270.6.$$

Therefore it requires 271 observations if the test is to reject the null hypothesis 95% of the time when in fact μ is as large as 69.

A similar procedure can be used to determine the sample size $n = n_1 = n_2$ required for a specific power of the test in which two population means are being compared. For example, suppose we wish to test the hypothesis

$$H_0: \quad \mu_1 = \mu_2$$
$$H_1: \quad \mu_1 \neq \mu_2$$

when σ_1 and σ_2 are known. For a specific alternative, say $\mu_1 - \mu_2 = \delta$, the power of our test is given by

$$1 - \beta = Pr\left[\frac{|\bar{X}_1 - \bar{X}_2|}{\sqrt{(\sigma_1^2 + \sigma_2^2)/n}} > z_{\alpha/2} \mid \mu_1 - \mu_2 = \delta\right].$$

Therefore

$$\beta = Pr\left[-z_{\alpha/2} < \frac{\bar{X}_1 - \bar{X}_2}{\sqrt{(\sigma_1^2 + \sigma_2^2)/n}} < z_{\alpha/2} \mid \mu_1 - \mu_2 = \delta\right]$$

$$= Pr\left[-z_{\alpha/2} - \frac{\delta}{\sqrt{(\sigma_1^2 + \sigma_2^2)/n}} < \frac{\bar{X}_1 - \bar{X}_2 - \delta}{\sqrt{(\sigma_1^2 + \sigma_2^2)/n}} < z_{\alpha/2}\right.$$

$$\left. - \frac{\delta}{\sqrt{(\sigma_1^2 + \sigma_2^2)/n}} \mid \mu_1 - \mu_2 = \delta\right].$$

Under the alternative hypothesis $\mu_1 - \mu_2 = \delta$, the statistic

$$\frac{\bar{X}_1 - \bar{X}_2 - \delta}{\sqrt{(\sigma_1^2 + \sigma_2^2)/n}}$$

is a standard normal variable. Therefore

$$\beta = Pr\left[-z_{\alpha/2} - \frac{\delta}{\sqrt{(\sigma_1^2 + \sigma_2^2)/n}} < Z < z_{\alpha/2} - \frac{\delta}{\sqrt{(\sigma_1^2 + \sigma_2^2)/n}}\right]$$

from which we conclude that

$$-z_\beta \simeq z_{\alpha/2} - \frac{\delta}{\sqrt{(\sigma_1^2 + \sigma_2^2)/n}}$$

and hence

$$n \simeq \frac{(z_{\alpha/2} + z_\beta)^2(\sigma_1^2 + \sigma_2^2)}{\delta^2}.$$

For the one-tailed test, the expression for the required sample size when $n = n_1 = n_2$ is given by

$$n = \frac{(z_\alpha + z_\beta)^2(\sigma_1^2 + \sigma_2^2)}{\delta^2}.$$

When the population variance (or variances in the two-sample situation) is unknown, the choice of sample size is not straightforward. In testing the hypothesis $\mu = \mu_0$ when the true value is $\mu = \mu_0 + \delta$, the statistic

$$\frac{\bar{X} - (\mu_0 + \delta)}{S/\sqrt{n}}$$

does not follow the t distribution as one might expect but instead follows the *noncentral t distribution*. However, tables or charts based on the non-central t distribution do exist for determining the appropriate sample size if some estimate of σ is available or if δ is a multiple of σ. Table XIII (see Statistical Tables) gives the sample sizes needed to control the values of α and β for various values of

$$\Delta = \frac{|\delta|}{\sigma} = \frac{|\mu - \mu_0|}{\sigma}$$

for both one-tailed and two-tailed tests. In the case of the two-sample t test in which the variances are unknown but assumed equal, we obtain the sample sizes $n = n_1 = n_2$ needed to control the values of α and β for various values of

$$\Delta = \frac{|\delta|}{\sigma} = \frac{|\mu_1 - \mu_2|}{\sigma}$$

from Table XIV.

Example 7.7 In comparing the performance of two catalysts on the effect of a reaction yield, a two-sample t test is to be conducted with $\alpha = 0.05$. The variances in the yields are considered to be the same for the two catalysts. How large a sample for each catalyst is needed to test the hypothesis

$$H_0: \quad \mu_1 = \mu_2$$
$$H_1: \quad \mu_1 \neq \mu_2$$

if it is essential to detect a difference of 0.8σ between the catalysts with probability 0.9?

Solution. From Table XIV, with $\alpha = 0.05$ for a two-tailed test, $\beta = 0.1$, and

$$\Delta = \frac{|0.8\sigma|}{\sigma} = 0.8,$$

we find the required sample size to be $n = 34$.

7.6 Wilcoxon Two-Sample Test

The procedures discussed in Section 7.5 for testing hypotheses about the difference between two means are valid only if the populations are approximately normal or if the samples are large. How then does one make such a test when small independent samples are selected from nonnormal populations? In 1945 Frank Wilcoxon proposed a very simple test that assumes no knowledge about the distribution and parameters of the population. A test performed without this information is called a *non-parametric* or *distribution-free* test. The particular nonparametric test that we shall consider here is referred to as the *Wilcoxon two-sample test*.

We shall test the null hypothesis H_0 that $\mu_1 = \mu_2$ against some suitable alternative. First we select a random sample from each of the populations. Let n_1 be the number of observations in the smaller sample and n_2 the number of observations in the larger sample. Arrange the $n_1 + n_2$ observations of the combined samples in ascending order and substitute a rank of $1, 2, \ldots, n_1 + n_2$ for each observation. In the case of ties (identical observations), we replace the observations by the mean of the ranks that the observations would have if they were distinguishable. For example, if the seventh and eighth observations are identical, we would assign a rank of 7.5 to each of the two observations.

The sum of the ranks corresponding to the n_1 observations in the smaller sample is denoted by w_1. Similarly, the value w_2 represents the sum of the n_2 ranks corresponding to the larger sample. When $n_1 = n_2$, we take w_1 to be the sum of the ranks corresponding to the observations in the sample with the smaller mean. The total $w_1 + w_2$ depends only on the number of observations in the two samples and is in no way affected by the results of the experiment. Hence, if $n_1 = 3$ and $n_2 = 4$, then $w_1 + w_2 = 1 + 2 + \cdots + 7 = 28$, regardless of the numerical values of the observations.

In choosing repeated samples of size n_1 and n_2, we would expect w_1 to vary. Thus we may think of w_1 as a value of some random variable W_1. The null hypothesis $\mu_1 = \mu_2$ will be rejected in favor of the alternative $\mu_1 < \mu_2$ only if $\bar{x}_1 < \bar{x}_2$ and w_1 is small. Likewise, the alternative $\mu_1 > \mu_2$ can be accepted only if $\bar{x}_1 > \bar{x}_2$ and w_1 is large. For a two-tailed test, we may reject H_0 in favor of H_1 if either $\bar{x}_1 < \bar{x}_2$ and w_1 is small or if $\bar{x}_1 > \bar{x}_2$ and w_1 is large. Upper tail probabilities are required only when $\bar{x}_1 > \bar{x}_2$. Owing to symmetry these probabilities may be obtained from the lower tail probabilities. However, it is usually simpler to rank the observations in descending order whenever $\bar{x}_1 > \bar{x}_2$ and assign a rank of 1 to the largest observation, a rank of 2 to the next largest, and so forth (see Example 7.9). Small values of w_1 will then be significant. Hence, no matter what the alternative hypothesis may be, we are interested in significantly small values of the statistic W_1. Knowing the distribution of W_1, we can determine $Pr(W_1 \leq w_1 \mid H_0$ is true$)$. If this probability is less than or equal to 0.05, our test is significant and we would reject H_0 in favor of the appropriate one-sided alternative. When the probability does not exceed 0.01 the test is highly significant. In the case of a two-tailed test, symmetry permits us to base our decision on the value of $2Pr(W_1 \leq w_1 \mid H_0$ is true$)$. Therefore, when $2Pr(W_1 \leq w_1 \mid H_0$ is true$) < 0.05$, the test is significant and we conclude that $\mu_1 \neq \mu_2$.

The distribution of W_1, when H_0 is true, is based on the fact that all the observations in the smaller sample could be assigned ranks at random as long as their sum is less than or equal to w_1. The total number of ways of assigning $n_1 + n_2$ ranks to n_1 observations so that the sum of the ranks does not exceed w_1 is denoted by $n(W_1 \leq w_1)$. There are $\binom{n_1 + n_2}{n_1}$ equally likely ways to assign the $n_1 + n_2$ ranks to n_1 observations giving all possible values of W_1. Hence

$$Pr(W_1 \leq w_1 \mid H_0 \text{ is true}) = \frac{n(W_1 \leq w_1)}{\dbinom{n_1 + n_2}{n_1}}, \qquad \text{for } n_1 \leq n_2.$$

It is possible to find $n(W_1 \leq w_1)$ for any given test by listing all the cases and counting them. Thus, when $n_1 = 3$ and $n_2 = 5$, the number of cases where the sum of the ranks in the smaller sample is less than or equal to 8 may be listed as follows:

$$1 + 2 + 3 = 6$$
$$1 + 2 + 4 = 7$$
$$1 + 3 + 4 = 8$$
$$1 + 2 + 5 = 8.$$

Therefore, there are 4 favorable cases out of a possible $\binom{8}{3} = 56$ equally likely cases. Hence,

$$Pr(W_1 \leq 8 \mid H_0 \text{ is true}) = \tfrac{4}{56} = 0.0714.$$

It is usually easier to use Table VIII to find $Pr(W_1 \leq w_1 \mid H_0 \text{ is true})$ when n_2 does not exceed 8. Table VIII is based on the statistic U, where

$$U = W_1 - \text{(minimum value of } W_1).$$

Clearly, the minimum value of W_1 is the arithmetic sum

$$1 + 2 + 3 + \cdots + n_1 = \frac{n_1(n_1 + 1)}{2}.$$

Hence we may write

$$U = W_1 - \frac{n_1(n_1 + 1)}{2},$$

and then

$$Pr(W_1 \leq w_1 \mid H_0 \text{ is true}) = Pr(U \leq u \mid H_0 \text{ is true}).$$

In the preceding illustration, where we had $n_1 = 3$, $n_2 = 5$, and $w_1 = 8$, we find $u = 8 - [(3)(4)/2] = 2$. Using Table VIII we have

$$Pr(W_1 \leq 8 \mid H_0 \text{ is true}) = Pr(U \leq 2 \mid H_0 \text{ is true}) = 0.071,$$

which agrees with the previous answer. If, for the same illustration, $w_1 = 7$ so that $u = 1$, we find

$$Pr(W_1 \leq 7 \mid H_0 \text{ is true}) = Pr(U \leq 1 \mid H_0 \text{ is true}) = 0.036,$$

which is significant for a one-tailed test at the 0.05 level but not at 0.01 level. For a two-tailed test, the probability that the sample means differ by an amount as great as or greater than that observed is

$$2Pr(W_1 \leq 7 \mid H_0 \text{ is true}) = 2(0.036) = 0.072,$$

from which we conclude that H_0 is true.

When n_2 is between 9 and 20, Table IX (see Statistical Tables) may be used. If the observed value of U is less than or equal to the tabled value, the null hypothesis may be rejected at the level of significance indicated by the table. Table IX gives critical values of U for levels of significance equal to 0.001, 0.01, 0.025, and 0.05 for a one-tailed test. In the case of a two-tailed test the critical values of U correspond to the 0.002, 0.02, 0.05, and 0.1 levels of significance. When n_1 and n_2 increase in size, the sampling distribution of U approaches the normal distribution with mean

$$\mu_U = \frac{n_1 n_2}{2}$$

and variance

$$\sigma_U^2 = \frac{n_1 n_2 (n_1 + n_2 + 1)}{12}.$$

Consequently, when n_2 is greater than 20 one could use the statistic $Z = (U - \mu_U)/\sigma_U$ to establish the critical region for the test.

To test the null hypothesis that the means of two nonnormal populations are equal when only small independent samples are available, we proceed by the following steps:

1. H_0: $\mu_1 = \mu_2$.
2. H_1: Alternatives are $\mu_1 < \mu_2$, $\mu_1 > \mu_2$, or $\mu_1 \neq \mu_2$.
3. Choose a level of significance equal to α.
4. Critical region:
   (a) All u values for which $Pr(U \leq u \mid H_0$ is true$) < \alpha$ when $n_2 \leq 8$ and the test is one-tailed.
   (b) All u values for which $2Pr(U \leq u \mid H_0$ is true$) < \alpha$ when $n_2 \leq 8$ and the test is two-tailed.
   (c) All u values less than or equal to the appropriate critical value in Table IX when $9 \leq n_2 \leq 20$.
5. Compute w_1 and $u = w_1 - [n_1(n_1 + 1)/2]$ from independent samples of size n_1 and n_2, where $n_1 \leq n_2$. Determine whether u falls in the acceptance or critical region.
6. Conclusion: Reject H_0 if u falls in the critical region; otherwise accept H_0.

Example 7.8 To find out whether a new serum will arrest leukemia, nine mice that have all reached an advanced stage of the disease are selected. Five mice receive the treatment and four do not. The survival times, in years, from the time the experiment commenced are

Treatment	2.1	5.3	1.4	4.6	0.9
No treatment	1.9	0.5	2.8	3.1	

At the 0.05 level of significance can the serum be said to be effective?

Solution. From the data we have $n_1 = 4$, $n_2 = 5$, $\bar{x}_1 = 2.075$, and $\bar{x}_2 = 2.86$. We follow the six-step procedure:

1. H_0: $\mu_1 = \mu_2$.
2. H_1: $\mu_1 < \mu_2$.
3. $\alpha = 0.05$.
4. Critical region: All u values for which $Pr(U \le u \mid H_0 \text{ is true}) < 0.05$.
5. Computations: Since $\bar{x}_1 < \bar{x}_2$, the observations are arranged in ascending order and ranks from 1 to 9 assigned.

Original data	0.5	0.9	1.4	1.9	2.1	2.8	3.1	4.6	5.3
Ranks	1	2	3	4	5	6	7	8	9

The treatment observations are underscored for identification purposes. Now

$$w_1 = 1 + 4 + 6 + 7 = 18$$

and

$$u = 18 - \frac{(4)(5)}{2} = 8.$$

Since $Pr(U \le 8 \mid H_0 \text{ is true}) = 0.365 > 0.05$, the value $u = 8$ falls in the acceptance region.

6. Conclusion: Accept H_0 and conclude that the serum does not prolong life by arresting leukemia.

Example 7.9 The nicotine content of two brands of cigarettes, measured in milligrams, was found to be as follows:

Brand A	22.1	24.0	26.3	25.4	24.8	23.7	26.1	23.3		
Brand B	24.1	20.6	23.1	22.5	24.0	26.2	21.6	22.2	21.9	25.4

Test the hypothesis, at the 0.05 level of significance, that the average nicotine contents of the two brands are equal against the alternative that they are unequal.

Solution. Examining the data, we see that $n_1 = 8$, $n_2 = 10$, $\bar{x}_1 = 24.5$, and $\bar{x}_2 = 23.2$. We proceed by the six-step rule:

1. H_0: $\mu_1 = \mu_2$.
2. H_1: $\mu_1 \ne \mu_2$.
3. $\alpha = 0.05$.
4. Critical region: $U \le 17$ (from Table IX).

5. Computations: Since $\bar{x}_1 > \bar{x}_2$, the observations are arranged in descending order and ranks from 1 to 18 assigned.

Original data	26.3	26.2	26.1	25.4	25.4	24.8	24.1	24.0	24.0
Ranks	1	2	3	4.5	4.5	6	7	8.5	8.5

Original data	23.7	23.3	23.1	22.5	22.2	22.1	21.9	21.6	20.6
Ranks	10	11	12	13	14	15	16	17	18

The ranks of the observations belonging to the smaller sample are underscored. Now

$$w_1 = 1 + 3 + 4.5 + 6 + 8.5 + 10 + 11 + 15 = 59$$

and

$$u = 59 - \frac{(8)(9)}{2} = 23.$$

6. Conclusion: Accept H_0 and conclude that there is no difference in the average nicotine contents of the two brands of cigarettes.

The use of the Wilcoxon two-sample test is not restricted to nonnormal populations. It can be used in place of the t test when the populations are normal, although the probability of committing a type II error will be larger. The Wilcoxon two-sample test is always superior to the t test for nonnormal populations.

7.7 Wilcoxon Test for Paired Observations

Assume that n pairs of observations are selected from two nonnormal populations. For large n the distribution of the mean of the differences in repeated sampling is approximately normal and tests of hypotheses concerning the means may be carried out using the statistic

$$T = (\bar{D} - d_0)/(S_d/\sqrt{n})$$

given in Table 7.1. However, if n is small and the population of differences is decidedly nonnormal, we must resort to a nonparametric test. Perhaps the easiest and quickest to perform is a test called the *sign test*. In testing the null hypothesis that $\mu_1 = \mu_2$ or $\mu_D = 0$, each difference d_i is assigned a plus or minus sign, depending on whether d_i is positive or negative. If the null hypothesis is true, the sum of the plus signs should be approximately equal to the sum of the minus signs. When one sign appears more frequently than it should, based on chance alone, we would reject the hypothesis that the population means are equal.

The sign test shows, by the assigned plus or minus sign, which member of a pair of observations is the larger, but it does not indicate the magnitude of the difference. A test utilizing both direction and magnitude was proposed in 1945 by Frank Wilcoxon and is now commonly referred to as the *Wilcoxon test for paired observations*. Wilcoxon's test is more sensitive than the sign test in detecting a difference in the population means and therefore will be considered in detail.

To test the hypothesis that $\mu_1 = \mu_2$ by the Wilcoxon test, first discard all differences equal to zero and then rank the remaining d_i's without regard to sign. A rank of 1 is assigned to the smallest d_i in absolute value, a rank of 2 to the next smallest, and so on. When the absolute value of two or more differences is the same, assign to each the average of the ranks that would have been assigned if the differences were distinguishable. If there is no difference between the two population means, the total of the ranks corresponding to the positive differences should be almost equal to the total of the ranks corresponding to the negative differences. Let us represent these totals by w_+ and w_-, respectively. We shall designate the smaller of the w_+ and w_- by w and find the probability of obtaining, by chance alone, a value less than or equal to w when H_0 is true.

In selecting repeated samples of paired observations, we would expect w to vary. Thus we may think of w as a value of some random variable W. Once the distribution of W is known, we can determine $Pr(W \leq w \mid H_0$ is true). For a level of significance equal to α, we reject H_0 when

$$Pr(W \leq w \mid H_0 \text{ is true}) < \alpha,$$

and accept the appropriate one-sided alternative. In the case of a two-tailed test, we reject H_0 at the α level of significance in favor of the alternate two-sided hypothesis $\mu_1 \neq \mu_2$ when

$$2Pr(W \leq w \mid H_0 \text{ is true}) < \alpha.$$

If we assume that there is no difference in the population means, each d_i is just as likely to be positive as it is to be negative. Thus there are two equally likely ways for a given rank to receive a sign. For n differences, there are 2^n equally likely ways for the n ranks to receive signs. Let $n(W \leq w)$ be the number of the 2^n ways of assigning signs to the n ranks such that the value of W does not exceed w. Then

$$Pr(W \leq w \mid H_0 \text{ is true}) = \frac{n(W \leq w)}{2^n}.$$

Consider, for example, the case of $n = 6$ matched pairs that yield a value $w = 5$. What is the probability that $W \leq 5$ when the two population means are equal? The sets of ranks whose total does not exceed 5 may be listed as follows:

Value of W	Sets of ranks totaling W
0	∅
1	{1}
2	{2}
3	{3}, {1,2}
4	{4}, {1,3}
5	{5}, {1,4}, {2,3}

Therefore $n(W \leq 5) = 10$ out of a possible $2^6 = 64$ equally likely cases. Hence

$$Pr(W \leq 5 \mid H_0 \text{ is true}) = \tfrac{10}{64} = 0.1563,$$

a result that is quite likely to occur when $\mu_1 = \mu_2$.

When $5 \leq n \leq 30$, Table X (see Statistical Tables) gives approximate critical values of W for levels of significance equal to 0.01, 0.025, and 0.05 for a one-tailed test and equal to 0.02, 0.05, and 0.10 for a two-tailed test. In the preceding example for which $n = 6$, Table X shows that a value of $W \leq 2$ is required for a one-tailed test to be significant at the 0.05 level. When $n > 30$ the sampling distribution of W approaches the normal distribution with mean

$$\mu_W = \frac{n(n + 1)}{4}$$

and variance

$$\sigma_W^2 = \frac{n(n + 1)(2n + 1)}{24}.$$

In this case, the statistic $Z = (W - \mu_W)/\sigma_W$ can be used to determine the critical region for our test.

The Wilcoxon test for paired observations may also be used to test the null hypothesis that $\mu_1 - \mu_2 = \mu_D = d_0$. We simply apply the same procedure as before after each d_i is adjusted by subtracting d_0. Therefore, to test a hypothesis about the difference between the means of two populations whose distributions are unknown, where the observations occur in pairs and the sample size is small, we proceed by the following six steps:

1. H_0: $\mu_1 - \mu_2 = \mu_D = d_0$.
2. H_1:   Alternatives are $\mu_1 - \mu_2 < d_0$, $\mu_1 - \mu_2 > d_0$, or $\mu_1 - \mu_2 \neq d_0$.

3. Choose a level of significance equal to α.
4. Critical region:
   (a) All w values for which $Pr(W \leq w \mid H_0$ is true) $< \alpha$ when $n < 5$ and the test is one-tailed.
   (b) All w values for which $2Pr(W \leq w \mid H_0$ is true) $< \alpha$ when $n < 5$ and the test is two-tailed.
   (c) All w values less than or equal to the appropriate critical value in Table X when $5 \leq n \leq 30$.
5. Rank the n differences, $d_i - d_0$, without regard to sign, and then compute w.
6. Conclusion: Reject H_0 if w falls in the critical region; otherwise accept H_0.

Example 7.10 It is claimed that a college senior can increase his score in the major field area of the graduate record examination by at least 50 points if he is provided sample problems in advance. To test this claim, 20 college seniors were divided into 10 pairs such that each matched pair had almost the same overall quality point average for their first 3 years in college. Sample problems and answers were provided at random to one member of each pair 1 week prior to the examination. The following examination scores were recorded:

	Pair									
	1	2	3	4	5	6	7	8	9	10
With sample problems	531	621	663	579	451	660	591	719	543	575
Without sample problems	509	540	688	502	424	683	568	748	530	524

Test the null hypothesis at the 0.05 level of significance that sample problems increase the scores by 50 points against the alternative hypothesis that the increase is less than 50 points.

Solution. Let μ_1 and μ_2 represent the mean score of all students taking the test in question with and without sample problems, respectively. We follow the six-step procedure already outlined:

1. H_0: $\mu_1 - \mu_2 = 50$.
2. H_1: $\mu_1 - \mu_2 < 50$.
3. $\alpha = 0.05$.
4. Critical region: Since $n = 10$, Table X shows the critical region to be $W \leq 11$.

5. Computations:

	Pair									
	1	*2*	*3*	*4*	*5*	*6*	*7*	*8*	*9*	*10*
d_i	22	81	−25	77	27	−23	23	−29	13	51
$d_i - d_0$	−28	31	−75	27	−23	−73	−27	−79	−37	1
Ranks	5	6	9	3.5	2	8	3.5	10	7	1

Now $w_+ = 10.5$ and $w_- = 44.5$, so that $w = 10.5$, the smaller of w_+ and w_-.

6. Conclusion: Reject H_0 and conclude that the sample problems do not, on the average, increase one's graduate record score by as much as 50 points.

7.8 Tests Concerning Proportions

Tests of hypotheses concerning proportions are required in many areas. The politician is certainly interested in knowing what fraction of the voters will favor him in the next election. All manufacturing firms are concerned about the proportion of defectives when a shipment is made. The gambler depends on a knowledge of the proportion of outcomes that he considers favorable.

We shall consider the problem of testing the hypothesis that the proportion of successes in a binomial experiment equals some specified value. That is, we are testing the null hypothesis H_0 that $p = p_0$, where p is the parameter of the binomial distribution. The alternative hypothesis may be one of the usual one-sided or two-sided alternatives, namely, $p < p_0$, $p > p_0$, or $p \neq p_0$.

The appropriate statistic on which we base our decision criterion is the binomial random variable X, although we could just as well use the statistic $\hat{P} = X/n$. Values of X that are far from the mean $\mu = np_0$ will lead to the rejection of the null hypothesis. To test the hypothesis

$$H_0: \quad p = p_0$$
$$H_1: \quad p < p_0$$

we use the binomial distribution with $p = p_0$ and $q = 1 - p_0$ to determine $Pr(X \leq x \mid H_0$ is true). The value x is the number of successes in our sample of size n. If $Pr(X \leq x \mid H_0$ is true) $< \alpha$, our test is significant at the α level and we reject H_0 in favor of H_1. Similarly, to test the hypothesis

$$H_0: \quad p = p_0$$
$$H_1: \quad p > p_0$$

we find $Pr(X \geq x \mid H_0$ is true) and reject H_0 in favor of H_1 if this probability is less than α. Finally, to test the hypothesis

$$H_0: \quad p = p_0$$
$$H_1: \quad p \neq p_0$$

at the α level of significance, we reject H_0 when $x < np_0$ and $Pr(X \leq x \mid H_0$ is true) $< \alpha/2$ or when $x > np_0$ and $Pr(X \geq x \mid H_0$ is true) $< \alpha/2$.

The steps for testing a hypothesis about a proportion against various alternatives are now summarized:

1. H_0: $\quad p = p_0$.
2. H_1: $\quad$ Alternatives are $p < p_0$, $p > p_0$, or $p \neq p_0$.
3. Choose a level of significance equal to α.
4. Critical region:
   (a) All x values such that $Pr(X \leq x \mid H_0$ is true) $< \alpha$ for the alternative $p < p_0$.
   (b) All x values such that $Pr(X \geq x \mid H_0$ is true) $< \alpha$ for the alternative $p > p_0$.
   (c) All x values such that $Pr(X \leq x \mid H_0$ is true) $< \alpha/2$ when $x < np_0$, and all x values such that $Pr(X \geq x \mid H_0$ is true) $< \alpha/2$ when $x > np_0$, for the alternative $p \neq p_0$.
5. Computations: Find x and compute the appropriate probability.
6. Conclusion: Reject H_0 if x falls in the critical region; otherwise accept H_0.

Example 7.11 A pheasant hunter claims that he hits 80% of the birds he shoots at. Would you agree with this claim if on a given day he brings down 9 of the 15 pheasants he shoots at? Use a 0.05 level of significance.

Solution. We follow the six-step procedure:

1. H_0: $\quad p = 0.8$.
2. H_1: $\quad p \neq 0.8$.
3. $\alpha = 0.05$.
4. Critical region: All x values such that $Pr(X \leq x \mid H_0$ is true) < 0.025.
5. Computations: We have $x = 9$ and $n = 15$. Therefore, using Table II,

$$Pr(X \leq 9 \mid p = 0.8) = \sum_{x=0}^{9} b(x; 15, 0.8)$$

$$= 0.0611 > 0.025.$$

6. Conclusion: Accept H_0 and conclude that there is no reason to doubt the hunter's claim.

In Section 4.3 we saw that binomial probabilities were obtainable from the actual binomial formula or from Table II when n is small. For large n, approximation procedures are required. When the hypothesized value p_0 is very close to zero or 1, the Poisson distribution, with parameter $\mu = np_0$, may be used. The normal curve approximation is usually preferred for large n and is very accurate as long as p_0 is not extremely close to zero or 1. Using the normal approximation, we base our decision criterion on the standard normal variable

$$Z = \frac{\hat{P} - p_0}{\sqrt{p_0 q_0/n}} = \frac{X - np_0}{\sqrt{np_0 q_0}}.$$

Hence, for a two-tailed test at the α level of significance, the critical region is $Z < -z_{\alpha/2}$ and $Z > z_{\alpha/2}$. For the one-sided alternative $p < p_0$, the critical region is $Z < -z_\alpha$, and for the alternative $p > p_0$, the critical region is $Z > z_\alpha$.

To test a hypothesis about a proportion using the normal curve approximation, we proceed as follows:

1. H_0: $p = p_0$.
2. H_1: Alternatives are $p < p_0$, $p > p_0$, or $p \neq p_0$.
3. Choose a level of significance equal to α.
4. Critical region:
   (a) $Z < -z_\alpha$ for the alternative $p < p_0$.
   (b) $Z > z_\alpha$ for the alternative $p > p_0$.
   (c) $Z < -z_{\alpha/2}$ and $Z > z_{\alpha/2}$ for the alternative $p \neq p_0$.
5. Computations: Find x from a sample of size n, and then compute

$$z = \frac{x - np_0}{\sqrt{np_0 q_0}}.$$

6. Conclusion: Reject H_0 if z falls in the critical region; otherwise accept H_0.

Example 7.12 A manufacturing company has submitted a claim that 90% of items produced by a certain process are nondefective. An improvement in the process is being considered that they feel will lower the proportion of defectives below the current 10%. In an experiment 100 items are produced with the new process and 5 are defective. Is this evidence sufficient to conclude that the method has been improved? Use a 0.05 level of significance.

Solution. As usual, we follow the six-step procedure:

1. H_0: $p = 0.9$.
2. H_1: $p > 0.9$.
3. $\alpha = 0.05$.
4. Critical region: $Z > 1.645$.
5. Computations: $x = 95$, $n = 100$, $np_0 = (100)(0.95) = 95$, and

$$z = \frac{95 - 90}{\sqrt{(100)(0.90)(0.10)}} = 1.67.$$

6. Conclusion: Reject H_0 and conclude that the improvement has reduced the proportion of defectives.

7.9 Testing the Difference Between Two Proportions

Situations often arise where we wish to test the hypothesis that two proportions are equal. For example, we might try to prove that the proportion of doctors who are pediatricians in one state is equal to the proportion of pediatricians in another state. A person may decide to give up smoking only if he is convinced that the proportion of smokers with lung cancer exceeds the proportion of nonsmokers with lung cancer.

In general we wish to test the null hypothesis

$$H_0: \quad p_1 = p_2 = p$$

against some suitable alternative. The parameters p_1 and p_2 are the two population proportions of the attribute under investigation. The statistic on which we base our decision criterion is the random variable $\hat{P}_1 - \hat{P}_2$. Independent samples of size n_1 and n_2 are selected at random from two binomial populations and the proportion of successes $\hat{P}_1$ and $\hat{P}_2$ for the two samples are computed. From Section 6.6 we know that the statistic

$$Z = \frac{\hat{P}_1 - \hat{P}_2}{\sqrt{(p_1 q_1/n_1) + (p_2 q_2/n_2)}} = \frac{\hat{P}_1 - \hat{P}_2}{\sqrt{pq[(1/n_1) + (1/n_2)]}}$$

has a standard normal distribution when H_0 is true and n_1 and n_2 are large. To compute a value of Z we must estimate the parameter p appearing in the radical. Pooling the data from both samples, we write

$$\hat{p} = \frac{x_1 + x_2}{n_1 + n_2}$$

where x_1 and x_2 are the number of successes in each of the two samples. Substituting $\hat{p}$ for p the statistic Z assumes the form

$$Z = \frac{\hat{P}_1 - \hat{P}_2}{\sqrt{\hat{p}\hat{q}[(1/n_1) + (1/n_2)]}},$$

where $\hat{q} = 1 - \hat{p}$. The critical regions for the appropriate alternative hypotheses are set up as before using the critical points of the standard normal curve.

To test the hypothesis that two proportions are equal when the samples are large, we proceed by the following six steps:

1. H_0: $p_1 = p_2$.
2. H_1: Alternatives are $p_1 < p_2$, $p_1 > p_2$, or $p_1 \neq p_2$.
3. Choose a level of significance equal to α.
4. Critical region:
   (a) $Z < -z_\alpha$ for the alternative $p_1 < p_2$.
   (b) $Z > z_\alpha$ for the alternative $p_1 > p_2$.
   (c) $Z < -z_{\alpha/2}$ and $Z > z_{\alpha/2}$ for the alternative $p_1 \neq p_2$.
5. Computations: Compute $\hat{p}_1 = x_1/n_1$, $\hat{p}_2 = x_2/n_2$, and $\hat{p} = (x_1 + x_2)/(n_1 + n_2)$ and then find

$$z = \frac{\hat{p}_1 - \hat{p}_2}{\sqrt{\hat{p}\hat{q}[(1/n_1) + (1/n_2)]}}.$$

6. Conclusion: Reject H_0 if z falls in the critical region; otherwise accept H_0.

Example 7.13 A vote is to be taken among the residents of a town and the surrounding county to determine whether a proposed chemical plant should be constructed. The construction site is within the town limits and for this reason many voters in the county feel that the proposal will pass because of the large proportion of town voters who favor the construction. To determine if there is a significant difference in the proportion of town voters and county voters favoring the proposal, a poll is taken. If 120 of 200 town voters favor the proposal and 240 of 500 county residents favor it, would you agree that the proportion of town voters favoring the proposal is higher than the proportion of county voters? Use a 0.025 level of significance.

Solution. Let p_1 and p_2 be the true proportion of voters in the town and county, respectively, favoring the proposal. We now follow the six-step procedure:

1. H_0: $p_1 = p_2$.
2. H_1: $p_1 > p_2$.
3. $\alpha = 0.025$.
4. Critical region: $Z > 1.96$.
5. Computations:

$$\hat{p}_1 = \frac{x_1}{n_1} = \frac{120}{200} = 0.60$$

$$\hat{p}_2 = \frac{x_2}{n_2} = \frac{240}{500} = 0.48$$

$$\hat{p} = \frac{x_1 + x_2}{n_1 + n_2} = \frac{120 + 240}{200 + 500} = 0.51.$$

Therefore

$$z = \frac{0.60 - 0.48}{\sqrt{(0.51)(0.49)(\frac{1}{200} + \frac{1}{500})}} = 2.9.$$

6. Conclusion: Reject H_0 and agree that the proportion of town voters favoring the proposal is higher than the proportion of county voters.

7.10 Goodness of Fit Test

Throughout this chapter we have been concerned with the testing of statistical hypotheses about single population parameters such as μ, σ^2, and p. Now we shall consider a test to determine if a population has a specified theoretical distribution. The test is based on how good a fit we have between the frequency of occurrence of observations in an observed sample and the expected frequencies obtained from the hypothesized distribution.

To illustrate, consider the tossing of a die. We hypothesize that the die is honest, which is equivalent to testing the hypothesis that the distribution of outcomes is uniform. Suppose the die is tossed 120 times and each outcome is recorded. Theoretically, if the die is balanced, we would expect each face to occur 20 times. The results are given in Table 7.2. By comparing the observed frequencies with the corresponding expected frequencies we must decide whether these discrepancies are likely to occur as a result of sampling fluctuations and the die is balanced or the die is not honest and the distribution of outcomes is not uniform. It is common practice to refer to each possible outcome of an experiment as a

Table 7.2 Observed and Expected Frequencies
of 120 Tosses of a Die

	Faces					
	1	*2*	*3*	*4*	*5*	*6*
Observed	20	22	17	18	19	24
Expected	20	20	20	20	20	20

cell. Hence in our illustration we have six cells. The appropriate statistic
on which we base our decision criterion for an experiment involving k
cells is defined by the following theorem.

THEOREM 7.1 *A* goodness of fit *test between observed and expected*
frequencies is based on the quantity

$$\chi^2 = \sum_{i=1}^{k} \frac{(o_i - e_i)^2}{e_i},$$

where χ^2 is a value of the random variable X^2 *whose sampling distribution*
is approximated very closely by the chi-square distribution. The symbols
o_i *and* e_i *represent the observed and expected frequencies, respectively, for*
the ith cell.

If the observed frequencies are close to the corresponding expected
frequencies, the χ^2 value will be small, indicating a good fit. If the observed
frequencies differ considerably from the expected frequencies, the χ^2 value
will be large and the fit is poor. A good fit leads to the acceptance of H_0,
whereas a poor fit leads to its rejection. The critical region will, therefore,
fall in the right tail of the chi-square distribution. For a level of significance equal to α, we find the critical value χ_α^2 from Table VI, and then
$X^2 > \chi_\alpha^2$ constitutes the critical region. The decision criterion described
here should not be used unless each of the expected frequencies is at least
equal to 5.

The number of degrees of freedom associated with the chi-square distribution used here depends on two factors: the number of cells in the
experiment, and the number of quantities obtained from the observed
data that are necessary in the calculation of the expected frequencies. We

arrive at this number by the following theorem:

THEOREM 7.2 *The number of* degrees of freedom *in a chi-square good-ness of fit test is equal to the number of cells minus the number of quan-tities obtained from the observed data that are used in the calculations of the expected frequencies.*

The only quantity provided by the observed data, in computing expected frequencies for the outcome when a die is tossed, is the total frequency. Hence, according to our definition, the computed x^2 value has $6 - 1 = 5$ degrees of freedom.

From Table 7.2 we find the x^2 value to be

$$x^2 = \frac{(20 - 20)^2}{20} + \frac{(22 - 20)^2}{20} + \frac{(17 - 20)^2}{20} + \frac{(18 - 20)^2}{20}$$
$$+ \frac{(19 - 20)^2}{20} + \frac{(24 - 20)^2}{20}$$

$$= 1.7.$$

Using Table VI, we find $x_{0.05}^2 = 11.070$ for $\nu = 5$ degrees of freedom. Since 1.7 is less than the critical value we fail to reject H_0 and conclude that the distribution is uniform. In other words, the die is balanced.

As a second illustration let us test the hypothesis that the frequency distribution of battery lives given in Chapter 2, Table 2.2, may be approximated by the normal distribution. The expected frequencies for each class (cell), listed in Table 7.3, are obtained from a normal curve

Table 7.3 Observed and Expected Frequencies
of Battery Lives Assuming Normality

Class boundaries	o_i		e_i	
1.45–1.95	2		0.6	
1.95–2.45	1	7	2.7	10.1
2.45–2.95	4		6.8	
2.95–3.45	15		10.6	
3.45–3.95	10		10.3	
3.95–4.45	5	8	6.1	8.3
4.45–4.95	3		2.2	

having the same mean and standard deviation as our sample. From the data of Table 2.1 we find that the sample of 40 batteries has a mean

$\bar{x} = 3.4125$ and a standard deviation $s = 0.703$. These values will be used for μ and σ in computing z values corresponding to the class boundaries. The z value corresponding to the boundaries of the fourth class, for example, are

$$z_1 = \frac{2.95 - 3.4125}{0.703} = -0.658$$

$$z_2 = \frac{3.45 - 3.4125}{0.703} = 0.053.$$

From Table IV we find the area between $z_1 = -0.658$ and $z_2 = 0.053$ to be

$$\begin{aligned} \text{Area} &= Pr(-0.658 < Z < 0.053) \\ &= Pr(Z < 0.053) - Pr(Z < -0.658) \\ &= 0.5211 - 0.2552 \\ &= 0.2659. \end{aligned}$$

Hence the expected frequency for the fourth class is

$$e_4 = (0.2659)(40) = 10.6.$$

The expected frequency for the first class interval is obtained by using the total area under the normal curve to the left of the boundary 1.95. For the last class interval, we use the total area to the right of the boundary 4.45. All other expected frequencies are determined by the method described for the fourth class. Note that we have combined adjacent classes in Table 7.3 where the expected frequencies are less than 5. Consequently, the total number of intervals is reduced from 7 to 4. The χ^2 value is then given by

$$\chi^2 = \frac{(7 - 10.1)^2}{10.1} + \frac{(15 - 10.6)^2}{10.6} + \frac{(10 - 10.3)^2}{10.3} + \frac{(8 - 8.3)^2}{8.3}$$

$$= 2.797.$$

The number of degrees of freedom for this test will be $4 - 3 = 1$ since three quantities, namely, the total frequency, mean, and standard deviation of the observed data, were required to find the expected frequencies. Since the computed χ^2 value is less than $\chi^2_{0.05} = 3.841$ for 1 degree of freedom, we have no reason to reject the null hypothesis and conclude that the normal distribution provides a good fit for the distribution of battery lives.

7.11 Test for Independence

The chi-square test procedure discussed in Section 7.10 can also be used to test the hypothesis of independence of two variables. Suppose we wish to study the relationship between religious affiliation and geographical region. Two groups of people are chosen at random, one from the east coast and one from the west coast of the United States, and each person is classified as Protestant, Catholic, or Jewish. The observed frequencies are presented in Table 7.4, which is known as a *contingency table*.

Table 7.4 2 × 3 Contingency Table

	Protestant	Catholic	Jewish	Totals
East coast	182	215	203	600
West coast	154	136	110	400
Totals	336	351	313	1000

A contingency table with r rows and c columns is referred to as an $r \times c$ table. The symbol $r \times c$ is read "r by c." The row and column totals in Table 7.4 are called *marginal frequencies*. To test the null hypothesis H_0 of independence between a person's religious faith and the region where he lives, we must first find the expected frequencies for each cell of Table 7.4 under the assumption that H_0 is true.

Let us define the following events:

P: an individual selected from our sample is Protestant.

C: an individual selected from our sample is Catholic.

J: an individual selected from our sample is Jewish.

E: an individual selected from our sample lives on the east coast.

W: an individual selected from our sample lives on the west coast.

Using the marginal frequencies, we can list the following probabilities:

$$Pr(P) = \frac{336}{1000}, \quad Pr(C) = \frac{351}{1000}, \quad Pr(J) = \frac{313}{1000}, \quad Pr(E) = \frac{600}{1000},$$

$$Pr(W) = \frac{400}{1000}.$$

Now, if H_0 is true and the two variables are independent, we should have

$$Pr(P \cap E) = Pr(P)Pr(E) = \left(\frac{336}{1000}\right)\left(\frac{600}{1000}\right)$$

$$Pr(P \cap W) = Pr(P)Pr(W) = \left(\frac{336}{1000}\right)\left(\frac{400}{1000}\right)$$

$$Pr(C \cap E) = Pr(C)Pr(E) = \left(\frac{351}{1000}\right)\left(\frac{600}{1000}\right)$$

$$Pr(C \cap W) = Pr(C)Pr(W) = \left(\frac{351}{1000}\right)\left(\frac{400}{1000}\right)$$

$$Pr(J \cap E) = Pr(J)Pr(E) = \left(\frac{313}{1000}\right)\left(\frac{600}{1000}\right)$$

$$Pr(J \cap W) = Pr(J)Pv(W) = \left(\frac{313}{1000}\right)\left(\frac{400}{1000}\right).$$

The expected frequencies are obtained by multiplying each cell probability by the total number of observations. Thus the expected number of Protestants living on the east coast in our sample will be

$$\left(\frac{336}{1000}\right)\left(\frac{600}{1000}\right)(1000) = \frac{(336)(600)}{1000} = 202$$

when H_0 is true. The general formula for obtaining the expected frequency of any cell is given by

$$e = \frac{RC}{T},$$

where R and C are the corresponding row and column totals and T is the grand total of all the observed frequencies. The expected frequencies for each cell are recorded in parentheses beside the actual observed value in Table 7.5. Note that the expected frequencies in any row or column add up to the appropriate marginal total. In our example, we need to compute only the two expected frequencies in the top row of Table 7.5 and then

Table 7.5 Observed and Expected Frequencies

	Protestant	Catholic	Jewish	Totals
East coast	182 (202)	215 (211)	203 (187)	600
West coast	154 (134)	136 (140)	110 (126)	400
Totals	336	351	313	1000

find the others by subtraction. By using three marginal totals and the grand total to arrive at the expected frequencies we have lost 4 degrees of freedom, leaving a total of 2. A simple formula providing the correct number of degrees of freedom is given by $\nu = (r-1)(c-1)$. Hence, for our example $\nu = (2-1)(3-1) = 2$ degrees of freedom.

To test the null hypothesis of independence, we use the following decision criterion:

TEST FOR INDEPENDENCE *Calculate*

$$\chi^2 = \sum_i \frac{(o_i - e_i)^2}{e_i},$$

where the summation extends over all cells in the $r \times c$ contingency table. If $\chi^2 > \chi^2_\alpha$, reject the null hypothesis of independence at the α level of significance; otherwise accept the null hypothesis. The number of degrees of freedom is

$$\nu = (r-1)(c-1).$$

Applying this criterion to our example, we find that

$$\chi^2 = \frac{(182-202)^2}{202} + \frac{(215-211)^2}{211} + \frac{(203-187)^2}{187}$$

$$+ \frac{(154-134)^2}{134} + \frac{(136-140)^2}{140} + \frac{(110-126)^2}{126}$$

$$= 8.556.$$

From Table VI we find that $\chi^2_{0.05} = 5.991$ for $\nu = (2-1)(3-1) = 2$ degrees of freedom. The null hypothesis is rejected at the 0.05 level of significance and we conclude that religious faith and the region where one lives are not independent.

The chi-square statistic for testing independence is also applicable when testing the hypothesis that k binomial populations have the same parameter p. This is, therefore, an extension of the test presented in Section 7.9 for the difference between two proportions to the differences among k proportions. Hence, we are interested in testing the hypothesis

$$H_0: \quad p_1 = p_2 = \cdots = p_k = p$$

against the alternative hypothesis that the population proportions are not all equal, which is equivalent to testing that the number of successes or failures is independent of the sample chosen. To perform this test we first select independent random samples of size n_1, n_2, ..., n_k from the k populations and arrange the data as in a $2 \times k$ contingency table (Table 7.6). The expected cell frequencies are calculated as before and substituted together with the observed frequencies into the preceding chi-square formula for independence with $\nu = k - 1$ degrees of freedom. By selecting an appropriate critical region, one can now reach a conclusion concerning H_0.

Table 7.6 k Independent Binomial Samples

	Sample			
	1	*2*	. . .	*k*
Successes	x_1	x_2	. . .	x_k
Failures	$n_1 - x_1$	$n_2 - x_2$	. . .	$n_k - x_k$

It is important to remember that the statistic on which we base our decision has a distribution that is only approximated by the chi-square distribution. The computed χ^2 values depend on the cell frequencies and consequently are discrete. The continuous chi-square distribution seems to approximate the discrete sampling distribution of X^2 very well, provided the number of degrees of freedom is greater than 1. In a 2×2 contingency table, where we have only 1 degree of freedom, a correction called *Yates' correction for continuity* is applied. The corrected formula then becomes

$$\chi^2(\text{corrected}) = \sum_i \frac{(|o_i - e_i| - 0.5)^2}{e_i}.$$

If the expected cell frequencies are large, the corrected and uncorrected results are almost the same. When the expected frequencies are between 5 and 10, Yates' correction should be applied. For expected frequencies less than 5, the Fisher-Irwin exact test should be used. A discussion of this test may be found in *Basic Concepts of Probability and Statistics* by Hodges and Lehmann. The Fisher-Irwin test may be avoided, however, by choosing a larger sample.

EXERCISES

1. The proportion of adults living in a small town who are college graduates is estimated to be $p = 0.3$. To test this hypothesis a random sample of 15 adults is selected. If the number of college graduates in our sample is

anywhere from 2 to 7, we shall accept the null hypothesis that $p = 0.3$; otherwise we shall conclude that $p \neq 0.3$. Evaluate α assuming $p = 0.3$. Evaluate β for the alternatives $p = 0.2$ and $p = 0.4$. Is this a good test procedure?

2. The proportion of families buying milk from company A in a certain city is believed to be $p = 0.6$. If a random sample of 10 families shows that 3 or less buy milk from company A, we shall reject the hypothesis that $p = 0.6$ in favor of the alternative $p < 0.6$. Find the probability of committing a type I error if the true proportion is $p = 0.6$. Evaluate the probability of committing a type II error for the alternatives $p = 0.3$, $p = 0.4$, and $p = 0.5$.

3. In a large experiment to determine the success of a new drug, 400 patients with a certain disease are to be given the drug. If more than 300 but less than 340 patients are cured, we shall conclude that the drug is 80% effective. Find the probability of committing a type I error. What is the probability of committing a type II error if the new drug is only 70% effective?

4. A new cure has been developed for a certain type of cement that results in a compressive strength of 5000 pounds per square inch and a standard deviation of 120. To test the hypothesis that $\mu = 5000$ against the alternative that $\mu < 5000$, a random sample of 50 pieces of cement are tested. The critical region is defined to be $\bar{X} < 4970$. Find the probability of committing a type I error. Evaluate β for the alternatives $\mu = 4970$ and $\mu = 4960$.

5. An electrical firm manufactures light bulbs that have a length of life that is approximately normally distributed with a mean of 800 hours and a standard deviation of 40 hours. Test the hypothesis that $\mu = 800$ hours against the alternative $\mu \neq 800$ hours if a random sample of 30 bulbs has an average life of 788 hours. Use a 0.04 level of significance.

6. A random sample of 36 drinks from a soft-drink machine has an average content of 7.4 ounces with a standard deviation of 0.48 ounce. Test the hypothesis that $\mu = 7.5$ ounces against the alternative hypothesis $\mu < 7.5$ at the 0.05 level of significance.

7. The average height of males in the freshman class of a certain college has been 68.5 inches, with a standard deviation of 2.7 inches. Is there reason to believe that there has been a change in the average height if a random sample of 50 males in the present freshman class has an average height of 69.7 inches? Use a 0.02 level of significance.

8. It is claimed that an automobile is driven on the average less than 12,000 miles per year. To test this claim a random sample of 100 automobile owners are asked to keep a record of the miles they travel. Would you agree with this claim if the random sample showed an average of 14,500 miles and a standard deviation of 2,400 miles? Use a 0.01 level of significance.

9. Test the hypothesis that the average weight of containers of a particular lubricant is 10 ounces if the weights of a random sample of 10 containers are 10.2, 9.7, 10.1, 10.3, 10.1, 9.8, 9.9, 10.4, 10.3, and 9.8 ounces. Use a 0.01 level of significance and assume that the distribution of weights is normal.

10. A random sample of size 20 from a normal distribution has a mean $\bar{x} = 32.8$

and a standard deviation $s = 4.51$. Does this suggest, at the 0.05 level of significance, that the population mean is greater than 30?

11. A random sample of eight cigarettes of a certain brand has an average nicotine content of 18.6 milligrams and a standard deviation of 2.4 milligrams. Is this in line with the manufacturer's claim that the average nicotine content does not exceed 17.5 milligrams? Use a 0.01 level of significance and assume the distribution of nicotine contents to be normal.

12. A male student will spend, on the average, $8.00 for a Saturday evening fraternity party. Test the hypothesis at the 0.1 level of significance that $\mu = \$8$ against the alternative $\mu \neq \$8$ if a random sample of 12 male students attending a homecoming party showed an average expenditure of $8.90 with a standard deviation of $1.75. Assume that the expenses are approximately normally distributed.

13. A random sample of size $n_1 = 25$, taken from a normal population with a standard deviation $\sigma_1 = 5.2$, has a mean $\bar{x}_1 = 81$. A second random sample of size $n_2 = 36$, taken from a different normal population with a standard deviation $\sigma_2 = 3.4$, has a mean $\bar{x}_2 = 76$. Test the hypothesis, at the 0.06 level of significance, that $\mu_1 = \mu_2$ against the alternative $\mu_1 \neq \mu_2$.

14. A farmer claims that the average yield of corn of variety A exceeds the average yield of variety B by at least 12 bushels per acre. To test this claim, 50 acres of each variety are planted and grown under similar conditions. Variety A yielded, on the average, 86.7 bushels per acre with a standard deviation of 6.28 bushels per acre, while variety B yielded, on the average, 77.8 bushels per acre with a standard deviation of 5.61 bushels per acre. Test the farmer's claim using a 0.05 level of significance.

15. A study was made to estimate the difference in salaries of college professors in the private and state colleges of Virginia. A random sample of 100 professors in private colleges showed an average 9-month salary of $13,000 with a standard deviation of $1,300. A random sample of 200 professors in state colleges showed an average salary of $13,900 with a standard deviation of $1,400. Test the hypothesis that the average salary for professors teaching in state colleges does not exceed the average salary for professors teaching in private colleges by more than $500. Use a 0.02 level of significance.

16. Given two random samples of size $n_1 = 11$ and $n_2 = 14$, from two independent normal populations, with $\bar{x}_1 = 75$, $\bar{x}_2 = 60$, $s_1 = 6.1$, and $s_2 = 5.3$, test the hypothesis at the 0.05 level of significance that $\mu_1 = \mu_2$ against the alternative that $\mu_1 \neq \mu_2$. Assume that the population variances are equal.

17. A study is made to see if increasing the substrate concentration has an appreciable effect on the velocity of a chemical reaction. With the substrate concentration of 1.5 moles per liter, the reaction was run 15 times with an average velocity of 7.5 micromoles per 30 minutes and a standard deviation of 1.5. With a substrate concentration of 2.0 moles per liter, 12 runs were made yielding an average velocity of 8.8 micromoles per 30 minutes and a sample standard deviation of 1.2. Would you say that the increase in substrate concentration increases the mean velocity by as much as 0.5 micromoles per 30 minutes? Use a 0.01 level of significance and assume the populations to be approximately normally distributed with equal variances.

18. A large automobile manufacturing company is trying to decide whether to purchase brand A or brand B tires for its new models. To help arrive at a decision an experiment is conducted using 12 of each brand. The tires are run until they wear out. The results are

$$\text{Brand A:} \quad \bar{x}_1 = 23{,}600 \text{ miles}, \quad s_1 = 3{,}200 \text{ miles}$$
$$\text{Brand B:} \quad \bar{x}_2 = 24{,}800 \text{ miles}, \quad s_2 = 3{,}700 \text{ miles}.$$

Test the hypothesis at the 0.05 level of significance that there is no difference in the two brands of tires. Assume the populations to be approximately normally distributed.

19. The following data represent the running times of films produced by two different motion-picture companies:

	Time (minutes)						
Company 1	102	86	98	109	92		
Company 2	81	165	97	134	92	87	114

Test the hypothesis that the average running time of films produced by company 2 exceeds the average running time of films produced by company 1 by 10 minutes against the one-sided alternative that the difference is more than 10 minutes. Use a 0.1 level of significance and assume the distributions of times to be approximately normal.

20. In Exercise 20, Chapter 6, use the t distribution to test the hypothesis, at the 0.05 level of significance, that the average yields of the two varieties of wheat are equal against the alternative hypothesis that they are unequal.

21. In Exercise 21, Chapter 6, use the t distribution to test the hypothesis, at the 0.01 level of significance, that $\mu_1 = \mu_2$ against the alternative hypothesis that $\mu_1 < \mu_2$.

22. In Exercise 22, Chapter 6, use the t distribution to test the hypothesis, at the 0.05 level of significance, that the diet reduces a person's weight by 10 pounds on the average against the alternative hypothesis that the mean difference in weight is less than 10 pounds.

23. Test the hypothesis that $\sigma^2 = 0.03$ against the alternative hypothesis that $\sigma^2 \neq 0.03$ in Exercise 9. Use a 0.01 level of significance.

24. Test the hypothesis that $\sigma = 6$ against the alternative that $\sigma < 6$ in Exercise 10. Use a 0.05 level of significance.

25. Test the hypothesis that $\sigma^2 = 2.3$ against the alternative that $\sigma^2 \neq 2.3$ in Exercise 11. Use a 0.05 level of significance.

26. Test the hypothesis that $\sigma = 1.40$ against the alternative that $\sigma > 1.40$ in Exercise 12. Use a 0.01 level of significance.

27. Test the hypothesis that $\sigma_1^2 = \sigma_2^2$ against the alternative that $\sigma_1^2 > \sigma_2^2$ in Exercise 16. Use a 0.01 level of significance.

28. Test the hypothesis that $\sigma_1 = \sigma_2$ against the alternative that $\sigma_1 < \sigma_2$ in Exercise 18. Use a 0.05 level of significance.

29. Test the hypothesis that $\sigma_1^2 = \sigma_2^2$ against the alternative that $\sigma_1^2 \neq \sigma_2^2$ in Exercise 19. Use a 0.10 level of significance.

30. How large a sample is required in Exercise 6 if the power of our test is to be 0.90 when the true mean is 7.2? Assume $\sigma = 0.48$.

31. How large a sample is required in Exercise 7 if the power of our test is to be 0.95 when the true average height differs from 68.5 by 1.2 inches?

32. How large should the samples be in Exercise 14 if the power of our test is to be 0.95 when the true difference between corn varieties A and B is 8 bushels per acre?

33. How large a sample is required in Exercise 11 if the power of our test is to be 0.8 when the true nicotine content exceeds the hypothesized value by 1.2σ?

34. On testing

$$H_0: \quad \mu = 14$$
$$H_1: \quad \mu \neq 14$$

an $\alpha = 0.05$ level t test is being considered. What sample size is necessary in order that the probability is 0.15 of falsely accepting H_0 when the true population mean differs from 14 by 0.5? From a preliminary sample we estimate σ to be 1.25.

35. In testing the hypothesis

$$H_0: \quad \sigma^2 = 2$$
$$H_1: \quad \sigma^2 > 2$$

at the 0.05 level of significance with a sample of size 25, find the power of the test when the true variance is actually $\sigma^2 = 3.5$?

36. The following data represent the weights of personal luggage carried on a large aircraft by the members of two baseball clubs:

Club A	34	39	41	28	33	
Club B	36	40	35	31	39	36

Use the Wilcoxon two-sample test with $\alpha = 0.05$ to test the hypothesis that the two clubs carry the same amount of luggage on the average against the alternative hypothesis that the average weight of luggage for club B is greater than that of club A.

37. A fishing line is being manufactured by two processes. To determine if there is a difference in the mean breaking strength of the lines, 10 pieces by each process are selected and tested for breaking strength. The results are as follows:

Process 1	10.4	9.8	11.5	10.0	9.9	9.6	10.9	11.8	9.3	10.7
Process 2	8.7	11.2	9.8	10.1	10.8	9.5	11.0	9.8	10.5	9.9

Use the Wilcoxon two-sample test with $\alpha = 0.1$ to determine if there is a difference between the mean breaking strengths of the lines manufactured by the two processes.

38. From a mathematics class of 12 equally capable students using programmed materials, 5 are selected at random and given additional instruction by the teacher. The results on the final examination were as follows:

	Grade						
Additional instruction	87	69	78	91	80		
No additional instruction	75	88	64	82	93	79	67

Use the Wilcoxon two-sample test with $\alpha = 0.05$ to determine if the additional instruction affects the average grade.

39. The weights of four people before they stopped smoking and 5 weeks after they stopped smoking are as follows:

	Individual			
	1	*2*	*3*	*4*
Before	148	176	153	116
After	154	179	151	121

Use the Wilcoxon test for paired observations to test the hypothesis, at the 0.05 level of significance, that giving up smoking has no effect on a person's weight against the alternative that one's weight increases if he quits smoking.

40. In Exercise 20, Chapter 6, use the Wilcoxon test for paired observations to test the hypothesis, at the 0.05 level of significance, that the average yields of the two varieties of wheat are equal against the alternative hypothesis that they are unequal. Compare your conclusion with that of Exercise 20.

41. In Exercise 21, Chapter 6, use the Wilcoxon test for paired observations to test the hypothesis, at the 0.01 level of significance, that $\mu_1 = \mu_2$ against the alternative hypothesis that $\mu_1 < \mu_2$. Compare your conclusions with that of Exercise 21.

42. In Exercise 22, Chapter 6, use the Wilcoxon test for paired observations to test the hypothesis, at the 0.05 level of significance, that the diet reduces a person's weight by 10 pounds on the average against the alternative hypothesis that the mean difference in weight is less than 10 pounds. Compare your conclusion with that of Exercise 22.

43. It is believed that at least 60% of the residents in a certain area favor an annexation suit by a neighboring city. What conclusion would you draw if only 110 in a sample of 200 voters favor the suit? Use a 0.04 level of significance.

44. A manufacturer of cigarettes claims that 20% of the cigarette smokers prefer brand X. To test this claim a random sample of 20 cigarette smokers is selected and the smokers are asked what brand they prefer. If 6 of the 20 smokers prefer brand X, what conclusion do we draw? Use a 0.01 level of significance.

45. A new radar device is being considered for a certain defense missile system. The system is checked by experimenting with actual aircraft in which a *kill* or a *no kill* is simulated. If in 300 trials, 250 kills occur, accept or reject, at the 0.04 level of significance, the claim that the probability of a kill with the new system exceeds 0.8.

46. At a certain college it is estimated that fewer than 25% of the students have cars on campus. Does this seem to be a valid estimate if in a random sample of 90 college students, 28 are found to have cars? Use a 0.05 level of significance.

47. In a study to estimate the proportion of housewives who own an automatic dryer, it is found that 63 of 100 urban residents have a dryer and 59 of 125 suburban residents own a dryer. Is there a significant difference between the proportion of urban and suburban housewives who own an automatic dryer? Use a 0.04 level of significance.

48. A cigarette manufacturing firm distributes two brands of cigarettes. If it is found that 56 of 200 smokers prefer brand A and that 29 of 150 smokers prefer brand B, can we conclude at the 0.06 level of significance that brand A outsells brand B?

49. A random sample of 100 men and 100 women at a southern college is asked if they have an automobile on campus. If 31 of the men and 24 of the women have cars, can we conclude that more men than women have cars on campus? Use a 0.01 level of significance.

50. A study is made to determine if a cold climate contributes more to absenteeism from school during a semester than a warmer climate. Two groups of students are selected at random, one group from Maine and the other from Alabama. Of the 300 students from Maine, 72 were absent at least 1 day during the semester, and of the 400 students from Alabama, 70 were absent 1 or more days. Can we conclude that a colder climate results in a greater number of students being absent from school at least 1 day during the semester? Use a 0.05 level of significance.

51. A die is tossed 180 times with the following results:

x	1	2	3	4	5	6
f	28	36	36	30	27	23

Is this a balanced die? Use a 0.01 level of significance.

52. In 100 tosses of a coin, 63 heads and 37 tails are observed. Is this a balanced coin? Use a 0.05 level of significance.

53. Three marbles are selected from an urn containing five red marbles and three green marbles. After recording the number X of red marbles, the marbles

are replaced in the urn and the experiment repeated 112 times. The results obtained are as follows:

x	0	1	2	3
f	1	31	55	25

Test the hypothesis at the 0.05 level of significance that the recorded data may be fitted by the hypergeometric distribution $h(x;8,3,5)$, $x = 0, 1, 2, 3$.

54. Three cards are drawn from an ordinary deck of playing cards, with replacement, and the number Y of spades is recorded. After repeating the experiment 64 times, the following outcomes were recorded:

y	0	1	2	3
f	21	31	12	0

Test the hypothesis at the 0.01 level of significance that the recorded data may be fitted by the binomial distribution $b(y;3,1/4)$, $y = 0, 1, 2, 3$.

55. The grades in a statistics course for a particular semester were as follows:

Grade	A	B	C	D	F
f	14	18	32	20	16

Test the hypothesis, at the 0.05 level of significance, that the distribution of grades is uniform.

56. A coin is thrown until a head occurs and the number X of tosses recorded. After repeating the experiment 256 times, we obtained the following results:

x	1	2	3	4	5	6	7	8
f	136	60	34	12	9	1	3	1

Test the hypothesis at the 0.05 level of significance that the observed distribution of X may be fitted by the geometric distribution $g(x;1/2)$, $x = 1, 2, 3 \ldots$.

57. In Exercise 15, Chapter 2, test the goodness of fit between the observed frequencies and the expected normal frequencies, using a 0.05 level of significance.

58. In an experiment to study the dependence of hypertension on smoking habits, the following data were taken on 180 individuals:

	Non-smokers	Moderate smokers	Heavy smokers
Hypertension	21	36	30
No hypertension	48	26	19

Test the hypothesis that the presence or absence of hypertension is independent of smoking habits. Use a 0.05 level of significance.

59. A random sample of 200 married men, all retired, were classified according to education and number of children:

	Number of children		
Education	0–1	2–3	Over 3
Elementary	14	37	32
Secondary	19	42	17
College	12	17	10

Test the hypothesis, at the 0.05 level of significance, that the size of a family is independent of the level of education attained by the father.

60. In a shop study, a set of data was collected to determine whether or not the proportion of defectives produced by workers was the same for the day, evening, or night shift worked. The following data were collected on the items produced:

	Shift		
	Day	Evening	Midnight
Defective	45	55	70
Nondefective	905	890	870

What is your conclusion? Use an $\alpha = 0.025$ level of significance.

8 LINEAR REGRESSION AND CORRELATION

8.1 Linear Regression

Often in practice one is called upon to solve problems involving sets of variables when it is known that there exists some inherent relationship among the variables. For example, in an industrial situation it may be known that the tar content in the outlet stream in a chemical process is related to the inlet temperature. It may be of interest to develop a method of prediction, that is, a procedure for estimating the tar content for various levels of the inlet temperature from experimental information. The statistical aspect of the problem then becomes one of arriving at the best estimate of the relationship between the variables.

For this example and most applications there is a clear distinction between the variables as far as their role in the experimental process is concerned. Quite often there is a single *dependent variable* or response Y, which is uncontrolled in the experiment. This response depends on one or more *independent variables*, say x_1, x_2, ..., x_k, which are measured with negligible error and indeed are often controlled in the experiment. Thus the independent variables x_1, x_2, ..., x_k are *not* random variables but are fixed quantities preselected by the investigator and have no distributional properties. In the example cited earlier, inlet temperature is the independent variable and tar content is the response Y. The relationship, fitted to a set of experimental data, is characterized by a prediction equation called a *regression equation*. In the case of a single Y and a single x, the situation becomes a regression of Y on x. For k independent variables, we speak in terms of a regression of Y on x_1, x_2, ..., x_k. A chemical engineer may, in fact, be concerned with the amount of hydrogen lost from samples of a particular metal when the material is placed in storage. In this case there may be two inputs, storage time x_1 in hours and storage temperature x_2 in degrees centigrade. The response would then be hydrogen loss Y in parts per million.

In this chapter we shall deal with the topic of *simple linear regression*, treating only the case of a single independent variable. For the case of more than one independent variable, the reader is referred to Chapter 9. Let us denote a random sample of size n by the set $\{(x_i, y_i); i = 1, 2, ..., n\}$. If additional samples were taken using exactly the same values of x, we would expect the y values to vary. Hence the value y_i in the ordered pair (x_i, y_i) is a value of some random variable Y_i. For convenience we define $Y \mid x$ to be the random variable Y corresponding to a fixed value x and denote its probability distribution by $f(y \mid x)$. Clearly then, if $x = x_i$, the symbol $Y \mid x_i$ represents the random variable Y_i.

The term *linear regression* implies that the mean of $Y \mid x$ is linearly related to x in the usual slope-intercept form; namely,

$$\mu_{Y \mid x} = \alpha + \beta x,$$

where α and β are parameters to be estimated from the sample data. Denoting their estimates by a and b, respectively, the estimated response $\hat{y}$ is obtained from the sample regression line

$$\hat{y} = a + bx.$$

Consider the experimental data of Table 8.1, which has been plotted in

Table 8.1 Experimental Data

i	1	2	3	4	5	6	7	8	9
x_i	1.5	1.8	2.4	3.0	3.5	3.9	4.4	4.8	5.0
y_i	4.8	5.7	7.0	8.3	10.9	12.4	13.1	13.6	15.3

Figure 8.1 to give a *scatter diagram*. The assumption of linearity appears to be reasonable.

The sample regression line and a hypothetical true regression line have been drawn on the scatter diagram of Figure 8.1. The agreement between the sample line and the unknown hypothetical line should be good if we have a large amount of data available.

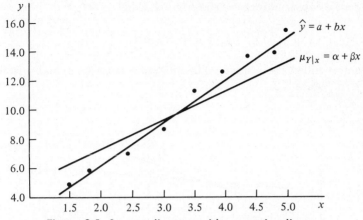

Figure 8.1 Scatter diagram with regression lines.

In the following sections we shall develop procedures for finding estimators of the *regression coefficients*, α and β, in order that the regression
equation can be used for predicting or estimating the mean response for
specific values of the independent variable x.

8.2 Simple Linear Regression

For the case where there is a single x and a single Y, the data take the
form of pairs of observations $\{(x_i, y_i); i = 1, 2, \ldots, n\}$. If the values for x
are controlled, that is, the experiment is *designed*, then the experimental
process is to fix or choose the x_i values in advance and observe the corresponding y_i values.

Assuming that all the means, $\mu_{Y|x}$, fall on a straight line, the random
variable $Y_i = Y \mid x_i$ may then be written

$$Y_i = \mu_{Y|x_i} + E_i = \alpha + \beta x_i + E_i,$$

where the random variable E_i must necessarily have a mean of zero. Each
observation (x_i, y_i) in our sample satisfies the relation

$$y_i = \alpha + \beta x_i + \epsilon_i.$$

Similarly, using the estimated regression line

$$\hat{y} = a + bx,$$

each pair of observations satisfies the relation

$$y_i = a + bx_i + e_i,$$

where e_i is called the *residual*. The difference between e_i and ϵ_i is clearly shown in Figure 8.2.

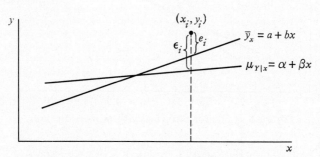

Figure 8.2 Comparing ϵ_i with the residual e_i.

We shall find a and b, the estimates of α and β, so that the sum of the squares of the residuals is a minimum. The residual sum of squares is often called the sum of squares of the errors about the regression line and denoted by SSE. This minimization procedure for estimating parameters is called *the method of least squares*. Hence, we shall find a and b so as to minimize

$$\text{SSE} = \sum_{i=1}^{n} e_i^2 = \sum_{i=1}^{n} (y_i - a - bx_i)^2.$$

Differentiating SSE with respect to a and b, we have

$$\frac{\partial(\text{SSE})}{\partial a} = -2 \sum_{i=1}^{n} (y_i - a - bx_i)$$

$$\frac{\partial(\text{SSE})}{\partial b} = -2 \sum_{i=1}^{n} (y_i - a - bx_i)x_i.$$

Setting the partial derivatives equal to zero and rearranging the terms, we obtain the equations (called the *normal equations*)

$$na + b \sum_{i=1}^{n} x_i = \sum_{i=1}^{n} y_i$$

$$a \sum_{i=1}^{n} x_i + b \sum_{i=1}^{n} x_i^2 = \sum_{i=1}^{n} x_i y_i,$$

which yield

$$b = \frac{n \sum\limits_{i=1}^{n} x_i y_i - \left(\sum\limits_{i=1}^{n} x_i\right)\left(\sum\limits_{i=1}^{n} y_i\right)}{n \sum\limits_{i=1}^{n} x_i^2 - \left(\sum\limits_{i=1}^{n} x_i\right)^2}.$$

From the first of these two normal equations we can write

$$a = \bar{y} - b\bar{x}.$$

ESTIMATION OF THE PARAMETERS α AND β *The regression line*

$$\mu_{Y|x} = \alpha + \beta x$$

is estimated from the sample $\{(x_i, y_i); \ i = 1, 2, \ldots, n\}$ by the line

$$\hat{y} = a + bx,$$

where

$$b = \frac{n \sum\limits_{i=1}^{n} x_i y_i - \left(\sum\limits_{i=1}^{n} x_i\right)\left(\sum\limits_{i=1}^{n} y_i\right)}{n \sum\limits_{i=1}^{n} x_i^2 - \left(\sum\limits_{i=1}^{n} x_i\right)^2}$$

$$a = \bar{y} - b\bar{x}.$$

The calculations of a and b using the data of Table 8.1 are illustrated by the following example.

Example 8.1 Estimate the regression line for the data of Table 8.1.

Solution. Using a desk calculator we find that

$$\sum_{i=1}^{9} x_i = 30.3, \quad \sum_{i=1}^{9} y_i = 91.1, \quad \sum_{i=1}^{9} x_i y_i = 345.09,$$

$$\sum_{i=1}^{9} x_i^2 = 115.11, \quad \bar{x} = 3.3667, \quad \bar{y} = 10.1222.$$

Therefore

$$b = \frac{(9)(345.09) - (30.3)(91.1)}{(9)(115.11) - (30.3)^2} = 2.9303$$

$$a = 10.1222 - (2.9303)(3.3667) = 0.2568.$$

Thus the estimated regression line is given by

$$\hat{y} = 0.2568 + 2.9303x.$$

By substituting any two of the given values of x into this equation, say $x_1 = 1.5$ and $x_2 = 5.0$, we obtain the ordinates $\hat{y}_1 = 4.7$ and $\hat{y}_2 = 14.9$. The sample regression line in Figure 8.1 was drawn by connecting these two points with a straight line.

8.3 Confidence Limits and Tests of Significance

Aside from merely estimating the linear relationship between x and Y for purposes of prediction, the experimenter may also be interested in drawing certain inferences about the slope, intercept, or the general quality of the estimated regression line. Too often, in fact, regression results are reported by the scientist without any reference to how well b estimates β or how well the regression line will eventually predict response.

In addition to the assumption that the error term in the model

$$Y_i = \alpha + \beta x_i + E_i$$

is a random variable with mean zero, suppose we make the further assumption that each E_i is normally distributed with the same variance σ^2 and that $E_1, E_2, \ldots, E_n$ are independent from run to run in the experiment. With this normality assumption on the E_i's, we have a procedure for constructing confidence intervals for α and β and for single points on the actual regression line.

It is important to remember that our values of a and b are only estimates of the true parameters α and β based on a given sample of n observations. The different estimates of α and β that could be computed by drawing several samples of size n may be thought of as values assumed by the random variables A and B.

Since the values of x remain fixed, the values of A and B depend on the variations in the values of y, or, more precisely, on the values of the random variables $Y_1, Y_2, \ldots, Y_n$. The distributional assumptions on the E_i's implies that the Y_i's, $i = 1, 2, \ldots, n$, are likewise independently distributed, each with probability distribution $n(y_i; \alpha + \beta x_i, \sigma)$, and since the estimator

$$B = \frac{n \sum_{i=1}^{n} x_i Y_i - \left(\sum_{i=1}^{n} x_i\right)\left(\sum_{i=1}^{n} Y_i\right)}{n \sum_{i=1}^{n} x_i^2 - \left(\sum_{i=1}^{n} x_i\right)^2}$$

$$= \frac{\sum_{i=1}^{n} (x_i - \bar{x})(Y_i - \bar{Y})}{\sum_{i=1}^{n} (x_i - \bar{x})^2}$$

$$= \frac{\sum_{i=1}^{n} (x_i - \bar{x}) Y_i}{\sum_{i=1}^{n} (x_i - \bar{x})^2}$$

is a linear function of the random variables Y_1, Y_2, ..., Y_n with coefficients

$$a_i = \frac{x_i - \bar{x}}{\sum_{i=1}^{n} (x_i - \bar{x})^2}, \qquad i = 1, 2, \ldots, n,$$

we may conclude from Theorem 5.11 that it is normally distributed with mean

$$\mu_B = E(B) = \frac{\sum_{i=1}^{n} (x_i - \bar{x}) E(Y_i)}{\sum_{i=1}^{n} (x_i - \bar{x})^2}$$

$$= \frac{\sum_{i=1}^{n} (x_i - \bar{x})(\alpha + \beta x_i)}{\sum_{i=1}^{n} (x_i - \bar{x})^2} = \beta$$

and variance

$$\sigma_B^2 = \frac{\sum_{i=1}^{n} (x_i - \bar{x})^2 \sigma_{Y_i}^2}{\left[\sum_{i=1}^{n} (x_i - \bar{x})^2\right]^2} = \frac{\sigma^2}{\sum_{i=1}^{n} (x_i - \bar{x})^2}.$$

It can be shown (Exercise 7) that the random variable A is also normally distributed with mean

$$\mu_A = \alpha$$

and variance

$$\sigma_A^2 = \left[\frac{\sum\limits_{i=1}^{n} x_i^2}{n \sum\limits_{i=1}^{n} (x_i - \bar{x})^2} \right] \sigma^2.$$

To be able to draw inferences on α and β or on the regression line itself, it becomes necessary to arrive at an estimator for σ^2. An unbiased estimator of σ^2 with $n - 2$ degrees of freedom, reflecting the variation about the regression line, is given by

$$s^2 = \frac{\text{SSE}}{n - 2}.$$

From a computational point of view it is advantageous to introduce the notation

$$S_{xx} = \sum_{i=1}^{n} (x_i - \bar{x})^2 = \sum_{i=1}^{n} x_i^2 - \frac{\left(\sum\limits_{i=1}^{n} x_i \right)^2}{n}$$

$$S_{yy} = \sum_{i=1}^{n} (y_i - \bar{y})^2 = \sum_{i=1}^{n} y_i^2 - \frac{\left(\sum\limits_{i=1}^{n} y_i \right)^2}{n}$$

$$S_{xy} = \sum_{i=1}^{n} (x_i - \bar{x})(y_i - \bar{y}) = \sum_{i=1}^{n} x_i y_i - \frac{\left(\sum\limits_{i=1}^{n} x_i \right) \left(\sum\limits_{i=1}^{n} y_i \right)}{n}.$$

Now we may write the error sum of squares as follows:

$$\begin{aligned}
\text{SSE} &= \sum_{i=1}^{n} (y_i - a - bx_i)^2 \\
&= \sum_{i=1}^{n} [(y_i - \bar{y}) - b(x_i - \bar{x})]^2 \\
&= \sum_{i=1}^{n} (y_i - \bar{y})^2 - 2b \sum_{i=1}^{n} (x_i - \bar{x})(y_i - \bar{y}) + b^2 \sum_{i=1}^{n} (x_i - \bar{x})^2 \\
&= S_{yy} - 2b S_{xy} + b^2 S_{xx} \\
&= S_{yy} - b S_{xy},
\end{aligned}$$

the final step following from the fact that $b = S_{xy}/S_{xx}$.

Since B is a normal random variable and $(n - 2)S^2/\sigma^2$ is a chi-square variable with $n - 2$ degrees of freedom, Theorem 5.17 assures us that the statistic

$$T = \frac{(B - \beta)/(\sigma/\sqrt{S_{xx}})}{S/\sigma}$$

$$= \frac{B - \beta}{S/\sqrt{S_{xx}}}$$

has a t distribution with $n - 2$ degrees of freedom. The statistic T can be used to construct a $(1 - \alpha)100\%$ confidence interval for the coefficient β.

CONFIDENCE INTERVAL FOR β *A $(1 - \alpha)100\%$ confidence interval for the parameter β in the regression line $\mu_{Y|x} = \alpha + \beta x$ is*

$$b - \frac{t_{\alpha/2}s}{\sqrt{S_{xx}}} < \beta < b + \frac{t_{\alpha/2}s}{\sqrt{S_{xx}}},$$

where $t_{\alpha/2}$ is a value of the t distribution with $n - 2$ degrees of freedom.

Example 8.2 Find a 95% confidence interval for β in the regression line $\mu_{Y|x} = \alpha + \beta x$ based on the data in Table 8.1.

Solution. In Example 8.1 we found that

$$\sum_{i=1}^{9} x_i = 30.3, \quad \sum_{i=1}^{9} x_i^2 = 115.11, \quad \sum_{i=1}^{9} y_i = 91.1, \quad \sum_{i=1}^{9} x_i y_i = 345.09.$$

Referring to the data in Table 8.1, we now find $\sum_{i=1}^{9} y_i^2 = 1036.65$. Therefore

$$S_{xx} = 115.11 - \frac{(30.3)^2}{9} = 13.10$$

$$S_{yy} = 1036.65 - \frac{(91.1)^2}{9} = 114.52$$

$$S_{xy} = 345.09 - \frac{(30.3)(91.1)}{9} = 38.39.$$

Recall that $b = 2.9303$. Hence

$$s^2 = \frac{S_{yy} - bS_{xy}}{n - 2}$$

$$= \frac{114.52 - (2.9303)(38.39)}{7} = 0.2894.$$

Therefore, taking square roots we obtain $\sqrt{S_{xx}} = 3.6194$ and $s = 0.5380$. Using Table V we find $t_{0.025} = 2.365$ for 7 degrees of freedom. Therefore a 95% confidence interval for β is given by

$$2.9305 - \frac{(2.365)(0.5380)}{(3.6194)} < \beta < 2.9305 + \frac{(2.365)(0.5380)}{(3.6194)},$$

which simplifies to

$$2.579 < \beta < 3.282.$$

To test the null hypothesis H_0 that $\beta = \beta_0$ against a suitable alternative, we again use the t distribution with $n - 2$ degrees of freedom to establish a critical region and then base our decision on the value of

$$t = \frac{b - \beta_0}{s/\sqrt{S_{xx}}}.$$

The method is illustrated in the following example.

Example 8.3 Using the estimated value $b = 2.9303$ of Example 8.1, test the hypothesis that $\beta = 2.5$ at the 0.01 level of significance against the alternative that $\beta > 2.5$.

Solution.

1. H_0: $\beta = 2.5$.
2. H_1: $\beta > 2.5$.
3. Choose a 0.01 level of significance.
4. Critical region: $T > 2.998$.
5. Computations:

$$t = \frac{2.9303 - 2.5}{0.5380/3.6194} = 2.8948.$$

6. Conclusion: Accept H_0 and conclude that β does not differ significantly from 2.5.

Confidence intervals and hypothesis testing on the coefficient α may be established from the fact that A is also normally distributed. It is not difficult to show that

$$T = \frac{A - \alpha}{S\sqrt{\sum\limits_{i=1}^{n} x_i^2/nS_{xx}}}$$

has a t distribution with $n - 2$ degrees of freedom from which we may construct a $(1 - \alpha)100\%$ confidence interval for α.

CONFIDENCE INTERVAL FOR α A $(1 - \alpha)100\%$ *confidence interval for the parameter* α *in the regression line* $\mu_{Y|x} = \alpha + \beta x$ *is*

$$a - \frac{t_{\alpha/2}s\sqrt{\sum\limits_{i=1}^{n} x_i^2}}{\sqrt{nS_{xx}}} < \alpha < a + \frac{t_{\alpha/2}s\sqrt{\sum\limits_{i=1}^{n} x_i^2}}{\sqrt{nS_{xx}}},$$

where $t_{\alpha/2}$ *is a value of the t distribution with* $n - 2$ *degrees of freedom.*

To test the null hypothesis H_0 that $\alpha = \alpha_0$ against a suitable alternative, we can use the t distribution with $n - 2$ degrees of freedom to establish a critical region and then base our decision on the value of

$$t = \frac{a - \alpha_0}{s\sqrt{\sum\limits_{i=1}^{n} x_i^2/nS_{xx}}}.$$

Quite often it is important for the experimenter to attach a confidence interval on the *mean response* at some fixed level of x not necessarily one of the prechosen values. Suppose we are interested in the mean of Y for $x = x_0$; we are then estimating $\mu_{Y|x_0} = \alpha + \beta x_0$ with the point estimator $\hat{Y}_0 = A + Bx_0$.

Since $E(\hat{Y}_0) = E(A + Bx_0) = \alpha + \beta x_0$, the estimator $\hat{Y}_0$ is unbiased with variance

$$\sigma_{\hat{Y}_0}^2 = \sigma_{A+Bx_0}^2 = \sigma_{\bar{Y}+B(x_0-\bar{x})}^2$$

$$= \sigma^2\left[\frac{1}{n} + \frac{(x_0 - \bar{x})^2}{S_{xx}}\right],$$

the latter following from the fact that $\text{cov}(\bar{Y},B) = 0$ (see Exercise 6).

Thus the $(1 - \alpha)100\%$ confidence interval on the mean response $\mu_{Y|x_0}$ can now be constructed from the statistic

$$T = \frac{\hat{Y}_0 - \mu_{Y|x_0}}{S \sqrt{(1/n) + [(x_0 - \bar{x})^2/S_{xx}]}},$$

which has a t distribution with $n - 2$ degrees of freedom.

CONFIDENCE INTERVAL FOR $\mu_{Y|x_0}$ A $(1 - \alpha)100\%$ *confidence interval for the mean response* $\mu_{Y|x_0}$ *is given by*

$$\hat{y}_0 - t_{\alpha/2}s \sqrt{\frac{1}{n} + \frac{(x_0 - \bar{x})^2}{S_{xx}}} < \mu_{Y|x_0} < \hat{y}_0 + t_{\alpha/2}s \sqrt{\frac{1}{n} + \frac{(x_0 - \bar{x})^2}{S_{xx}}},$$

where $t_{\alpha/2}$ *is a value of the t distribution with* $n - 2$ *degrees of freedom.*

Example 8.4 Using the data of Table 8.1, construct 95% confidence limits for the mean response $\mu_{Y|x}$.

Solution. From the regression equation we find for $x_0 = 2$, say,

$$\hat{y}_0 = 0.2568 + (2.9303)(2) = 6.1174.$$

Previously we had $\bar{x} = 3.3667$, $S_{xx} = 13.10$, $s = 0.5380$, and $t_{0.025} = 2.365$ for 7 degrees of freedom. Therefore a 95% confidence interval for $\mu_{Y|2}$ is given by

$$6.1174 - (2.365)(0.5380) \sqrt{\frac{1}{9} + \frac{(2 - 3.3667)^2}{13.10}} < \mu_{Y|2} < 6.1174$$

$$+ (2.365)(0.5380) \sqrt{\frac{1}{9} + \frac{(2 - 3.3667)^2}{13.10}}$$

or simply

$$5.4765 < \mu_{Y|2} < 6.7583.$$

Repeating the previous calculations for each of several different values of x_0, one can obtain the corresponding confidence limits on each $\mu_{Y|x_0}$. Figure 8.3 displays the data points, the estimated regression line, and the upper and lower confidence limits on the mean of $Y \mid x$.

Another type of interval that is often misinterpreted and confused with that given for $\mu_{Y|x}$ is the prediction interval on a future observed response. Actually, in many instances the prediction interval is more relevant to the

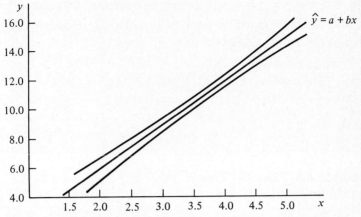

Figure 8.3 Confidence limits for the mean value of $Y \mid x$.

scientist or engineer than the confidence interval on the mean. In the tar content-inlet temperature example, there would certainly be interest not only in estimating the mean tar content at a specific temperature but also in constructing a confidence interval for predicting the actual amount of tar content at the given temperature for some future measurement.

Suppose we begin with the distributional properties of the differences between the ordinates $\hat{y}_0$, obtained from the computed regression line in repeated sampling, and the corresponding true ordinate y_0 at $x = x_0$. We may think of the difference $\hat{y}_0 - y_0$ as a value of the random variable $\hat{Y}_0 - Y_0$, whose sampling distribution can be shown to be normal with mean

$$\mu_{\hat{Y}_0 - Y_0} = E(\hat{Y}_0 - Y_0)$$
$$= E[A + Bx_0 - (\alpha + \beta x_0 + E_0)]$$
$$= 0$$

and variance

$$\sigma^2_{\hat{Y}_0 - Y_0} = \sigma^2_{A + Bx_0 - E_0}$$
$$= \sigma^2_{\bar{Y} + B(x_0 - x) - E_0}$$
$$= \sigma^2 \left[1 + \frac{1}{n} + \frac{(x_0 - \bar{x})^2}{S_{xx}} \right].$$

Thus the $(1 - \alpha)100\%$ confidence interval for a single predicted value y_0 can be constructed from the statistic

$$T = \frac{\hat{Y}_0 - Y_0}{S \sqrt{1 + (1/n) + [(x_0 - \bar{x})^2/S_{xx}]}},$$

which has a t distribution with $n - 2$ degrees of freedom.

CONFIDENCE INTERVAL FOR y_0 A $(1 - \alpha)100\%$ *confidence interval for a single response y_0 is given by*

$$\hat{y}_0 - t_{\alpha/2}s \sqrt{1 + \frac{1}{n} + \frac{(x_0 - \bar{x})^2}{S_{xx}}} < y_0 < \hat{y}_0 + t_{\alpha/2}s \sqrt{1 + \frac{1}{n} + \frac{(x_0 - \bar{x})^2}{S_{xx}}},$$

where $t_{\alpha/2}$ is a value of the t distribution with $n - 2$ degrees of freedom.

Example 8.5 Using the data of Table 8.1, construct a 95% confidence interval for y_0 when $x_0 = 2$.

Solution. We have $n = 9, x_0 = 2, \bar{x} = 3.3667, \hat{y}_0 = 6.1171, S_{xx} = 13.10$, $s = 0.5380$, and $t_{0.025} = 2.365$ for 7 degrees of freedom. Therefore a 95% confidence interval for y_0 is given by

$$6.1174 - (2.365)(0.5380) \sqrt{1 + \frac{1}{9} + \frac{(2 - 3.3667)^2}{13.10}} < y_0 < 6.1174$$
$$+ (2.365)(0.5380) \sqrt{1 + \frac{1}{9} + \frac{(2 - 3.3667)^2}{13.10}},$$

which simplifies to

$$4.6927 < y_0 < 7.5421.$$

8.4 Choice of a Regression Model

Much of what has been presented to this point on regression involving a single independent variable depends on the assumption that the model chosen is correct, namely, the presumption that $\mu_{Y|x}$ is related to x linearly in the parameters. Certainly one would not expect the prediction of the response to be good if there are several independent variables, not considered in the model, that are affecting the response and are varying in the system. In addition the prediction would certainly be inadequate if the true structure relating $\mu_{Y|x}$ to x is extremely nonlinear in the range of the variables considered.

Often the simple linear regression model is used even though it is known that the model is something other than linear or that the true structure is unknown. This approach is often sound, particularly when the range of x is narrow. Thus the model used becomes an approximating function that one hopes is an adequate representation of the true picture in the region of interest. One should note, however, the effect of an inadequate model on the results presented thus far. For example, if the true model, unknown to the experimenter, is linear in more than one x, say,

$$\mu_{Y|x_1,x_2} = \alpha + \beta x_1 + \gamma x_2,$$

then the ordinary least squares estimate $b = S_{xy}/S_{xx}$, calculated by only considering x_1 in the experiment, is, under general circumstances, a biased estimate of the coefficient β, the bias being a function of the additional coefficient γ (see Exercise 16). Also, the estimate s^2 for σ^2 is biased due to the additional variable.

8.5 Analysis-of-Variance Approach

Often the problem of analyzing the quality of the estimated regression line is handled through an *analysis-of-variance* approach. This is merely a procedure whereby the total variation in the dependent variable is subdivided into meaningful components that are then observed and treated in a systematic fashion. The analysis of variance, discussed extensively in Chapter 11, is a powerful tool that is used in many applications.

Suppose we have n experimental data points in the usual form (x_i, y_i) and that the regression line is estimated. In our estimation of σ^2 in Section 8.3 we established the identity

$$S_{yy} = bS_{xy} + \text{SSE}.$$

So we have achieved a partitioning of the *total corrected sum of squares of y* into two components that should reflect particular meaning to the experimenter. We shall indicate this partitioning symbolically as

$$\text{SST} = \text{SSR} + \text{SSE}.$$

The first component on the right is called the *regression sum of squares* and it reflects the amount of variation in the y values explained by the model, in this case the postulated straight line. The second component is just the familiar error sum of squares, which reflects variation about the regression line.

Since SSR/σ^2 and SSE/σ^2 are values of independent chi-square variables with 1 and $n-2$ degrees of freedom, respectively, it follows that SST/σ^2 is also a value of a chi-square variable with $n-1$ degrees of freedom. To test the hypothesis

$$H_0: \quad \beta = 0$$

$$H_1: \quad \beta \neq 0,$$

where the null hypothesis essentially says that variation in Y is not explained by the straight line but rather by chance or random fluctuations, we compute

$$f = \frac{\text{SSR}/1}{\text{SSE}/(n-2)} = \frac{\text{SSR}}{s^2}$$

and reject H_0 at the α level of significance when $f > f_\alpha(1, n-2)$.

In practice, one first computes

$$\text{SST} = S_{yy}$$
$$\text{SSR} = bS_{xy}$$

and then, making use of the previous sum of squares identity, obtains

$$\text{SSE} = \text{SST} - \text{SSR}.$$

The computations are usually summarized by means of an *analysis-of-variance table* as indicated in Table 8.2.

Table 8.2 Analysis of Variance for Testing $\beta = 0$

Source of variation	Sum of squares	Degrees of freedom	Mean square	Computed f
Regression	SSR	1	SSR	SSR/s^2
Error	SSE	$n-2$	$s^2 = \dfrac{\text{SSE}}{n-2}$	
Total	SST	$n-1$		

When the null hypothesis is rejected, that is, when the computed F statistic exceeds the critical value $f_\alpha(1, n-2)$, we conclude that there is a significant amount of variation in the response accounted for by the postulated model, the straight-line function. If the F statistic is in the acceptance region, we conclude that the data did not reflect sufficient evidence to support the model postulated.

In Section 8.3 a procedure was given whereby the statistic

$$T = \frac{B - \beta_0}{S/\sqrt{S_{xx}}}$$

was used to test the hypothesis

$$H_0: \quad \beta = \beta_0$$
$$H_1: \quad \beta \neq \beta_0,$$

where T follows the t distribution with $n - 2$ degrees of freedom. The hypothesis is rejected if $|T| > t_{\alpha/2}$ for an α level of significance. It is interesting to note that in the special case in which we are testing

$$H_0: \quad \beta = 0$$
$$H_1: \quad \beta \neq 0$$

the value of our T statistic becomes

$$t = \frac{b}{s/\sqrt{S_{xx}}}$$

and the hypothesis is identical to that being tested in Table 8.2. Namely, the null hypothesis states that the variation in the response is due merely to chance. The analysis of variance uses the F distribution rather than the t distribution but *the two procedures are identical*. This we can see by writing

$$t^2 = \frac{b^2 S_{xx}}{s^2} = \frac{b S_{xy}}{s^2} = \frac{\text{SSR}}{s^2},$$

which is identical to the f value used in the analysis of variance. The basic relationship between the t distribution with ν degrees of freedom and the F distribution with 1 and ν degrees of freedom is given by

$$t^2_{\alpha/2} = f_\alpha(1,\nu).$$

8.6 Repeated Measurements on the Response

Quite often it is advantageous for the experimenter to obtain repeated observations for each value of x. While it is not necessary to have these repetitions in order to estimate α and β, nevertheless it does enable the experimenter to obtain quantitative information concerning the appropriateness of the model. In fact if repeated observations are at his disposal, the experimenter can make a significance test to aid in determining whether or not the model is adequate.

Let us select a random sample of n observations using k distinct values of x, say $x_1, x_2, \ldots, x_k$, such that the sample contains n_1 observed values of the random variable Y_1 corresponding to x_1, n_2 observed values of Y_2 corresponding to $x_2, \ldots, n_k$ observed values of Y_k corresponding to x_k. Of necessity, $n = \sum\limits_{i=1}^{k} n_i$. We define

$$y_{ij} = \text{the } j\text{th value of the random variable } Y_i$$

$$T_i. = \sum_{i=1}^{n_i} y_{ij}.$$

Hence, if $n_4 = 3$ measurements of Y are made corresponding to $x = x_4$, we would indicate these observations by y_{41}, y_{42}, and y_{43}. Then

$$T_4. = y_{41} + y_{42} + y_{43}.$$

The error sum of squares consists of two parts: the amount due to the variation between the values of Y within given values of x and a component that is normally called the *lack of fit* contribution. The first component reflects mere random variation or *pure experimental error*, while the second component is a measure of the systematic variation brought about by higher-order terms. In our case these are terms in x other than the linear or *first-order contribution*. Note that in choosing a linear model we are essentially assuming that this second component does not exist and hence our error sum of squares is completely due to random errors. If this should be the case, then $s^2 = \text{SSE}/(n - 2)$ is an unbiased estimate of σ^2. However, if the model does not adequately fit the data, then the error sum of squares is inflated and produces a biased estimate of σ^2. The procedure for separating the error sum of squares into two components representing pure error and lack of fit is as follows:

1. Compute the pure error sum of squares

$$\sum_{i=1}^{k} \sum_{j=1}^{n_i} (y_{ij} - \bar{y}_i.)^2 = \sum_{i=1}^{k} \sum_{j=1}^{n_i} y_{ij}^2 - \sum_{i=1}^{k} \frac{T_i.^2}{n_i}.$$

   This sum of squares has $n - k$ degrees of freedom associated with it and the resulting mean square is our unbiased estimator of σ^2 and thus will be denoted by s^2.

2. Subtract the pure error sum of squares from the error sum of squares, SSE, thereby obtaining the sum of squares owing to lack of fit. The degrees of freedom for lack of fit are also obtained by simply subtracting $(n - 2) - (n - k) = k - 2$.

The computations required for testing hypotheses in a regression problem with repeated measurements on the response may be summarized as shown in Table 8.3.

Table 8.3 Analysis of Variance for Repeated Measurements on the Response

Source of variation	Sum of squares	Degrees of freedom	Mean square	Computed f
Regression	SSR	1	SSR	$\dfrac{\text{SSR}}{s^2}$
Error	SSE	$n-2$		
Lack of fit	$\{$ SSE $-$ SSE(pure)	$\{$ $k-2$	$\dfrac{\text{SSE}-\text{SSE(pure)}}{k-2}$	$\dfrac{\text{SSE}-\text{SSE(pure)}}{s^2(k-2)}$
Pure error	$\quad$ SSE(pure)	$\quad$ $n-k$	$s^2 = \dfrac{\text{SSE(pure)}}{n-k}$	
Total	SST	$n-1$		

The concept of lack of fit is extremely important in applications of regression analysis. In fact the need to construct or design an experiment that will account for lack of fit becomes more critical as the problem and the underlying mechanism involved become more complicated. Surely one cannot always be certain that his postulated structure, in this case the linear regression model, is correct or even an adequate representation. The following example shows how the error sum of squares is partitioned into the two components representing pure error and lack of fit. The adequacy of the model is tested at the α level of significance by comparing the lack-of-fit mean square divided by s^2 with $f_\alpha(k-2, n-k)$.

Example 8.6 Observations on the yield of a chemical reaction taken at various temperatures were recorded as follows:

$y(\%)$	$x(°C)$	$y(\%)$	$x(°C)$
77.4	150	88.9	250
76.7	150	89.2	250
78.2	150	89.7	250
84.1	200	94.8	300
84.5	200	94.7	300
83.7	200	95.9	300

Estimate the linear model $\mu_{Y|x} = \alpha + \beta x$ and test for lack of fit.

Solution. We have $n_1 = n_2 = n_3 = n_4 = 3$. Therefore

$$S_{yy} = \sum_{i=1}^{4} \sum_{j=1}^{3} y_{ij}^2 - \frac{\left(\sum_{i=1}^{4} \sum_{j=1}^{3} y_{ij} \right)^2}{12}$$
$$= 90{,}265.5200 - 89{,}752.4033$$
$$= 513.1167$$

$$S_{xx} = \sum_{i=1}^{4} n_i x_i^2 - \frac{\left(\sum_{i=1}^{4} n_i x_i \right)^2}{12}$$
$$= 645{,}000 - 607{,}500$$
$$= 37{,}500$$

$$S_{xy} = \sum_{i=1}^{4} \sum_{j=1}^{3} x_i y_{ij} - \frac{\left(\sum_{i=1}^{4} n_i x_i \right) \left(\sum_{i=1}^{4} \sum_{j=1}^{3} y_{ij} \right)}{12}$$
$$= 237{,}875 - 233{,}505$$
$$= 4{,}370,$$
$$\bar{y} = 86.4833 \quad \text{and} \quad \bar{x} = 225.$$

The regression coefficients are then given by

$$b = \frac{4{,}370}{37{,}500} = 0.1165$$

and

$$a = 86.4833 - (0.1165)(225) = 60.2708.$$

Hence, our estimated regression line is

$$\hat{y} = 60.2708 + 0.1165x.$$

To test for lack of fit, we proceed in the usual manner:

1. H_0: the regression is linear in x.
2. H_1: the regression is nonlinear in x.
3. Choose a 0.05 level of significance.
4. Critical region: $F > 4.46$ with 2 and 8 degrees of freedom.
5. Computations: We have

$$\text{SST} = S_{yy} = 513.1167$$
$$\text{SSR} = bS_{xy} = (0.1165)(4370) = 509.1050$$
$$\text{SSE} = S_{yy} - bS_{xy} = 4.0117.$$

To compute the pure error sum of squares we first write

$$x_1 = 150 \quad T_1. = 232.3$$
$$x_2 = 200 \quad T_2. = 252.3$$
$$x_3 = 250 \quad T_3. = 267.8$$
$$x_4 = 300 \quad T_4. = 285.4.$$

Therefore

$$\text{SSE(pure)} = 90{,}265.52 - \frac{232.3^2 + 252.3^2 + 267.8^2 + 285.4^2}{3}$$

$$= 2.66.$$

These results and the remaining computations are exhibited in Table 8.4.

Table 8.4 Analysis of Variance on Yield-Temperature Data

Source of variation	Sum of squares	Degrees of freedom	Mean square	Computed f
Regression	509.1050	1	509.1050	1531.60
Error	4.0117	10		
Lack of fit	⎰1.3517	⎰2	0.6758	2.03
Pure error	⎱2.6600	⎱8	0.3324	
Total	513.1167	11		

6. Conclusion: The partitioning of the total variation in this manner reveals a significant variation accounted for by the linear model and an insignificant amount of variation due to lack of fit. Thus the experimental data do not seem to suggest the need to consider terms higher than first order in the model and the null hypothesis is accepted.

8.7 Correlation

Up to this point we have assumed that the independent variable x is controlled and therefore is not a random variable. In fact, in this context, x is often called a *mathematical variable*, which, in the sampling process, is measured with negligible error. In many applications of regression techniques, it is more realistic to assume that both X and Y are random variables and the measurements $\{(x_i, y_i); i = 1, 2, \ldots, n\}$ are observations from a joint density function $f(x, y)$. For example, in an archeological study, it might be assumed that two measurements on a particular kind of bone in the adult body are both random variables and follow a bivariate distribution.

It is often assumed that the conditional distribution $f(y \mid x)$ of Y, for fixed values of X, is normal with mean $\mu_{Y|x} = \alpha + \beta x$ and variance $\sigma_{Y|x}^2 = \sigma^2$ and that X is likewise normally distributed with mean μ_X and variance σ_X^2. The joint density of X and Y is then given by

$$f(x,y) = n(y \mid x; \alpha + \beta x, \sigma)n(x; \mu_X, \sigma_X)$$

$$= \frac{1}{2\pi\sigma_X\sigma} e^{-\left(\frac{1}{2}\right)\left(\left\{\frac{[y-(\alpha+\beta x)]}{\sigma}\right\}^2 + \left[\frac{(x-\mu_X)}{\sigma_X}\right]^2\right)},$$

for $-\infty < x < \infty$ and $-\infty < y < \infty$.

Writing the random variable Y in the form

$$Y = \alpha + \beta X + E,$$

where X is now a random variable independent of the random error E, we have

$$\mu_Y = \alpha + \beta\mu_X$$
$$\sigma_Y^2 = \sigma^2 + \beta^2\sigma_X^2.$$

Substituting into the above expression for $f(x,y)$, we obtain the *bivariate normal distribution*

$$f(x,y) =$$

$$\frac{1}{2\pi\sigma_X\sigma_Y\sqrt{1-\rho^2}} e^{-\frac{1}{2(1-\rho^2)}\left[\left(\frac{x-\mu_X}{\sigma_X}\right)^2 - 2\rho\left(\frac{x-\mu_X}{\sigma_X}\right)\left(\frac{y-\mu_Y}{\sigma_Y}\right) + \left(\frac{y-\mu_Y}{\sigma_Y}\right)^2\right]},$$

for $-\infty < x < \infty$ and $-\infty < y < \infty$, where

$$\rho^2 = 1 - \frac{\sigma^2}{\sigma_Y^2} = \beta^2\frac{\sigma_X^2}{\sigma_Y^2}.$$

The constant ρ is called the *correlation coefficient* and plays a major role in many bivariate data analysis problems. It is important for the reader to understand the physical interpretation of this correlation coefficient and the distinction between correlation and regression. The term *regression* still has meaning here. In fact the straight line given by $\mu_{Y|x} = \alpha + \beta x$ is called the regression line as before. The value of ρ is zero when $\beta = 0$, which results when there essentially is no linear regression; that is, the regression line is horizontal and any knowledge of X is useless in predicting Y. Since $\sigma_Y^2 \geq \sigma^2$, we must have $-1 \leq \rho \leq 1$. Values of $\rho = \pm 1$ only occur when $\sigma^2 = 0$, in which case we have a perfect linear

relationship between the two variables. Thus a value of ρ equal to $+1$ implies a perfect linear relationship with a positive slope, while a value of ρ equal to -1 results from a perfect linear relationship with a negative slope. It might be said then that sample estimates of ρ close to unity in magnitude imply good correlation or *linear association* between X and Y, while values near zero indicate little or no correlation. So, given observations from the bivariate normal density, we can use ordinary least squares procedures to yield estimates for α and β, and therefore $\mu_{Y|x}$, that are identical to those given in Section 8.2. This, of course, is the standard regression analysis. In addition we can draw inferences on the correlation coefficient ρ to gain insight on the degree of linear relationship between X and Y.

CORRELATION COEFFICIENT *The measure ρ of linear relationship between two variables X and Y is estimated by the* sample correlation coefficient r, *where*

$$r = \frac{S_{xy}}{\sqrt{S_{xx}S_{yy}}} = b\sqrt{\frac{S_{xx}}{S_{yy}}}.$$

To show that r must also range from -1 to $+1$, recall from Section 8.3 that

$$\text{SSE} = S_{yy} - bS_{xy}.$$

Dividing both sides of this equation by S_{yy} and replacing b by S_{xy}/S_{xx}, we obtain the relation

$$r^2 = 1 - \frac{\text{SSE}}{S_{yy}},$$

and since $S_{yy} \geq \text{SSE}$, we conclude that r^2 must be between zero and 1. Consequently r must range from -1 to $+1$. A value of -1 or $+1$ will occur when $\text{SSE} = 0$, but this is the case where all points lie in a straight line. Hence a perfect linear relationship exists between X and Y when $r = \pm 1$. For values of r between -1 and $+1$ we must be careful in our interpretation. For example, values of r equal to 0.3 and 0.6 only mean that we have two positive correlations, one somewhat stronger than the other. It is wrong to conclude that $r = 0.6$ indicates a linear relationship twice as good as that indicated by the value $r = 0.3$. On the other hand, if we consider r^2, then $100 \times r^2\%$ of the variation in the values of Y may be accounted for by the linear relationship with the variable X. Thus a correlation of 0.6 means that 36% of the variation of the random variable Y is accounted for by differences in the variable X.

Example 8.7 In a study of the correlation between the amount of rainfall and the quantity of air pollution removed, the following data were collected:

x, daily rainfall (0.01 inch)	y, particulate removed (micrograms per cubic meter)
4.3	126
4.5	121
5.9	116
5.6	118
6.1	114
5.2	118
3.8	132
2.1	141
7.5	108

Compute and interpret the sample correlation coefficient.

Solution. From the data we find

$$S_{xx} = 19.2600, \quad S_{yy} = 804.2222, \quad S_{xy} = -121.8000.$$

Therefore

$$r = \frac{-121.8000}{\sqrt{(19.2600)(804.2222)}} = -0.9786.$$

A correlation coefficient of -0.9786 indicates a very good linear relationship between X and Y. Since $r^2 = 0.9581$, we can say that approximately 96% of the variation in the values of Y is accounted for by a linear relationship with X.

A test of the hypothesis

$$H_0: \quad \rho = \rho_0$$
$$H_1: \quad \rho \neq \rho_0$$

is easily conducted from the sample information. For observations from the bivariate normal distribution, the quantity

$$\frac{1}{2} \ln \left(\frac{1 + r}{1 - r} \right)$$

is a value of a random variable that follows approximately the normal distribution with mean $(1/2) \ln [(1 + \rho)/(1 - \rho)]$ and variance $1/(n - 3)$. Thus the test procedure is to compute

$$z = \frac{\sqrt{n - 3}}{2} \left[\ln \left(\frac{1 + r}{1 - r} \right) - \ln \left(\frac{1 + \rho_0}{1 - \rho_0} \right) \right]$$

$$= \frac{\sqrt{n - 3}}{2} \ln \left[\frac{(1 + r)(1 - \rho_0)}{(1 - r)(1 + \rho_0)} \right]$$

and compare to the critical points of the standard normal distribution.

Example 8.8 For the data of Example 8.7, test the null hypothesis that there is no linear association between the variables. Use a 0.05 level of significance.

Solution.

1. H_0: $\rho = 0$.
2. H_1: $\rho \neq 0$.
3. $\alpha = 0.05$.
4. Critical region: $Z < -1.96$ and $Z > 1.96$.
5. Computations:

$$Z = \frac{\sqrt{6}}{2} \ln \left(\frac{0.0214}{1.9786} \right) = -5.55.$$

6. Conclusion: Reject the hypothesis of no linear relationship.

It should be pointed out that in correlation studies, as in linear regression problems, the results obtained are only as good as the model that

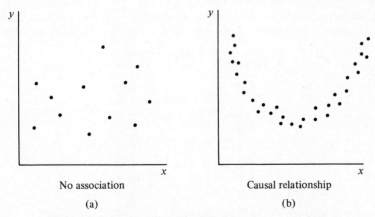

No association   Causal relationship

(a)   (b)

Figure 8.4 Scatter diagrams showing zero correlation.

is assumed. In the correlation techniques studied here, a bivariate normal density is assumed for the variables X and Y, with the mean value of Y at each x value being linearly related to x. To observe the suitability of the linearity assumption, a preliminary plotting of the experimental data is often helpful. A value of the sample correlation coefficient close to zero will result from data that display a strictly random effect as in Figure 8.4(a), thus implying little or no causal relationship. It is important to remember that the correlation coefficient between two variables is a measure of their linear relationship, and that a value of $r = 0$ implies a lack of linearity and not a lack of association. Hence if a strong quadratic relationship exists between X and Y as indicated in Figure 8.4(b), we shall still obtain a zero correlation indicating a nonlinear relationship.

EXERCISES

1. Consider the following experimental data:

x	y	x	y
10.0	18.7	13.5	22.4
10.5	21.5	14.0	23.3
11.0	18.5	14.5	19.6
11.5	19.6	15.0	23.8
12.0	18.2	15.5	21.7
12.5	20.8	16.0	23.2
13.0	21.6		

(a) Estimate the linear regression line.
(b) Graph the line on a scatter diagram.
(c) Find a point estimate of $\mu_{Y|12}$.

2. A study was made on the amount of converted sugar in a certain process at various temperatures. The data were coded and recorded as follows:

x, temperature	y, converted sugar
1.0	8.1
1.1	7.8
1.2	8.5
1.3	9.8
1.4	9.5
1.5	8.9
1.6	8.6
1.7	10.2
1.8	9.3
1.9	9.2
2.0	10.5

(a) Estimate the linear regression line.

(b) Estimate the amount of converted sugar produced when the coded temperature is 1.75.

3. In a certain type of test specimen, the normal stress on a specimen is known to be functionally related to the shear resistance. The following is a set of coded experimental data on the two variables:

x, normal stress	y, shear resistance
26.8	26.5
25.4	27.3
28.9	24.2
23.6	27.1
27.7	23.6
23.9	25.9
24.7	26.3
28.1	22.5
26.9	21.7
27.4	21.4
22.6	25.8
25.6	24.9

(a) Estimate the regression line $\mu_{Y|x} = \alpha + \beta x$.

(b) Estimate the shear resistance for a normal stress of 24.5 pounds per square inch.

4. The following data are the amounts, y, of an unconverted substance from six similar chemical reactions after x minutes:

x (minutes)	y (milligrams)
1	23.5
2	16.9
2	17.5
3	14.0
5	9.8
5	8.9

(a) Fit a curve of the form $\mu_{Y|x} = \gamma \delta^x$ by means of the nonlinear sample regression equation $\hat{y} = cd^x$. [Hint: Write

$$\log \hat{y} = \log c + (\log d)x$$
$$= a + bx,$$

where $a = \log c$ and $b = \log d$, and then estimate a and b by the formulas of Section 8.2 using the sample points $(x_i, \log y_i)$.]

(b) Estimate the amount of unconverted substance in this type of reaction after 4 minutes have elapsed.

5. The pressure P of a gas corresponding to various volumes V was recorded as follows:

V (cubic inches)	50	60	70	90	100
P (pounds per square inch)	64.7	51.3	40.5	25.9	7.8

The ideal gas law is given by the equation $PV^\gamma = C$, where γ and C are constants.

(a) Following the suggested procedure of Exercise 4, find the least squares estimates of γ and C from the given data.

(b) Estimate P when $V = 80$ cubic inches.

6. For a simple linear regression model

$$Y_i = \alpha + \beta x_i + E_i, \qquad i = 1, 2, \ldots, n,$$

where the E_i's are independent and normally distributed with zero means and equal variances σ^2, show that $\bar{Y}$ and

$$B = \frac{\sum_{i=1}^{n} (x_i - \bar{x})Y_i}{\sum_{i=1}^{n} (x_i - \bar{x})^2}$$

are uncorrelated.

7. Show that A, the least squares estimator of the intercept α in the equation $\mu_{Y|x} = \alpha + \beta x$, is normally distributed with mean α and variance

$$\sigma_A^2 = \left[\frac{\sum_{i=1}^{n} x_i^2}{n \sum_{i=1}^{n} (x_i - \bar{x})^2} \right] \sigma^2.$$

8. Test the hypothesis that $\beta = 0$ in Exercise 1, using a 0.05 level of significance, against the alternative that $\beta > 0$.

9. Construct 95% confidence intervals for α and β in Exercise 2.

10. Construct 99% confidence intervals for α and β in Exercise 3.

11. Construct 95% confidence intervals for the mean response $\mu_{Y|x}$ and for the single predicted value y_0, corresponding to $x = 24.5$ in Exercise 3.

12. Graph the regression line and the 95% confidence bands for the mean response $\mu_{Y|x}$ for the data of Exercise 2.

13. Construct a 95% confidence interval for the amount of converted sugar corresponding to $x = 1.6$ in Exercise 2.

14. (a) Find the least squares estimate for the parameter β in the linear equation $\mu_{Y|x} = \beta x$.

(b) Estimate the regression line passing through the origin for the following data:

x	0.5	1.5	3.2	4.2	5.1	6.5
y	1.3	3.4	6.7	8.0	10.0	13.2

15. Suppose it is not known in Exercise 14 whether or not the true regression should pass through the origin. Estimate the general linear model $\mu_{Y|x} = \alpha + \beta x$ and test the hypothesis that $\alpha = 0$ at the 0.10 level of significance against the alternative that $\alpha \neq 0$.

16. Suppose an experimenter postulates a model of the type

$$Y_i = \alpha + \beta x_{1i} + E_i, \qquad i = 1, 2, \ldots, n,$$

when in fact, an additional variable, say x_2, also contributes linearly to the response. The true model is then given by

$$Y_i = \alpha + \beta x_{1i} + \gamma x_{2i} + E_i, \qquad i = 1, 2, \ldots, n.$$

Compute the expected value of the estimator

$$B = \frac{\sum\limits_{i=1}^{n} (x_{1i} - \bar{x}_1) Y_i}{\sum\limits_{i=1}^{n} (x_{1i} - \bar{x}_1)^2}.$$

17. Use an analysis-of-variance approach to test the hypothesis that $\beta = 0$ against the alternative hypothesis $\beta \neq 0$ in Exercise 2 at the 0.05 level of significance.

18. The amounts of a chemical compound y, which were dissolved in 100 grams of water at various temperatures, x, were recorded as follows:

x (°C)	y (grams)		
0	8	6	8
15	12	10	14
30	25	21	24
45	31	33	28
60	44	39	42
75	48	51	44

(a) Find the equation of the regression line.
(b) Estimate the amount of chemical that will dissolve in 100 grams of water at 50°C.
(c) Test the hypothesis that $\alpha = 6$, using a 0.01 level of significance, against the alternative that $\alpha \neq 6$.
(d) Test whether the linear model is adequate.

19. The amounts of solids removed from a particular material when exposed
to drying periods of different lengths are as follows:

x (hours)	y (grams)	
4.4	13.1	14.2
4.5	9.0	11.5
4.8	10.4	11.5
5.5	13.8	14.8
5.7	12.7	15.1
5.9	9.9	12.7
6.3	13.8	16.5
6.9	16.4	15.7
7.5	17.6	16.9
7.8	18.3	17.2

(a) Estimate the linear regression line.

(b) Test whether the linear model is adequate.

(c) Construct a 90% confidence interval for the slope β.

20. Compute and interpret the correlation coefficient for the following data:

x	4	5	9	14	18	22	24
y	16	22	11	16	7	3	17

21. Compute and interpret the correlation coefficient for the following grades
of six students selected at random:

Math grade	70	92	80	74	65	83
English grade	74	84	63	87	78	90

22. Compute the correlation coefficient for the random variables of Exercise 1
and test the hypothesis that $\rho = 0$ against the alternative that $\rho \neq 0$.
Use a 0.05 level of significance.

9 MULTIPLE LINEAR REGRESSION

9.1 Introduction

In most research problems where regression analysis is applied, more than one independent variable is needed in the regression model. The complexity of most scientific mechanisms is such that in order to be able to predict an important response, a *multiple regression* model is needed. When this model is linear in the coefficients it is called a *multiple linear regression model*. For the case of k independent variables x_1, x_2, ..., x_k, the mean of $Y \mid x_1$, x_2, ..., x_k is given by the multiple linear regression model

$$\mu_{Y|x_1, x_2, \ldots, x_k} = \beta_0 + \beta_1 x_1 + \cdots + \beta_k x_k$$

and the estimated response is obtained from the sample regression equation

$$\hat{y} = b_0 + b_1 x_1 + \cdots + b_k x_k,$$

where each regression coefficient β_i is estimated by b_i from the sample data.

As is the case of a single independent variable, the model is often used as an approximation of a more complicated structure and, in fact, can be an adequate representation in certain ranges of the independent variables.

9.2 Least Squares Estimators

Suppose the experimenter has data of the type

$$\{(x_{1i}, x_{2i}, \ldots, x_{ki}, y_i);\ i = 1, 2, \ldots, n \text{ and } n > k\},$$

where y_i is the observed response to the values x_{1i}, x_{2i}, $\ldots$, x_{ki} of the k independent variables x_1, x_2, $\ldots$, x_k. Each observation $(x_{1i}, x_{2i}, \ldots, x_{ki}, y_i)$ satisfies the equation

$$\begin{aligned}
y_i &= \beta_0 + \beta_1 x_{1i} + \beta_2 x_{2i} + \cdots + \beta_k x_{ki} + \epsilon_i \\
&= b_0 + b_1 x_{1i} + b_2 x_{2i} + \cdots + b_k x_{ki} + e_i,
\end{aligned}$$

where ϵ_i and e_i are the random error and residual, respectively, associated with the response y_i. We can use the concept of least squares again to arrive at the estimates $b_0, b_1, b_2, \ldots, b_k$. Consider the expression

$$\begin{aligned}
\text{SSE} &= \sum_{i=1}^{n} e_i^2 \\
&= \sum_{i=1}^{n} (y_i - b_0 - b_1 x_{1i} - b_2 x_{2i} - \cdots - b_k x_{ki})^2,
\end{aligned}$$

which is to be minimized. Differentiating SSE in turn with respect to $b_0, b_1, \ldots, b_k$ and equating to zero, we generate the set of normal equations

$$nb_0 + b_1 \sum_{i=1}^{n} x_{1i} + b_2 \sum_{i=1}^{n} x_{2i} + \cdots + b_k \sum_{i=1}^{n} x_{ki} = \sum_{i=1}^{n} y_i$$

$$b_0 \sum_{i=1}^{n} x_{1i} + b_1 \sum_{i=1}^{n} x_{1i}^2 + b_2 \sum_{i=1}^{n} x_{1i} x_{2i} + \cdots + b_k \sum_{i=1}^{n} x_{1i} x_{ki} = \sum_{i=1}^{n} x_{1i} y_i$$

$$\vdots \qquad \vdots \qquad \vdots \qquad \qquad \vdots \qquad \qquad \vdots$$

$$b_0 \sum_{i=1}^{n} x_{ki} + b_1 \sum_{i=1}^{n} x_{ki} x_{1i} + b_2 \sum_{i=1}^{n} x_{ki} x_{2i} + \cdots + b_k \sum_{i=1}^{n} x_{ki}^2 = \sum_{i=1}^{n} x_{ki} y_i.$$

The normal equations can be put in the matrix form

$$\mathbf{Ab} = \mathbf{g},$$

where

$$
\mathbf{A} =
\begin{bmatrix}
n & \sum_{i=1}^{n} x_{1i} & \sum_{i=1}^{n} x_{2i} & \cdots & \sum_{i=1}^{n} x_{ki} \\
\sum_{i=1}^{n} x_{1i} & \sum_{i=1}^{n} x_{1i}^2 & \sum_{i=1}^{n} x_{1i}x_{2i} & \cdots & \sum_{i=1}^{n} x_{1i}x_{ki} \\
\cdot & \cdot & \cdot & & \cdot \\
\cdot & \cdot & \cdot & & \cdot \\
\cdot & \cdot & \cdot & & \cdot \\
\sum_{i=1}^{n} x_{ki} & \sum_{i=1}^{n} x_{ki}x_{1i} & \sum_{i=1}^{n} x_{ki}x_{2i} & \cdots & \sum_{i=1}^{n} x_{ki}^2
\end{bmatrix},
$$

$$
\mathbf{b} =
\begin{bmatrix}
b_0 \\ b_1 \\ \cdot \\ \cdot \\ \cdot \\ b_k
\end{bmatrix},
\quad
\mathbf{g} =
\begin{bmatrix}
g_0 = \sum_{i=1}^{n} y_i \\
g_1 = \sum_{i=1}^{n} x_{1i}y_i \\
\cdot \\ \cdot \\ \cdot \\
g_k = \sum_{i=1}^{n} x_{ki}y_i
\end{bmatrix}.
$$

If the matrix $\mathbf{A}$ is nonsingular, we can write the solution for the regression coefficients as

$$\mathbf{b} = \mathbf{A}^{-1}\mathbf{g}.$$

Thus one can obtain the prediction equation or regression equation by solving a set of $k+1$ equations in a like number of unknowns. This involves the inversion of the $k+1$ by $k+1$ matrix $\mathbf{A}$. Techniques for inverting this matrix are explained in most textbooks on elementary determinants and matrices. There are many high-speed computer programs available for multiple regression problems, programs that not only give estimates of the regression coefficients but also obtain other information relevant in making inferences concerning the regression equation.

Example 9.1 The percent survival of a certain type of animal semen after storage was measured at various combinations of concentrations of three materials used to increase chance of survival. The data are as follows:

$y(\%\ survival)$	$x_1(weight\ \%)$	$x_2(weight\ \%)$	$x_3(weight\ \%)$
25.5	1.74	5.30	10.80
31.2	6.32	5.42	9.40
25.9	6.22	8.41	7.20
38.4	10.52	4.63	8.50
18.4	1.19	11.60	9.40
26.7	1.22	5.85	9.90
26.4	4.10	6.62	8.00
25.9	6.32	8.72	9.10
32.0	4.08	4.42	8.70
25.2	4.15	7.60	9.20
39.7	10.15	4.83	9.40
35.7	1.72	3.12	7.60
26.5	1.70	5.30	8.20

Estimate the multiple linear regression model for the given data.

Solution. From the experimental data we list the following sums of squares and products:

$$\sum_{i=1}^{13} y_i = 377.5 \qquad \sum_{i=1}^{13} y_i^2 = 11{,}400.15 \qquad \sum_{i=1}^{13} x_{1i} = 59.43$$

$$\sum_{i=1}^{13} x_{2i} = 81.82 \qquad \sum_{i=1}^{13} x_{3i} = 115.40 \qquad \sum_{i=1}^{13} x_{1i}^2 = 394.7255$$

$$\sum_{i=1}^{13} x_{2i}^2 = 576.7264 \qquad \sum_{i=1}^{13} x_{3i}^2 = 1035.9600 \qquad \sum_{i=1}^{13} x_{1i}y_i = 1877.567$$

$$\sum_{i=1}^{13} x_{2i}y_i = 2246.661 \qquad \sum_{i=1}^{13} x_{3i}y_i = 3337.780 \qquad \sum_{i=1}^{13} x_{1i}x_{2i} = 360.6621$$

$$\sum_{i=1}^{13} x_{1i}x_{3i} = 522.0780 \qquad \sum_{i=1}^{13} x_{2i}x_{3i} = 728.3100 \qquad n = 13$$

The least squares estimating equations, $\mathbf{Ab} = \mathbf{g}$, are given by

$$\begin{bmatrix} 13 & 59.43 & 81.82 & 115.40 \\ 59.43 & 394.7255 & 360.6621 & 522.0780 \\ 81.82 & 360.6621 & 576.7264 & 728.3100 \\ 115.40 & 522.0780 & 728.3100 & 1035.9600 \end{bmatrix} \begin{bmatrix} b_0 \\ b_1 \\ b_2 \\ b_3 \end{bmatrix} = \begin{bmatrix} 377.5 \\ 1877.567 \\ 2246.661 \\ 3337.780 \end{bmatrix}.$$

From a computer readout we obtain the elements of the inverse matrix

$\mathbf{A}^{-1}$, and then using the relation $\mathbf{b} = \mathbf{A}^{-1}\mathbf{g}$ we may write

$$
\begin{bmatrix} b_0 \\ b_1 \\ b_2 \\ b_3 \end{bmatrix} = \begin{bmatrix} 8.064801 & -0.082595 & -0.094197 & -0.790526 \\ -0.082595 & 0.008480 & 0.001717 & 0.003720 \\ -0.094197 & 0.001717 & 0.016629 & -0.002063 \\ -0.790526 & 0.003720 & -0.002063 & 0.088601 \end{bmatrix}
$$

$$
\times \begin{bmatrix} 377.5 \\ 1877.567 \\ 2246.661 \\ 3337.780 \end{bmatrix} = \begin{bmatrix} 39.1574 \\ 1.0161 \\ -1.8616 \\ -0.3433 \end{bmatrix}.
$$

Hence

$$
b_0 = 39.1574, \quad b_1 = 1.0161, \quad b_2 = -1.8616, \quad b_3 = -0.3433
$$

and our estimated regression equation is

$$
\hat{y} = 39.1574 + 1.0161x_1 - 1.8616x_2 - 0.3433x_3.
$$

The means and variances of the estimators B_0, B_1, ..., B_k are easily obtained under certain assumptions on the random errors E_1, E_2, ..., E_k that are identical to those made in the case of simple linear regression. Suppose we assume these errors to be independent, each with zero mean and variance σ^2. It can then be shown that B_0, B_1, ..., B_k are, respectively, unbiased estimators of the regression coefficients β_0, β_1, ..., β_k. In addition the variances of the B's are obtained through the elements of the inverse of the $\mathbf{A}$ matrix. We can write this $\mathbf{A}$ matrix in rather conventional notation by initially considering the matrix

$$
\mathbf{X} = \begin{bmatrix} 1 & x_{11} & x_{21} & \cdots & x_{k1} \\ 1 & x_{12} & x_{22} & \cdots & x_{k2} \\ \vdots & \vdots & \vdots & & \vdots \\ 1 & x_{1n} & x_{2n} & \cdots & x_{kn} \end{bmatrix},
$$

where the ith row, apart from the initial element, represents the x values that give rise to the response y_i. One will note that the relationship between the $\mathbf{X}$ matrix and the $\mathbf{A}$ matrix is given by

$$
\mathbf{X}'\mathbf{X} = \mathbf{A}.
$$

In other words, the off diagonal elements of $\mathbf{A}$ represent sums of products of elements in the columns of $\mathbf{X}$, while the diagonal elements of $\mathbf{A}$ represent sums of squares of elements in the columns of $\mathbf{X}$. The inverse matrix, $\mathbf{A}^{-1}$, apart from the multiple σ^2, represents the *variance-covariance matrix* of the estimated regression coefficients. That is, the elements of the matrix $\mathbf{A}^{-1}\sigma^2$ display the variances of $B_0, B_1, \ldots, B_k$ on the main diagonal and covariances on the off diagonal. For example, in a $k = 2$ multiple linear regression problem, we might write

$$\mathbf{A}^{-1} = \begin{bmatrix} c_{00} & c_{01} & c_{02} \\ c_{10} & c_{11} & c_{12} \\ c_{20} & c_{21} & c_{22} \end{bmatrix}$$

with the elements below the main diagonal determined through the symmetry of the matrix. Then we can write

$$\sigma^2_{B_i} = c_{ii}\sigma^2, \qquad\qquad i = 0, 1, 2$$
$$\sigma_{B_iB_j} = \text{cov}\ (B_i, B_j) = c_{ij}\sigma^2, \qquad i \neq j.$$

Of course, the estimates of the variances and hence the standard errors of these estimators are obtained by replacing σ^2 with the appropriate estimate obtained through experimental data. The estimate of σ^2 is derived in Section 9.3.

9.3 Estimating the Error Variance

The partitioning of the total variation in the y's in the multiple linear regression case is identical to that given in Section 8.5 for the case of a single independent variable, namely,

$$\text{SST} = \text{SSR} + \text{SSE}.$$

Following the procedure outlined in Section 8.3, we may write

$$\text{SSE} = \sum_{i=1}^{n} (y_i - \hat{y}_i)^2$$
$$= \sum_{i=1}^{n} [y_i - (b_0 + b_1 x_{1i} + b_2 x_{2i} + \cdots + b_k x_{ki})]^2$$
$$= \sum_{i=1}^{n} [y_i - \mathbf{b}' \mathbf{X}_i]^2,$$

where $\mathbf{X} = [1, x_{1i}, x_{2i}, \ldots, x_{ki}]$. Expanding SSE, we have

$$\text{SSE} = \sum_{i=1}^{n} y_i^2 - 2\mathbf{b}' \sum_{i=1}^{n} y_i \mathbf{X}_i + \sum_{i=1}^{n} (\mathbf{b}' \mathbf{X}_i)^2.$$

Now

$$\sum_{i=1}^{n} y_i \mathbf{X}_i = y_1 \mathbf{X}_1 + y_2 \mathbf{X}_2 + \cdots + y_n \mathbf{X}_n$$

$$= \begin{bmatrix} y_1 \\ y_1 x_{11} \\ \cdot \\ \cdot \\ \cdot \\ y_1 x_{k1} \end{bmatrix} + \begin{bmatrix} y_2 \\ y_2 x_{12} \\ \cdot \\ \cdot \\ \cdot \\ y_2 x_{k2} \end{bmatrix} + \cdots + \begin{bmatrix} y_n \\ y_n x_{1n} \\ \cdot \\ \cdot \\ \cdot \\ y_n x_{kn} \end{bmatrix} = \begin{bmatrix} \sum_{i=1}^{n} y_i \\ \sum_{i=1}^{n} x_{1i} y_i \\ \cdot \\ \cdot \\ \cdot \\ \sum_{i=1}^{n} x_{ki} y_i \end{bmatrix} = \mathbf{g}$$

and

$$\sum_{i=1}^{n} (\mathbf{b}' \mathbf{X}_i)^2 = (\mathbf{b}' \mathbf{X}_1)^2 + (\mathbf{b}' \mathbf{X}_2)^2 + \cdots + (\mathbf{b}' \mathbf{X}_n)^2$$

$$= \mathbf{b}' \mathbf{X}_1 \mathbf{X}_1' \mathbf{b} + \mathbf{b}' \mathbf{X}_2 \mathbf{X}_2' \mathbf{b} + \cdots + \mathbf{b}' \mathbf{X}_n \mathbf{X}_n' \mathbf{b}$$

since

$$\mathbf{b}' \mathbf{X}_i = \mathbf{X}_i' \mathbf{b} = b_0 + b_1 x_{1i} + b_2 x_{2i} + \cdots + b_k x_{ki}, \qquad i = 1, 2, \cdots, n.$$

Therefore

$$\sum_{i=1}^{n} (\mathbf{b}' \mathbf{X}_i)^2 = \mathbf{b}' \sum_{i=1}^{n} (\mathbf{X}_i \mathbf{X}_i') \mathbf{b} = \mathbf{b}' \mathbf{X}' \mathbf{X} \mathbf{b}$$

$$= \mathbf{b}' \mathbf{A} \mathbf{b}$$

$$= \mathbf{b}' \mathbf{g}.$$

Thus the error sum of squares is written

$$\text{SSE} = \sum_{i=1}^{n} y_i^2 - \mathbf{b}' \mathbf{g}$$

$$= \text{SST} - \left[b_0 g_0 + b_1 g_1 + \cdots + b_k g_k - \frac{\left(\sum_{i=1}^{n} y_i \right)^2}{n} \right],$$

which implies that the regression sum of squares with k degrees of freedom is given by

$$\text{SSR} = \sum_{j=0}^{k} b_j g_j - \frac{\left(\sum_{i=1}^{n} y_i\right)^2}{n}.$$

There are $n - k - 1$ degrees of freedom associated with error in this general case and, as before, an unbiased estimate of σ^2 is obtained by dividing the error sum of squares by the corresponding degrees of freedom. Thus we have as our estimate

$$s^2 = \frac{\text{SSE}}{n - k - 1}.$$

Example 9.2 For the data of Example 9.1 estimate s^2.

Solution. The error sum of squares with $n - k - 1 = 9$ degrees of freedom can be found by writing

$$\text{SSE} = \text{SST} - \text{SSR},$$

where

$$\text{SST} = S_{yy} = 11,400.15 - \frac{(377.5)^2}{13} = 438.13$$

and

$$\text{SSR} = \sum_{j=0}^{3} b_j g_j - \frac{\left(\sum_{i=1}^{13} y_i\right)^2}{13}$$

$$= (39.1574)(377.5) + (1.0161)(1877.567)$$

$$\quad + (-1.8616)(2246.6610) + (-0.3433)(3337.780) - \frac{(377.5)^2}{13}$$

$$= 399.45,$$

with 3 degrees of freedom. Our estimate of σ^2 is then given by

$$s^2 = \frac{438.13 - 399.45}{9} = 4.298.$$

The error and regression sums of squares take on the same meaning that they did in the special case of a single independent variable. While these two portions give an intuitive indication of how adequate the model is, no form of hypothesis testing or confidence interval estimation can be accomplished without making a distributional assumption on the errors E_i. In Section 9.4, we outline certain methods for finding confidence intervals and testing hypotheses that can be helpful in the multiple linear regression analysis.

9.4 Inferences in Multiple Linear Regression

One of the most useful inferences that can be made regarding the ability of the regression equation to predict the response $\hat{y}_0$ corresponding to the values $x_{10}, x_{20}, \ldots, x_{k0}$ is the confidence interval on the mean response $\mu_{Y|x_{10},x_{20},\ldots,x_{k0}}$. In vector notation we are then interested in constructing a confidence interval on the mean response for the set of conditions given by $\mathbf{x}_0' = [1,x_{10},x_{20},\ldots,x_{k0}]$. We augment the conditions on the x's by the number 1 in order to facilitate using matrix notation. As in the $k = 1$ case, if we make the additional assumption that the errors are independent and normally distributed, then the B_j's are normal, with mean, variances, and covariances as indicated in Section 9.3, and hence

$$\hat{Y}_0 = B_0 + \sum_{j=1}^{k} B_j x_{j0}$$

is likewise normally distributed and is, in fact, an unbiased estimator for the *mean response* on which we are attempting to attach confidence intervals. The variance of $\hat{Y}_0$ written in matrix notation simply as a function of σ^2, $\mathbf{A}^{-1}$, and the condition vector $\mathbf{x}_0'$ is

$$\sigma_{\hat{Y}_0}^2 = \sigma^2 \mathbf{x}_0' \mathbf{A}^{-1} \mathbf{x}_0.$$

If this expression is expanded for a given case, say $k = 2$, it is easily seen that it appropriately accounts for the variances and covariances of the B_j's. Replacing σ^2 by s^2 as given in Section 9.3, the $100(1 - \alpha)\%$ confidence interval on $\mu_{Y|x_{10},x_{20},\ldots,x_{k0}}$ can be constructed from the statistic

$$T = \frac{\hat{Y}_0 - \mu_{Y|x_{10},x_{20},\ldots,x_{k0}}}{S \sqrt{\mathbf{x}_0' \mathbf{A}^{-1} \mathbf{x}_0}},$$

which has a t distribution with $n - k - 1$ degrees of freedom.

CONFIDENCE INTERVAL FOR $\mu_{Y|x_{10},x_{20},\ldots,x_{k0}}$ *A* $(1-\alpha)100\%$ *confidence interval for the mean response* $\mu_{Y|x_{10},x_{20},\ldots,x_{k0}}$ *is given by*

$$\hat{y}_0 - t_{\alpha/2}s\sqrt{\mathbf{x}_0'\mathbf{A}^{-1}\mathbf{x}_0} < \mu_{Y|x_{10},x_{20},\ldots,x_{k0}} < \hat{y}_0 + t_{\alpha/2}s\sqrt{\mathbf{x}_0'\mathbf{A}^{-1}\mathbf{x}_0},$$

where $t_{\alpha/2}$ *is a value of the t distribution with* $n-k-1$ *degrees of freedom.*

Example 9.3 Using the data of Example 9.1, construct a 95% confidence interval for the mean response when $x_1 = 3$, $x_2 = 8$, and $x_3 = 9$.

Solution. From the regression equation of Example 9.1 the estimated response when $x_1 = 3$, $x_2 = 8$, and $x_3 = 9$ is

$$\hat{y}_0 = 39.1574 + (1.0161)(3) - (1.8616)(8) - (0.3433)(9) = 24.2322.$$

Next we find

$$\mathbf{x}_0'\mathbf{A}^{-1}\mathbf{x}_0 = [1,3,8,9]\begin{bmatrix} 8.064 & -0.0826 & -0.0942 & -0.7905 \\ -0.0826 & 0.0085 & 0.0017 & 0.0037 \\ -0.0942 & 0.0017 & 0.0166 & -0.0021 \\ -0.7905 & 0.0037 & -0.0021 & 0.0886 \end{bmatrix}\begin{bmatrix} 1 \\ 3 \\ 8 \\ 9 \end{bmatrix}$$

$$= 0.1267.$$

Previously we found $s^2 = 4.298$ or $s = 2.073$, and using Table V we see that $t_{0.025} = 2.262$ for 9 degrees of freedom. Therefore a 95% confidence interval for $\mu_{Y|3,8,9}$ is given by

$$24.2322 - (2.262)(2.073)\sqrt{0.1267} < \mu_{Y|3,8,9} < 24.2322$$
$$+ (2.262)(2.073)\sqrt{0.1267}$$

or simply

$$22.5633 < \mu_{Y|3,8,9} < 25.9011.$$

A confidence interval for a single predicted response $\hat{y}_0$ is once again established by considering the differences $\hat{y}_0 - y_0$ of the random variable $\hat{Y}_0 - Y_0$. The sampling distribution can be shown to be normal with mean

$$\mu_{\hat{Y}_0-Y_0} = 0$$

and variance

$$\sigma^2_{\hat{Y}_0-Y_0} = \sigma^2[1 + \mathbf{x}_0'\mathbf{A}^{-1}\mathbf{x}_0].$$

Thus the $(1 - \alpha)100\%$ confidence interval for a single predicted value y_0 can be constructed from the statistic

$$T = \frac{\hat{Y}_0 - Y_0}{S \sqrt{1 + \mathbf{x}_0' \mathbf{A}^{-1} \mathbf{x}_0}},$$

which has a t distribution with $n - k - 1$ degrees of freedom.

CONFIDENCE INTERVAL FOR y_0 *A* $(1 - \alpha)100\%$ *confidence interval for a single response y_0 is given by*

$$\hat{y}_0 - t_{\alpha/2}s \sqrt{1 + \mathbf{x}_0' \mathbf{A}^{-1} \mathbf{x}_0} < y_0 < \hat{y}_0 + t_{\alpha/2}s \sqrt{1 + \mathbf{x}_0' \mathbf{A}^{-1} \mathbf{x}_0},$$

where $t_{\alpha/2}$ is a value of the t distribution with $n - k - 1$ degrees of freedom.

Example 9.4 Using the data of Example 9.1, construct a 95% confidence interval for the predicted response when $x_1 = 3$, $x_2 = 8$, and $x_3 = 9$.

Solution. Referring to the results of Example 9.3 the 95% confidence interval for the response y_0 when $x_1 = 3$, $x_2 = 8$, and $x_3 = 9$ is

$$24.2322 - (2.262)(2.073)\sqrt{1.1267} < y_0 < 24.2322$$
$$- (2.262)(2.073)\sqrt{1.1267},$$

which reduces to

$$19.2547 < y_0 < 29.2097.$$

A knowledge of the distributions of the individual coefficient estimators enables the experimenter to construct confidence intervals for the coefficients and to test hypotheses about them. One recalls that the B_j's $(j = 0, 1, 2, \ldots, k)$ are normally distributed with mean β_j and variance $c_{jj}\sigma^2$. Thus we can use the statistic

$$T = \frac{B_j - \beta_j}{S \sqrt{c_{jj}}}$$

with $n - k - 1$ degrees of freedom to test hypotheses and construct confidence intervals on β_j. For example, if we wish to test

$$H_0: \quad \beta_j = \beta_{j0}$$
$$H_1: \quad \beta_j \neq \beta_{j0},$$

we compute the statistic

$$t = \frac{b_j - \beta_{j0}}{s \sqrt{c_{jj}}}$$

and accept H_0 if

$$-t_{\alpha/2} < t < t_{\alpha/2},$$

where $t_{\alpha/2}$ has $n - k - 1$ degrees of freedom.

Example 9.5 For the model of Example 9.1, test the hypothesis that $\beta_2 = -2.5$ at the 0.05 level of significance against the alternative that $\beta_2 > -2.5$.

Solution.

1. H_0: $\beta_2 = -2.5$.
2. H_1: $\beta_2 > -2.5$.
3. Choose a 0.05 level of significance.
4. Critical region: $T > 1.833$.
5. Computations:

$$t = \frac{b_2 - \beta_{20}}{s \sqrt{c_{22}}} = \frac{-1.8616 + 2.5}{2.073 \sqrt{0.0166}} = 2.391.$$

6. Conclusion: Reject H_0 and conclude that $\beta_2 > -2.5$.

9.5 Adequacy of the Model

In many regression situations, individual coefficients are of importance to the experimenter. For example, in an economics application, β_1, β_2, etc., might have some particular significance, and thus confidence intervals and tests of hypotheses on these parameters are of interest to the economist. However, consider an industrial chemical situation in which the postulated model assumes that reaction yield is dependent linearly on reaction temperature and concentration of a certain catalyst. It is probably known that this is not the true model, but an adequate representation, so the interest is likely to be not in the individual parameters, but rather in the ability of the entire function to predict the true response in the range of the variables considered. Therefore in this situation, one would put more emphasis on $\sigma_{\hat{Y}}^2$, confidence intervals on the mean response, and so forth, and likely de-emphasize inferences on individual parameters.

The experimenter using regression analysis is also interested in deletion of variables when the situation dictates that, in addition to arriving at a workable prediction equation, he must find the "best regression" involving only variables that are useful predictors. There are a number of computer programs available for the practitioner that sequentially arrive at the so-called best regression equation depending on certain criteria. We shall not attempt, in this text, to cover these procedures. The best account may be found in *Applied Regression Analysis* by Draper and Smith.

One criterion that is commonly used to illustrate the adequacy of a fitted regression model is the *coefficient of multiple determination:*

$$R^2 = \frac{\text{SSR}}{\text{SST}} = \frac{\sum_{j=0}^{k} b_j g_j - \left(\sum_{i=1}^{n} y_i\right)^2/n}{S_{yy}}.$$

This quantity merely indicates what proportion of the total variation in the response Y is explained by the fitted model. Often an experimenter will report $R^2 \times 100\%$ and interpret the result as percentage variation explained by the postulated model. The square root of R^2 is called the *multiple correlation coefficient* between Y and the set $x_1, x_2, \ldots, x_k$. In Example 9.1, the value of R^2 indicating the proportion of variation explained by the three independent variables x_1, x_2, and x_3 is found to be

$$R^2 = \frac{\text{SSR}}{\text{SST}} = \frac{399.45}{438.13} = 0.9117,$$

which means that 91.17% of the variation in Y has been explained by the linear regression model.

The regression sum of squares can be used to give some indication concerning whether or not the model is an adequate explanation of the true situation. One can test the significance of regression for Example 9.1 by merely forming the ratio

$$f = \frac{\text{SSR}/3}{s^2} = \frac{133.15}{4.2981} = 30.978.$$

This value of F exceeds the tabulated critical point of the F distribution for 3 and 9 degrees of freedom at the $\alpha = 0.01$ level. The result here should not be misinterpreted. While it does indicate that the regression explained by the model is significant, this does not rule out the possibility that

1. The linear regression model in this set of x's is not the only model that can be used to explain the data; indeed there might be other models with transformations on the x's that might give a larger value of the F statistic.
2. The model might have been more effective with the inclusion of other variables in addition to x_1, x_2, and x_3 or perhaps with the deletion of one or more of the variables in the model.

The addition of any single variable to a regression system will increase the regression sum of squares and thus reduce the error sum of squares. Consequently we must decide whether the increase in regression is sufficient to warrant using it in the model. As one might expect, the use of unimportant variables can reduce the effectiveness of the prediction equation by increasing the variance of the estimated response. We shall pursue this point further by considering the importance of x_3 in Example 9.1. Initially we can test

$$H_0: \quad \beta_3 = 0$$
$$H_1: \quad \beta_3 \neq 0$$

by using the t distribution with 9 degrees of freedom. We have

$$t = \frac{b_3 - 0}{s \sqrt{c_{33}}} = \frac{-0.3413}{\sqrt{4.298 c_{33}}} = \frac{-0.3413}{2.073 \sqrt{0.0886}} = -0.556,$$

which indicates that β_3 does not differ significantly from zero and hence one may very well feel justified in removing x_3 from the model. Suppose we consider the regression of Y on the set (x_1, x_2), the least squares normal equations now reducing to

$$\begin{bmatrix} 13 & 59.43 & 81.82 \\ 59.43 & 394.7255 & 360.6621 \\ 81.82 & 360.6621 & 576.7264 \end{bmatrix} \begin{bmatrix} b_0 \\ b_1 \\ b_2 \end{bmatrix} = \begin{bmatrix} 377.75 \\ 1877.5670 \\ 2246.6610 \end{bmatrix}.$$

The estimated regression coefficients for this reduced model are given by

$$b_0 = 36.094, \quad b_1 = 1.031, \quad b_2 = -1.870,$$

and the resulting regression sum of squares with 2 degrees of freedom is as follows:

$$R(\beta_1, \beta_2) = \sum_{j=0}^{2} b_j g_j - \frac{\left(\sum_{i=1}^{13} y_i\right)^2}{13}$$
$$= 398.12.$$

Here we use the notation $R(\beta_1, \beta_2)$ to indicate the regression sum of squares of the restricted model and it is not to be confused with SSR, the regression sum of squares of the original model with 3 degrees of freedom. The new error sum of squares is then given by

$$SST - R(\beta_1, \beta_2) = 438.13 - 398.12$$
$$= 40.01,$$

and the resulting error mean square with 10 degrees of freedom becomes

$$s^2 = \frac{40.01}{10} = 4.001.$$

The amount of variation in the response, the percent survival, which is attributed to x_3, the weight percent of the third additive, in the presence of the variables x_1 and x_2, is given by

$$\begin{aligned} R(\beta_3 \mid \beta_1, \beta_2) &= \text{SSR} - R(\beta_1, \beta_2) \\ &= 399.45 - 398.12 \\ &= 1.33, \end{aligned}$$

which represents a small proportion of the entire regression variation. This amount of added regression is statistically insignificant as indicated by our previous test on β_3. An equivalent test involves the formation of the ratio

$$\begin{aligned} f &= \frac{R(\beta_3 \mid \beta_1, \beta_2)}{s^2} \\ &= \frac{1.33}{4.298} = 0.309, \end{aligned}$$

which is a value of the F distribution with 1 and 9 degrees of freedom. Recall that the basic relationship between the t distribution with ν degrees of freedom and the F distribution with 1 and ν degrees of freedom is given by

$$t_{\alpha/2}^2 = f_\alpha(1, \nu)$$

and we note that the f value of 0.309 is indeed the square of the t value of 0.556.

We can provide additional support for deleting x_3 from the model by considering $\sigma_{\hat{Y}}^2$ under both the full and reduced regression equation. We first note that the estimate of σ^2 was reduced from 4.298 to 4.001 by deleting x_3 from the model. In the case of the full model the variance of the estimated response $\hat{Y}_0$ at some point of interest, say

$$\mathbf{x}_0' = [1, 4.00, 5.5, 8.90],$$

is

$$\sigma_{\hat{Y}_0}^2 = s^2 \mathbf{x}_0' \mathbf{A}^{-1} \mathbf{x}_0$$
$$= 4.298[1,\ 4.00,\ 5.5,\ 8.90]$$

$$\begin{bmatrix} 8.064 & -0.0826 & -0.0942 & -0.7905 \\ -0.0826 & 0.0085 & 0.0017 & 0.0037 \\ -0.0942 & 0.0017 & 0.0166 & -0.0021 \\ -0.7905 & 0.0037 & -0.0021 & 0.0886 \end{bmatrix} \begin{bmatrix} 1 \\ 4.00 \\ 5.5 \\ 8.90 \end{bmatrix}$$

$$= 0.3936.$$

For the reduced model, our point of interest becomes $\mathbf{x}_0' = [1,\ 4.00,\ 5.5]$, our $\mathbf{A}$ matrix has been reduced to a 3×3 matrix with inverse given by

$$\mathbf{A}^{-1} = \begin{bmatrix} 1.0114 & -0.0494 & -0.1126 \\ -0.0494 & 0.0083 & 0.0018 \\ -0.1126 & 0.0018 & 0.0166 \end{bmatrix},$$

and the variance of the estimated response is now

$$\sigma_{\hat{Y}_0}^2 = 4.001[1,\ 4.00,\ 5.5] \begin{bmatrix} 1.0114 & -0.0494 & -0.1126 \\ -0.0494 & 0.0083 & 0.0018 \\ -0.1126 & 0.0018 & 0.0166 \end{bmatrix} \begin{bmatrix} 1 \\ 4.00 \\ 5.5 \end{bmatrix}$$

$$= 0.3668,$$

which represents a reduction over that found for the complete model.

To generalize the above concepts one can assess the work of an independent variable x_i in the general multiple linear regression model

$$\mu_Y|_{x_1, x_2, \ldots, x_k} = \beta_0 + \beta_1 x_1 + \cdots + \beta_k x_k$$

by observing the amount of regression attributed to x_i over that attributed to the other variables, that is, the regression on x_i *adjusted* for the other variables. This is computed by subtracting the regression sum of squares for a model with x_i removed, from SSR. For example, we say that x_1 is assessed by calculating

$$R(\beta_1 \mid \beta_2, \beta_3, \ldots, \beta_k) = \text{SSR} - R(\beta_2, \beta_3, \ldots, \beta_k),$$

where $R(\beta_2, \beta_3, \ldots, \beta_k)$ is the regression sum of squares with $\beta_1 x_1$ removed from the model. To test the hypothesis

$$H_0\colon \beta_1 = 0$$
$$H_1\colon \beta_1 \neq 0$$

compute

$$f = \frac{R(\beta_1 \mid \beta_2, \beta_3, \ldots, \beta_k)}{s^2}$$

and compare with $f_\alpha(1, n - k - 1)$.

In a similar manner we can test for the significance of a *set* of the variables. For example, to investigate simultaneously the importance of including x_1 and x_2 in the model, we test the hypothesis

$$H_0: \quad \beta_1 = \beta_2 = 0$$

$$H_1: \quad \beta_1 \text{ and } \beta_2 \text{ are not both zero}$$

by computing

$$f = \frac{[R(\beta_1, \beta_2 \mid \beta_3, \beta_4, \ldots, \beta_k)]/2}{s^2}$$

$$= \frac{[\text{SSR} - R(\beta_3, \beta_4, \ldots, \beta_k)]/2}{s^2}$$

and comparing with $f_\alpha(2, n - k - 1)$. The number of degrees of freedom associated with the numerator, in this case 2, equals the number of variables in the set.

9.6 Special Case of Orthogonality

Prior to our original development of the general linear regression problem, the assumption was made that the independent variables were measured without error and are often controlled by the experimenter. Quite often they occur as a result of an elaborately *designed experiment*. In fact, one can increase the effectiveness of the resulting prediction equation with the use of a suitable experimental plan.

Suppose we once again consider the $\mathbf{X}$ matrix as defined in Section 9.2. We can rewrite it to read

$$\mathbf{X} = [\mathbf{1}, \mathbf{x}_1, \mathbf{x}_2, \ldots, \mathbf{x}_k],$$

where $\mathbf{1}$ represents a column of ones and $\mathbf{x}_j$ is a column vector representing the levels of x_j. If

$$\mathbf{x}_p' \mathbf{x}_q = 0, \qquad p \neq q,$$

the variables x_p and x_q are said to be *orthogonal* to each other. There are certain obvious advantages to having a completely orthogonal situation whereby $\mathbf{x}'_p\mathbf{x}_q = 0$ for all possible p and q, $p \neq q$, and, in addition,

$$\sum_{i=1}^{n} x_{ji} = 0, \qquad j = 1, 2, \ldots, k.$$

The resulting $\mathbf{X}'\mathbf{X}$ is a diagonal matrix and the normal equations in Section 9.2 reduce to

$$nb_0 = \sum_{i=1}^{n} y_i$$

$$b_1 \sum_{i=1}^{n} x_{1i}^2 = \sum_{i=1}^{n} x_{1i}y_i$$

$$b_k \sum_{i=1}^{n} x_{ki}^2 = \sum_{i=1}^{n} x_{ki}y_i.$$

The most important advantage is that one is easily able to partition SSR into *single-degree-of-freedom components*, each of which corresponds to the amount of variation in Y accounted for by a given controlled variable. Solving the normal equations for $b_0, b_1, \ldots, b_k$ we can write

$$\text{SSR} = b_0 g_0 + b_1 g_1 + \cdots + b_k g_k - \frac{\left(\sum_{i=1}^{n} y_i\right)^2}{n}$$

$$= b_1 g_1 + b_2 g_2 + \cdots + b_k g_k$$

$$= \frac{\left(\sum_{i=1}^{n} x_{1i}y_i\right)^2}{\sum_{i=1}^{n} x_{1i}^2} + \frac{\left(\sum_{i=1}^{n} x_{2i}y_i\right)^2}{\sum_{i=1}^{n} x_{2i}^2} + \cdots + \frac{\left(\sum_{i=1}^{n} x_{ki}y_i\right)^2}{\sum_{i=1}^{n} x_{ki}^2}$$

$$= R(\beta_1) + R(\beta_2) + \cdots + R(\beta_k).$$

The quantity $R(\beta_i)$ is the amount of regression sum of squares associated with a model involving a single independent variable x_i.

To test simultaneously for the significance of a set of m variables in an orthogonal situation the regression sum of squares becomes

$$R(\beta_1,\beta_2,\ldots,\beta_m \mid \beta_{m+1},\beta_{m+2},\ldots,\beta_k) = R(\beta_1) + R(\beta_2) + \cdots + R(\beta_m)$$

and simplifies to

$$R(\beta_1 \mid \beta_2,\beta_3,\ldots,\beta_k) = R(\beta_1)$$

when evaluating a single independent variable. Therefore, the contribution of a given variable or set of variables is essentially found by *ignoring* the other variables in the model. Independent evaluations of the worth of the individual variables are accomplished using analysis-of-variance techniques as given in Table 9.1. The total variation in the response is

Table 9.1 Analysis of Variance for Orthogonal Variables

Source of variation	Sum of squares	Degrees of freedom	Mean square	Computed f
β_1	$R(\beta_1) = \dfrac{\left(\sum\limits_{i=1}^{n} x_{1i}y_i\right)^2}{\sum\limits_{i=1}^{n} x_{1i}^2}$	1	$R(\beta_1)$	$\dfrac{R(\beta_1)}{s^2}$
β_2	$R(\beta_2) = \dfrac{\left(\sum\limits_{i=1}^{n} x_{2i}y_i\right)^2}{\sum\limits_{i=1}^{n} x_{2i}^2}$	1	$R(\beta_2)$	$\dfrac{R(\beta_2)}{s^2}$
.	.	.	.	.
.	.	.	.	.
.	.	.	.	.
β_k	$R(\beta_k) = \dfrac{\left(\sum\limits_{i=1}^{n} x_{ki}y_i\right)^2}{\sum\limits_{i=1}^{n} x_{ki}^2}$	1	$R(\beta_k)$	$\dfrac{R(\beta_k)}{s^2}$
Error	SSE	$n - k - 1$	$s^2 = \dfrac{\text{SSE}}{n - k - 1}$	
Total	$\text{SST} = S_{yy}$	$n - 1$		

partitioned into single-degree-of-freedom components plus the error term with $n - k - 1$ degrees of freedom. Each computed f value is used to test one of the hypotheses

$$H_0: \quad \beta_i = 0 \left.\vphantom{\begin{matrix}a\\b\end{matrix}}\right\}$$
$$H_1: \quad \beta_i \neq 0 \qquad i = 1, 2, \ldots, k$$

by comparing with the critical point $f_\alpha (1, n - k - 1)$.

Example 9.6 Suppose a scientist takes experimental data on the radius of a propellant grain Y as a function of powder temperature x_1, extrusion rate x_2, and die temperature x_3. Fit a linear regression model for predicting grain radius and determine the effectiveness of each variable in the model. The data are given as follows:

Grain radius	Powder temperature	Extrusion rate	Die temperature
82	150 (−1)	12 (−1)	220 (−1)
93	190 (1)	12 (−1)	220 (−1)
114	150 (−1)	24 (1)	220 (−1)
124	150 (−1)	12 (−1)	250 (1)
111	190 (1)	24 (1)	220 (−1)
129	190 (1)	12 (−1)	250 (1)
157	150 (−1)	24 (1)	250 (1)
164	190 (1)	24 (1)	250 (1)

Solution. Note that each variable is controlled at two levels and the experiment represents each of the eight possible combinations. The data on the independent variables are coded for convenience by means of the following formulas:

$$x_1 = \frac{\text{powder temperature} - 170}{20}$$

$$x_2 = \frac{\text{extrusion rate} - 18}{6}$$

$$x_3 = \frac{\text{die temperature} - 235}{15}.$$

The resulting levels of x_1, x_2, and x_3 take on the values -1 and $+1$ as indicated in the table of data. This particular *experimental design* affords the orthogonality that we are illustrating here. A more thorough treatment of this type of experimental layout will be given in Chapter 13. The **X** matrix is given by

$$\mathbf{X} = \begin{bmatrix} 1 & -1 & -1 & -1 \\ 1 & 1 & -1 & -1 \\ 1 & -1 & 1 & -1 \\ 1 & -1 & -1 & 1 \\ 1 & 1 & 1 & -1 \\ 1 & 1 & -1 & 1 \\ 1 & -1 & 1 & 1 \\ 1 & 1 & 1 & 1 \end{bmatrix}$$

and it is easy to verify the orthogonality conditions.

One can now compute the coefficients

$$b_0 = \frac{\sum\limits_{i=1}^{8} y_i}{8} = 121.75$$

$$b_1 = \frac{\sum\limits_{i=1}^{8} x_{1i} y_i}{\sum\limits_{i=1}^{8} x_{1i}^2} = \frac{20}{8} = 2.5$$

$$b_2 = \frac{\sum\limits_{i=1}^{8} x_{2i} y_i}{\sum\limits_{i=1}^{8} x_{2i}^2} = \frac{118}{8} = 14.75$$

$$b_3 = \frac{\sum\limits_{i=1}^{8} x_{3i} y_i}{\sum\limits_{i=1}^{8} x_{3i}^2} = \frac{174}{8} = 21.75,$$

so in terms of the *coded* variables, the prediction equation is given by

$$y = 121.75 + 2.5x_1 + 14.75x_2 + 21.75x_3.$$

The analysis-of-variance table showing independent contributions to SSR for each variable is given in Table 9.2. The results, when compared to the

Table 9.2 Analysis of Variance on Grain Radius Data

Source of variation	Sum of squares	Degrees of freedom	Mean square	Computed f
β_1	$\dfrac{(20)^2}{8} = 50$	1	50	2.16
β_2	$\dfrac{(118)^2}{8} = 1740.50$	1	1740.50	75.26
β_3	$\dfrac{(174)^2}{8} = 3784.50$	1	3784.50	163.65
Error	92.5	4	23.1250	
Total	5667.50	7		

$f_{0.05}(1,4)$ critical point of 7.71, indicate that x_1 does not contribute significantly at the 0.05 level, while variables x_2 and x_3 are significant. In this example the estimate for σ^2 is 23.1250. As in the single independent variable case, it should be pointed out that this estimate does not solely contain experimental error variation unless the postulated model is correct. Otherwise the estimate is "contaminated" by lack of fit in addition to pure error, and the lack of fit can be separated only if one obtains multiple experimental observations at the various (x_1,x_2,x_3) combinations.

EXERCISES

1. Given the data

y	2	5	7	8	5
x_1	8	8	6	5	3
x_2	0	1	1	3	4

estimate the multiple linear regression equation.

2. The following data represent a set of 10 experimental runs in which two independent variables x_1 and x_2 are controlled and values of a response, y, are observed:

y	x_1	x_2
61.5	2400	54.5
61.2	2450	56.4
32.0	2500	43.2
52.5	2700	65.2
31.5	2750	45.5
22.5	2800	47.5
53.0	2900	65.0
56.8	3000	66.5
34.8	3100	57.3
52.7	3200	68.0

Estimate the multiple linear regression equation for the given data.

3. A set of experimental runs was made to determine a way of predicting coking time y at various levels of oven width x_1 and flue temperature x_2.

The coded data were recorded as follows:

y	x_1	x_2
6.40	1.32	1.15
15.05	2.69	3.40
18.75	3.56	4.10
30.25	4.41	8.75
44.85	5.35	14.82
48.94	6.20	15.15
51.55	7.12	15.32
61.50	8.87	18.18
100.44	9.80	35.19
111.42	10.65	40.40

Estimate the multiple regression equation.

4. The following data resulted from 15 experimental runs made on four independent variables and a single response y.

y	x_1	x_2	x_3	x_4
14.8	7.8	4.3	11.5	6.3
12.1	6.9	3.9	14.3	7.4
19.0	9.3	8.4	9.4	5.9
14.5	6.8	10.3	15.2	8.7
16.6	11.7	6.4	8.8	9.1
17.2	8.5	5.7	9.8	5.6
17.5	12.6	6.8	11.2	6.8
14.1	7.5	4.2	10.9	7.4
13.8	8.4	7.3	14.7	8.2
14.7	11.3	8.8	15.1	9.2
17.7	10.7	3.6	8.7	4.7
17.0	7.3	4.9	8.6	5.5
17.6	8.4	7.3	9.3	6.6
16.3	6.7	9.7	10.8	8.7
18.2	9.6	8.4	11.9	5.4

Estimate the multiple linear regression model relating y to x_1, x_2, x_3, and x_4.

5. For the data of Exercise 2 estimate σ^2.

6. For the data of Exercise 4 estimate σ^2.

7. Obtain estimates of the variances and the covariance of the estimators B_1 and B_2 of Exercise 2.

8. Referring to Exercise 4, find estimates of
   (a) $\sigma_{B_2}^2$.   (b) cov (B_1, B_4).

9. Using the data of Exercise 2 and the estimate of σ from Exercise 5, construct 95% confidence intervals for the predicted response and the mean response when $x_1 = 2500$ and $x_2 = 48.0$.

10. Using the data of Exercise 4 and the estimate of σ^2 from Exercise 6, construct 95% confidence intervals for the predicted response and the mean response when $x_1 = 8.2$, $x_2 = 6.0$, $x_3 = 10.3$, and $x_4 = 5.8$.

11. For the model of Exercise 2, use the T statistic to test the hypothesis that $\beta_1 = 0$ at the 0.05 level of significance against the alternative that $\beta_1 \neq 0$.

12. For the model of Exercise 3, test the hypothesis that $\beta_1 = 2$ at the 0.05 level of significance against the alternative that $\beta_1 \neq 2$.

13. Compute and interpret the coefficient of multiple determination for the variables of Exercise 3.

14. Test whether the regression explained by the model of Exercise 3 is significant at the 0.01 level of significance.

15. Test whether the regression explained by the model of Exercise 4 is significant at the 0.05 level of significance.

16. For the model of Exercise 4, test the hypothesis

$$H_0: \quad \beta_1 = \beta_2 = 0$$
$$H_1: \quad \beta_1 \text{ and } \beta_2 \text{ are not both zero.}$$

17. Repeat Exercise 11 using an F statistic.

18. A small experiment is conducted to fit a multiple regression equation relating the yield y to temperature x_1, reaction time x_2, and concentration of one of the reactants x_3. Two levels of each variable were chosen and measurements corresponding to the coded independent variables were recorded as follows:

y	x_1	x_2	x_3
7.6	-1	-1	-1
8.4	1	-1	-1
9.2	-1	1	-1
10.3	-1	-1	1
9.8	1	1	-1
11.1	1	-1	1
10.2	-1	1	1
12.6	1	1	1

(a) Estimate a multiple linear regression in the coded variables.

(b) Partition SSR, the regression sum of squares, into three single-degree-of-freedom components attributable to x_1, x_2, and x_3, respectively. Show an analysis-of-variance table, indicating significance tests on each variable.

10 POLYNOMIAL REGRESSION

In Chapters 8 and 9 we were concerned with problems of estimation and prediction, where it is postulated that a response is related to a single independent controlled variable x, or perhaps to a set $(x_1, x_2, \ldots, x_k)$ through an equation in which the x's occur linearly. Similar least squares techniques can be applied where the model involves, say, powers and products in the x's. This type of model is called a *polynomial regression model*. For example, when $k = 1$, the experimenter may feel that the means $\mu_{Y|x}$ do not fall in a straight line but are more appropriately described by the equation

$$\mu_{Y|x} = \beta_0 + \beta_1 x + \beta_2 x^2,$$

and the estimated response is obtained from the sample regression equation

$$\hat{y} = b_0 + b_1 x + b_2 x^2.$$

The purpose of the experiment in this case might be to estimate β_0, β_1, and β_2 and to perform tests of hypotheses on these parameters.

Confusion arises occasionally concerning what is meant by the term *linear model*. Normally, statisticians will refer to a linear model as one in

333

which the parameters occur linearly. By this definition the above equation for $\mu_{Y|x}$ represents a linear model. An example of a nonlinear model is the *exponential relationship* given by

$$\mu_{Y|x} = \alpha\beta^x,$$

which is estimated by the regression equation

$$\hat{y} = ab^x.$$

There are many phenomena in science and engineering that are inherently nonlinear in nature and, when the true structure is known, an attempt should certainly be made to fit the actual model. The literature on estimation by least squares of nonlinear models is voluminous and, except for Exercises 4 and 5 of Chapter 8, we do not attempt to cover the subject in this text. A student who wants a good account of some of the aspects of this subject should consult *Applied Regression Analysis* by Draper and Smith.

Often a good empirical representation of a scientific mechanism can be obtained through a polynomial model even though it is known that the true situation is, in fact, nonlinear. We treat polynomial regression here in the same order as we considered linear regression by initially considering the single independent variable case.

10.2 Single Independent Variable

For the case of a single independent variable, the *degree* of the best fitting polynomial can often be determined by plotting a scatter diagram of the data obtained from an experiment that yields n pairs of observations of the form $\{(x_i,y_i); i = 1, 2, \ldots, n\}$. Suppose a model is postulated so that each observation satisfies the equation

$$y_i = \beta_0 + \beta_1 x_i + \beta_2 x_i^2 + \cdots + \beta_r x_i^r + \epsilon_i$$
$$= b_0 + b_1 x_i + b_2 x_i^2 + \cdots + b_r x_i^r + e_i,$$

where r is the degree of the polynomial ϵ_i and e_i are again the random error and residual associated with the response y_i. Here, the number of pairs, n, must be at least as large as $r + 1$, the number of parameters to be estimated. We treat and interpret the ϵ_i term in the same fashion as we did in the case of linear regression. In fact, we notice that the polynomial model can be considered a special case of the more general multiple linear regression model where we set $x_1 = x$, $x_2 = x^2$, $\ldots$, $x_r = x^r$. We estimate $\beta_0, \beta_1, \beta_2, \ldots, \beta_r$ by again minimizing the sum of the squares of the residuals $e_i = y_i - \hat{y}_i$, where $\hat{y}_i$ is the point on the estimated regression curve at $x = x_i$. Therefore, we differentiate

$$
\begin{aligned}
\text{SSE} &= \sum_{i=1}^{n} e_i^2 \\
&= \sum_{i=1}^{n} (y_i - b_0 - b_1 x_i - b_2 x_i^2 \cdots - b_r x_i^r)^2
\end{aligned}
$$

with respect to $b_0, b_1, \ldots, b_r$, equate each partial derivative to zero, and obtain the set of normal equations given by

$$
\begin{bmatrix}
n & \sum_{i=1}^{n} x_i & \sum_{i=1}^{n} x_i^2 & \cdots & \sum_{i=1}^{n} x_i^r \\
\sum_{i=1}^{n} x_i & \sum_{i=1}^{n} x_i^2 & \sum_{i=1}^{n} x_i^3 & \cdots & \sum_{i=1}^{n} x_i^{r+1} \\
\sum_{i=1}^{n} x_i^2 & \sum_{i=1}^{n} x_i^3 & \sum_{i=1}^{n} x_i^4 & \cdots & \sum_{i=1}^{n} x_i^{r+2} \\
\cdot & \cdot & \cdot & & \cdot \\
\cdot & \cdot & \cdot & & \cdot \\
\cdot & \cdot & \cdot & & \cdot \\
\sum_{i=1}^{n} x_i^r & \sum_{i=1}^{n} x_i^{r+1} & \sum_{i=1}^{n} x_i^{r+2} & \cdots & \sum_{i=1}^{n} x_i^{2r}
\end{bmatrix}
\begin{bmatrix}
b_0 \\ b_1 \\ b_2 \\ \cdot \\ \cdot \\ \cdot \\ b_r
\end{bmatrix}
=
\begin{bmatrix}
\sum_{i=1}^{n} y_i \\ \sum_{i=1}^{n} x_i y_i \\ \cdot \\ \cdot \\ \cdot \\ \sum_{i=1}^{n} x_i^r y_i
\end{bmatrix}
$$

or, in more compact matrix notation, by $(\mathbf{X'X})\mathbf{b} = \mathbf{g}$. Solving these $r + 1$ equations we obtain the estimates $b_0, b_1, \ldots, b_r$ and thereby generate the polynomial regression prediction equation

$$
\hat{y} = b_0 + b_1 x + b_2 x^2 + \cdots + b_r x^r.
$$

10.3 Tests of Hypotheses

The hypothesis testing procedures for polynomial regression follow very closely those of linear regression. To make the standard tests, the assumptions of independence, common variance, and normality are made on the errors. As before the estimator of σ^2, the variance about the regression curve, is given by

$$
s^2 = \frac{\text{SSE}}{n - r - 1} = \frac{\text{SST} - \text{SSR}}{n - r - 1},
$$

with $n - r - 1$ degrees of freedom, where

$$\text{SSR} = \sum_{j=0}^{r} b_j g_j - \frac{\left(\sum_{i=1}^{n} y_i\right)^2}{n}.$$

We can test the hypothesis

$$H_0: \quad \beta_j = \beta_{j0}$$
$$H_1: \quad \beta_j \neq \beta_{j0}$$

by using the same techniques described in Section 9.4. In addition, a sequential testing procedure can be used for determining the proper degree of the polynomial to be fitted. For example, if we fit a polynomial of degree r, the test to determine whether or not the term $\beta_r x^r$ contributed significantly is given by comparing

$$f = \frac{R(\beta_r \mid \beta_1, \ldots, \beta_{r-1})}{s^2}$$

with the critical value $f_\alpha(1, n - r - 1)$. Here the numerator, representing the amount $\beta_r x^r$ contributed to SSR, is calculated using the identical steps discussed in Section 9.5. If this amount is not found to be significant, we drop $\beta_r x^r$ from the model and then conduct a similar investigation on the appropriateness of including $\beta_{r-1} x^{r-1}$ in the model. This process of eliminating unnecessary terms is continued until the experimenter is satisfied that he has the best polynomial representation.

10.4 Use of Orthogonal Polynomials

Before we proceed to the case of more than one independent variable (which is again handled by using general regression techniques discussed previously) it is important that we introduce the reader to the use of orthogonal polynomials. The advantage of these polynomials should be obvious after the student has been exposed to the discussion of orthogonality in Section 9.6. At that point we dwelled on the advantage of orthogonal columns in the so-called **X** matrix.

For the general polynomial model the **X** matrix is written

$$\mathbf{X} = \begin{bmatrix} 1 & x_1 & x_1^2 & \cdots & x_1^r \\ 1 & x_2 & x_2^2 & \cdots & x_2^r \\ \cdot & \cdot & \cdot & & \cdot \\ \cdot & \cdot & \cdot & & \cdot \\ \cdot & \cdot & \cdot & & \cdot \\ 1 & x_n & x_n^2 & \cdots & x_n^r \end{bmatrix}.$$

The subscript, of course, indicates the level of the controlled independent variable. Suppose we rewrite the model to read

$$y_i = \gamma_0 + \gamma_1 p_1(x_i) + \gamma_2 p_2(x_i) + \cdots + \gamma_r p_r(x_i) + \epsilon_i,$$

where the term $p_j(x_i)$ is a jth degree polynomial in x_i and the γ_j's are new parameters but obviously related to the β_j's. In addition we require that

$$\sum_{i=1}^{n} p_j(x_i) p_k(x_i) = 0, \qquad j \neq k$$

$$\sum_{i=1}^{n} p_j(x_i) = 0, \qquad j = 1, 2, \ldots, r$$

$$\sum_{i=1}^{n} p_j^2(x_i) \neq 0, \qquad j = 1, 2, \ldots, r.$$

Polynomials possessing these properties are called *orthogonal polynomials*. For the present we shall not be concerned about the exact structure of these polynomials. Now, the "new" X matrix is given by

$$X = \begin{bmatrix} 1 & p_1(x_1) & p_2(x_1) & \cdots & p_r(x_1) \\ 1 & p_1(x_2) & p_2(x_2) & \cdots & p_r(x_2) \\ \cdot & \cdot & \cdot & & \cdot \\ \cdot & \cdot & \cdot & & \cdot \\ \cdot & \cdot & \cdot & & \cdot \\ 1 & p_1(x_n) & p_2(x_n) & \cdots & p_r(x_n) \end{bmatrix}$$

and $X'X$ is the diagonal matrix

$$X'X = \begin{bmatrix} n & 0 & 0 & \cdots & 0 \\ 0 & \sum_{i=1}^{n} p_1^2(x_i) & 0 & \cdots & 0 \\ 0 & 0 & \sum_{i=1}^{n} p_2^2(x_i) & \cdots & 0 \\ \cdot & \cdot & \cdot & & \cdot \\ \cdot & \cdot & \cdot & & \cdot \\ \cdot & \cdot & \cdot & & \cdot \\ 0 & 0 & 0 & \cdots & \sum_{i=1}^{n} p_r^2(x_i) \end{bmatrix}$$

Clearly, then, we have orthogonal columns in the X matrix and all of the advantages of this condition are at our disposal. In fact we can very simply solve the normal equations, which, owing to the orthogonality conditions, reduce to

$$n\hat{\gamma}_0 = \sum_{i=1}^{n} y_i$$

$$\hat{\gamma}_1 \sum_{i=1}^{n} p_1^2(x_i) = \sum_{i=1}^{n} p_1(x_i)y_i$$

$$\hat{\gamma}_r \sum_{i=1}^{n} p_r^2(x_i) = \sum_{i=1}^{n} p_r(x_i)y_i.$$

Here the $\hat{\gamma}_i$'s, $i = 0, 1, \ldots, r$, are the least squares estimates, and the prediction equation relating Y to x is given by

$$\hat{y} = \hat{\gamma}_0 + \hat{\gamma}_1 p_1(x) + \hat{\gamma}_2 p_2(x) + \cdots + \hat{\gamma}_r p_r(x).$$

Tests of hypotheses on the individual coefficients are conducted with a minimum of effort. Suppose we make the usual normality, independence, and common variance assumptions on the random errors. The regression sum of squares may be simply written as

$$\text{SSR} = \hat{\gamma}_1 \sum_{i=1}^{n} p_1(x_i)y_i + \hat{\gamma}_2 \sum_{i=1}^{n} p_2(x_i)y_i + \cdots + \hat{\gamma}_r \sum_{i=1}^{n} p_r(x_i)y_i$$

with r degrees of freedom. In the expression for SSR, the individual terms represent meaningful single-degree-of-freedom components that can be used in testing the significance of individual coefficients. The orthogonality condition enables us to write

$$\text{SSR} = R(\gamma_1) + R(\gamma_2) + \cdots + R(\gamma_r),$$

where

$$R(\gamma_j) = \frac{\left[\sum_{i=1}^{n} p_j(x_i)y_i\right]^2}{\sum_{i=1}^{n} p_j^2(x_i)}.$$

Notice the analogy between the polynomial regression model using orthogonal polynomials and the ordinary multiple regression model with orthogonal columns in the $\mathbf{X}$ matrix. The $p_j(x_i)$'s have merely taken on the roles played by the independent variables in the multiple regression model. As always, the error sum of squares is just the difference between the corrected sum of squares of the response observations and the regression sum of squares. An analysis-of-variance table indicating the significance tests on the individual γ's is given in Table 10.1.

Table 10.1 Analysis of Variance for a Polynomial Model of Degree n Using Orthogonal Polynomials

Source of variation	Sum of squares	Degrees of freedom	Mean square	Computed f
γ_1	$R(\gamma_1) = \dfrac{\left[\sum\limits_{i=1}^{n} p_1(x_i)y_i\right]^2}{\sum\limits_{i=1}^{n} p_1^2(x_i)}$	1	$R(\gamma_1)$	$\dfrac{R(\gamma_1)}{s^2}$
γ_2	$R(\gamma_2) = \dfrac{\left[\sum\limits_{i=1}^{n} p_2(x_i)y_i\right]^2}{\sum\limits_{i=1}^{n} p_2^2(x_i)}$	1	$R(\gamma_2)$	$\dfrac{R(\gamma_2)}{s^2}$
$\vdots$	$\vdots$	$\vdots$	$\vdots$	$\vdots$
γ_r	$R(\gamma_r) = \dfrac{\left[\sum\limits_{i=1}^{n} p_r(x_i)y_i\right]^2}{\sum\limits_{i=1}^{n} p_r^2(x_i)}$	1	$R(\gamma_r)$	$\dfrac{R(\gamma_r)}{s^2}$
Error	SSE	$n - r - 1$	$s^2 = \dfrac{\text{SSE}}{n - r - 1}$	
Total	$\text{SST} = S_{yy}$	$n - 1$		

The F tests performed in Table 10.1 with 1 and $n - r - 1$ degrees of freedom, indicate the contribution (whether significant or not) of the individual polynomials. Of course, if we have $n > 1$ observations at each level, then SSE can be partitioned as usual into pure error and lack of fit so that the adequacy of the polynomial may be tested. The tests given in Table 10.1 are appropriate when the model hypothesized by the experimenter is correct.

10.5 Experiments with Equally Spaced Observations

The most important application of orthogonal polynomials in polynomial regression occurs when the values of the independent variable are *evenly spaced*. Hence, the experiment is designed in such a way that the x levels, determined by a controlled process, are set at $x_1, x_1 + d, x_1 + 2d, \ldots$. The term d will denote the *spacing* of the variable. In this kind of situation the polynomials are very simple to compute and are well tabulated. The stipulation of evenly spaced x values represents a rather elementary example of how a choice of an *experimental design* (in this case the mere spacing of the values of the independent variable) can aid the experimenter.

From *Biometrika Tables* by E. S. Pearson and H. O. Hartley, the first three polynomials, for m levels of the variable x, are seen to be

$$p_1(x) = \frac{x - \bar{x}}{d} \lambda_1$$

$$p_2(x) = \left[\left(\frac{x - \bar{x}}{d} \right)^2 - \frac{m^2 - 1}{12} \right] \lambda_2$$

$$p_3(x) = \left[\left(\frac{x - \bar{x}}{d} \right)^3 - \left(\frac{3m^2 - 7}{20} \right) \left(\frac{x - \bar{x}}{d} \right) \right] \lambda_3.$$

The λ values are actually not necessary but can be applied in order that the $p_j(x_i)$'s are integers. To explain more fully the use of orthogonal polynomials let us consider the following example.

Example 10.1 Suppose our experiment yields the following data:

x	y	x	y
2.00	14.5	3.00	19.34
2.00	14.7	3.50	19.51
2.00	15.8	3.50	19.43
2.50	16.93	3.50	18.97
2.50	17.77	4.00	18.45
2.50	17.92	4.00	19.43
3.00	19.15	4.00	18.78
3.00	18.75		

Fit a quadratic polynomial to the data. Perform an analysis of variance to determine the relative contributions of the linear and quadratic components, and test for lack of fit of the model.

Solution. Since three observations were made at each of five equally spaced levels of the independent variable x, the use of orthogonal polynomials is appropriate. We shall initially form the $\mathbf{X}$ matrix in order to illustrate the orthogonality. Note that $\bar{x} = 3.0$ and $d = 0.50$. Hence, the values $p_1(2.00)$ and $p_2(2.00)$ are given by

$$p_1(2.00) = \frac{2.00 - 3.00}{0.5} = -2$$

$$p_2(2.00) = \left[(-2)^2 - \frac{5^2 - 1}{12} \right] = +2.$$

In this example, λ_1 and λ_2 are both taken to be unity. The quantities p_1 and p_2 can be evaluated similarly for the remaining four levels of x. We may then write

$$
\mathbf{X} = \begin{bmatrix}
1 & -2 & 2 \\
1 & -2 & 2 \\
1 & -2 & 2 \\
\\
1 & -1 & -1 \\
1 & -1 & -1 \\
1 & -1 & -1 \\
\\
1 & 0 & -2 \\
1 & 0 & -2 \\
1 & 0 & -2 \\
\\
1 & 1 & -1 \\
1 & 1 & -1 \\
1 & 1 & -1 \\
\\
1 & 2 & 2 \\
1 & 2 & 2 \\
1 & 2 & 2
\end{bmatrix}, \quad
\mathbf{X}'\mathbf{X} = \begin{bmatrix}
15 & 0 & 0 \\
0 & 30 & 0 \\
0 & 0 & 42
\end{bmatrix},
$$

and

$$
\mathbf{g} = \begin{bmatrix}
g_0 = \sum_{i=1}^{15} y_i = 269.43 \\
g_1 = \sum_{i=1}^{15} p_1(x_i)y_i = 28.61 \\
g_2 = \sum_{i=1}^{15} p_2(x_i)y_i = -21.69
\end{bmatrix}.
$$

Solving the normal equations we obtain the estimates

$$
\hat{\gamma}_0 = \frac{269.43}{15} = 17.962
$$

$$
\hat{\gamma}_1 = \frac{28.61}{30} = 0.9537
$$

$$
\hat{\gamma}_2 = \frac{-21.69}{42} = -0.5164.
$$

The least squares quadratic prediction equation is then written

$$
\hat{y} = 17.962 + 0.9537\,\frac{x - 3.00}{0.5} - 0.5164\left[\left(\frac{x - 3.00}{0.5}\right)^2 - 2.00\right].
$$

In carrying out the analysis of variance we first compute

$$R(\gamma_1) = \frac{(28.61)^2}{30} = 27.284$$

$$R(\gamma_2) = \frac{(-21.69)^2}{42} = 11.201$$

$$S_{yy} = 40.9644,$$

and

$$SSE = 40.9644 - 27.284 - 11.201$$
$$= 2.4794.$$

To compute the pure error sum of squares we write

$$x_1 = 2.00 \quad T_1 = 45.00$$
$$x_2 = 2.50 \quad T_2 = 52.62$$
$$x_3 = 3.00 \quad T_3 = 57.24$$
$$x_4 = 3.50 \quad T_4 = 57.91$$
$$x_5 = 4.00 \quad T_5 = 56.66.$$

Then

$$SSE(pure) = 4880.4661 - \frac{45.00^2 + 52.62^2 + 57.24^2 + 57.91^2 + 56.66^2}{3}$$

$$= 2.3980.$$

These results and the remaining computations are exhibited in Table 10.2. The results indicate that essentially all of the variation is explained by the linear and quadratic terms.

Table 10.2 Analysis of Variance for Quadratic Regression

Source of variation	Sum of squares	Degrees of freedom	Mean square	Computed f
γ_1	27.284	1	27.284	113.78
γ_2	11.201	1	11.201	46.71
Error	2.4794	12		
Lack of fit	⎰0.0814	⎰2	0.0407	0.17
Pure error	⎱2.3980	⎱10	0.2398	
Total	40.9644	14		

In this example, three observations were taken at each level of x. If multiple observations are used, it is necessary that equal numbers be taken at each level if we are to maintain orthogonality. This does not, of course, rule out the use of the general procedure described earlier in this chapter.

10.6 Several Independent Variables

The procedure for fitting a polynomial regression model can be generalized to the case of more than one independent variable. In fact, the student of regression analysis should at this stage have the facility for fitting any linear model in, say, k independent variables. Suppose, for example, we have a response Y with $k = 2$ independent variables and a quadratic model is postulated of the type

$$y_i = \beta_0 + \beta_1 x_{1i} + \beta_2 x_{2i} + \beta_{11} x_{1i}^2 + \beta_{22} x_{2i}^2 + \beta_{12} x_{1i} x_{2i} + \epsilon_i,$$

where y_i, $i = 1, 2, \ldots, n$, is the response to the combination (x_{1i}, x_{2i}) of the independent variables in the experiment. In this situation n must be at least 6 since there are six parameters to estimate by the least squares procedure. In addition, since the model contains quadratic terms in both variables, at least three levels of each variable must be used. The reader should easily verify that the least squares normal equations $(\mathbf{X'X})\mathbf{b} = \mathbf{g}$ are given by

$$
\begin{bmatrix}
n & \sum_{i=1}^{n} x_{1i} & \sum_{i=1}^{n} x_{2i} & \sum_{i=1}^{n} x_{1i}^2 & \sum_{i=1}^{n} x_{2i}^2 & \sum_{i=1}^{n} x_{1i}x_{2i} \\
\sum_{i=1}^{n} x_{1i} & \sum_{i=1}^{n} x_{1i}^2 & \sum_{i=1}^{n} x_{1i}x_{2i} & \sum_{i=1}^{n} x_{1i}^3 & \sum_{i=1}^{n} x_{1i}x_{2i}^2 & \sum_{i=1}^{n} x_{1i}^2 x_{2i} \\
\sum_{i=1}^{n} x_{2i} & \sum_{i=1}^{n} x_{1i}x_{2i} & \sum_{i=1}^{n} x_{2i}^2 & \sum_{i=1}^{n} x_{1i}^2 x_{2i} & \sum_{i=1}^{n} x_{2i}^3 & \sum_{i=1}^{n} x_{1i}x_{2i}^2 \\
\sum_{i=1}^{n} x_{1i}^2 & \sum_{i=1}^{n} x_{1i}^3 & \sum_{i=1}^{n} x_{1i}^2 x_{2i} & \sum_{i=1}^{n} x_{1i}^4 & \sum_{i=1}^{n} x_{1i}^2 x_{2i}^2 & \sum_{i=1}^{n} x_{1i}^3 x_{2i} \\
\sum_{i=1}^{n} x_{2i}^2 & \sum_{i=1}^{n} x_{1i}x_{2i}^2 & \sum_{i=1}^{n} x_{2i}^3 & \sum_{i=1}^{n} x_{1i}^2 x_{2i}^2 & \sum_{i=1}^{n} x_{2i}^4 & \sum_{i=1}^{n} x_{1i}x_{2i}^3 \\
\sum_{i=1}^{n} x_{1i}x_{2i} & \sum_{i=1}^{n} x_{1i}^2 x_{2i} & \sum_{i=1}^{n} x_{1i}x_{2i}^2 & \sum_{i=1}^{n} x_{1i}^3 x_{2i} & \sum_{i=1}^{n} x_{1i}x_{2i}^3 & \sum_{i=1}^{n} x_{1i}^2 x_{2i}^2
\end{bmatrix}
\times
\begin{bmatrix}
b_0 \\
b_1 \\
b_2 \\
b_{11} \\
b_{22} \\
b_{12}
\end{bmatrix}
=
\begin{bmatrix}
\sum_{i=1}^{n} y_i \\
\sum_{i=1}^{n} x_{1i}y_i \\
\sum_{i=1}^{n} x_{2i}y_i \\
\sum_{i=1}^{n} x_{1i}^2 y_i \\
\sum_{i=1}^{n} x_{2i}^2 y_i \\
\sum_{i=1}^{n} x_{1i}x_{2i}y_i
\end{bmatrix}.
$$

Orthogonal polynomials can be used in the $k > 1$ case under conditions similar to those described for the single variable case. Evenly spaced levels of the variables are again required in order that any real advantage be attained. Let us consider an example as already described with $k = 2$ independent variables, where the postulated model is quadratic.

Example 10.2 The following data represent the percent of impurities that occurs at various temperatures and sterilizing times in a reaction associated with the manufacturing of a certain beverage.

Sterilizing time, x_2 (minutes)	Temperature, x_1 (°F)		
	75	100	125
15	14.05	10.55	7.55
	14.93	9.48	6.59
20	16.56	13.63	9.23
	15.85	11.75	8.78
25	22.41	18.55	15.93
	21.66	17.98	16.44

Use orthogonal polynomials to estimate the coefficients in the model

$$\mu_{Y|x_1, x_2} = \gamma_0 + \gamma_1 p_1(x_1) + \gamma_2 p_1(x_2) + \gamma_{11} p_2(x_1)$$
$$+ \gamma_{22} p_2(x_2) + \gamma_{12} p_1(x_1) p_1(x_2).$$

Solution. Using the formulas for orthogonal polynomials given in Section 10.5, we have

$$p_1(x_{1i}) = \frac{x_{1i} - 100}{25}$$

$$p_1(x_{2i}) = \frac{x_{2i} - 20}{5}$$

$$p_2(x_{1i}) = \left[\left(\frac{x_{1i} - 100}{25} \right)^2 - \frac{2}{3} \right] 3$$

$$p_2(x_{2i}) = \left[\left(\frac{x_{2i} - 20}{5} \right)^2 - \frac{2}{3} \right] 3$$

with the coefficient 3 being the value of λ_2 that generates integral values when we evaluate p_2 at the three levels given in the experiment. The values of p_1 turn out to be -1, 0, and $+1$ at the low, medium, and high levels, respectively, for both independent variables, while the corresponding values of p_2 are 1, -2, and 1. We may then write

$$
\mathbf{X} = \begin{array}{cccccc}
& p_1(x_1) & p_1(x_2) & p_2(x_1) & p_2(x_2) & p_1(x_1)p_1(x_2) \\
\begin{bmatrix}
1 & -1 & -1 & 1 & 1 & 1 \\
1 & -1 & -1 & 1 & 1 & 1 \\
1 & -1 & 0 & 1 & -2 & 0 \\
1 & -1 & 0 & 1 & -2 & 0 \\
1 & -1 & 1 & 1 & 1 & -1 \\
1 & -1 & 1 & 1 & 1 & -1 \\
1 & 0 & -1 & -2 & 1 & 0 \\
1 & 0 & -1 & -2 & 1 & 0 \\
1 & 0 & 0 & -2 & -2 & 0 \\
1 & 0 & 0 & -2 & -2 & 0 \\
1 & 0 & 1 & -2 & 1 & 0 \\
1 & 0 & 1 & -2 & 1 & 0 \\
1 & 1 & -1 & 1 & 1 & -1 \\
1 & 1 & -1 & 1 & 1 & -1 \\
1 & 1 & 0 & 1 & -2 & 0 \\
1 & 1 & 0 & 1 & -2 & 0 \\
1 & 1 & 1 & 1 & 1 & 1 \\
1 & 1 & 1 & 1 & 1 & 1
\end{bmatrix}
\end{array},
$$

$$
\mathbf{X'X} = \begin{bmatrix}
18 & 0 & \cdot & \cdot & \cdot & 0 \\
0 & 12 & \cdot & \cdot & \cdot & \cdot \\
\cdot & & 12 & & & \cdot \\
\cdot & & & 36 & & \cdot \\
\cdot & & & & 36 & 0 \\
0 & \cdot & \cdot & \cdot & 0 & 8
\end{bmatrix},
$$

and

$$
\mathbf{g} = \begin{bmatrix}
\sum_{i=1}^{18} y_i = 251.92 \\[2mm]
\sum_{i=1}^{18} y_i p_1(x_{1i}) = -40.94 \\[2mm]
\sum_{i=1}^{18} y_i p_1(x_{2i}) = 49.82 \\[2mm]
\sum_{i=1}^{18} y_i p_2(x_{1i}) = 6.10 \\[2mm]
\sum_{i=1}^{18} y_i p_2(x_{2i}) = 24.52 \\[2mm]
\sum_{i=1}^{18} y_i p_1(x_{1i}) p_1(x_{2i}) = 3.14
\end{bmatrix}.
$$

The estimates of the coefficients are given by

$$\hat{\gamma}_0 = \frac{251.92}{18} = 13.9956$$

$$\hat{\gamma}_1 = \frac{-40.94}{12} = -3.4117$$

$$\hat{\gamma}_2 = \frac{49.82}{12} = 4.1517$$

$$\hat{\gamma}_{11} = \frac{6.10}{36} = 0.1694$$

$$\hat{\gamma}_{22} = \frac{24.52}{36} = 0.6811$$

$$\hat{\gamma}_{12} = \frac{3.14}{8} = 0.3982,$$

and our estimated regression line in terms of the orthogonal polynomials takes the form

$$\hat{y} = 13.9956 - 3.4117p_1(x_1) + 4.1517p_1(x_2) + 0.1694p_2(x_1)$$
$$+ 0.6811p_2(x_2) + 0.3982p_1(x_1)p_1(x_2).$$

Tests of hypotheses can be constructed on the parameters of Example 10.2 under the usual normality, independence, and equal variance assumptions on the errors. The format for these tests is the usual analysis-of-variance table for which the orthogonality allows the regression sum of squares to be partitioned into independent single-degree-of-freedom components much like that described previously where we had orthogonal columns in the **X** matrix.

Many of the principles and procedures associated with the estimation of polynomial regression functions fall into the category of *response surface methodology*, a collection of techniques that have been used quite successfully by scientists and engineers in many fields. Such problems as selecting a proper experimental design, particularly in cases where a large number of variables are in the model, and choosing "optimum" operating conditions on x_1, x_2, ..., x_k are often approached through the use of these methods. For an extensive exposure the reader is referred to *Response Surface Methodology* by Myers.

EXERCISES

1. Given the data

x	0	1	2	3	4	5	6	7	8	9
y	9.1	7.3	3.2	4.6	4.8	2.9	5.7	7.1	8.8	10.2

(a) Fit a regression curve of the form $\mu_{Y|x} = \beta_0 + \beta_1 x + \beta_2 x^2$.
(b) Estimate Y when $x = 2$.

2. An experiment was conducted in order to determine if cerebral blood flow in human beings can be predicted from arterial oxygen tension (millimeters of mercury). Fifteen patients were used in the study and the following data were observed:

Blood flow, y	Arterial oxygen tension, x
84.33	603.40
87.80	582.50
82.20	556.20
78.21	594.60
78.44	558.90
80.01	575.20
83.53	580.10
79.46	451.20
75.22	404.00
76.58	484.00
77.90	452.40
78.80	448.40
80.67	334.80
86.60	320.30
78.20	350.30

Estimate the quadratic regression equation

$$\mu_{Y|x} = \alpha + \beta_1 x + \beta_2 x^2.$$

3. For the model of Exercise 2, test the hypothesis that $\beta_2 = 0$ at the 0.05 level of significance against the alternative that $\beta_2 \neq 0$.

4. The following is a set of coded experimental data on the compressive strength of a particular alloy at various values of the concentration of some additive:

Concentration, x	Compressive strength, y		
10.0	25.2	27.3	28.7
15.0	29.8	31.1	27.8
20.0	31.2	32.6	29.7
25.0	31.7	30.1	32.3
30.0	29.4	30.8	32.8

(a) Use orthogonal polynomials to estimate a quadratic regression function.
(b) Perform an analysis of variance to determine the relative contributions of the linear and quadratic components. Use the error mean square in testing for significance.
(c) Test for lack of fit of the model. Were you justified in using the error mean square in part (b)?

5. For Exercise 4, using the quadratic model, construct a 90% confidence interval for the mean compressive strength when the concentration is $x = 19.5$.

6. Consider the following two-way table of experimental data:

	x_2		
x_1	0.5	1.0	1.5
1.0	18.0	20.9	19.8
	18.5	21.3	18.9
2.0	17.5	18.8	18.2
	16.8	18.5	17.9

(a) Use orthogonal polynomials to estimate the coefficients in the model

$$\mu_{Y|x_1, x_2} = \gamma_0 + \gamma_1 p_1(x_1) + \gamma_2 p_1(x_2) + \gamma_{22} p_2(x_2) + \gamma_{12} p_1(x_1) p_1(x_2).$$

(b) Conduct individual significance tests on each coefficient.

11 ANALYSIS OF VARIANCE

11.1 Introduction

In the estimation and hypotheses testing material covered in Chapters 6 and 7, we were restricted in each case to considering no more than two parameters. Such was the case, for example, in testing for the equality of two population means using independent samples from normal populations with common but unknown variance, where it was necessary to obtain a pooled estimate of σ^2. It would seem obvious and desirable that the reader be able to extend the techniques developed so far to cover tests of hypotheses in which there are, say, k population means.

Analysis of variance is certainly not a new technique to the reader if he has followed the material on regression theory. We used the analysis-of-variance approach to partition the total sum of squares into a portion due to regression and a portion due to error. Further we were able, in some cases, to conveniently partition SSR into meaningful components for the purpose of testing relevant hypotheses on the parameters in the model. The term *analysis of variance* describes a technique whereby the total variation is being analyzed or divided into meaningful components.

Regression problems in which the model contains *quantitative variables* (like those we have discussed to this point) are not the only type in which the analysis of variance plays an important role. In this chapter we shall present and study other types of models in which this technique is used. The degree of difficulty of the analysis depends on the complexity of the problem.

Suppose in an industrial experiment that an engineer is interested in how the mean absorption of moisture in concrete varies among five different concrete aggregates. The samples are exposed to moisture for 48 hours. It is decided that 6 samples are to be tested for each aggregate, requiring a total of 30 samples to be tested. The data are recorded in Table 11.1.

Table 11.1 Absorption of Moisture in Concrete Aggregates

	Aggregate (weight %)					
	1	2	3	4	5	
	551	595	639	417	563	
	457	580	615	449	631	
	450	508	511	517	522	
	731	583	573	438	613	
	499	633	648	415	656	
	632	517	677	555	679	
Total	3,320	3,416	3,663	2,791	3,664	16,854
Mean	553.33	569.33	610.50	465.17	610.67	561.80

The model for this situation may be considered as follows. There are six observations taken from each of five populations with means $\mu_1, \mu_2, \ldots, \mu_5$, respectively. We might wish to test

$$H_0: \quad \mu_1 = \mu_2 = \cdots = \mu_5$$

$H_1:$ at least two of the means are not equal.

In addition we might be interested in making individual comparisons among these five population means.

In the analysis-of-variance procedure, it is assumed that whatever variation exists between the aggregate averages is attributed to (1) variation in absorption among observations within aggregate types and (2) variation due to aggregate types, that is, due to differences in the chemical composition of the aggregates. The *within-aggregate* variation is, of course, brought about by various causes. Perhaps humidity and temperature conditions were not kept entirely constant throughout the experiment. It is possible that there was a certain amount of heterogeneity in the batches of raw materials that were used. At any rate, we shall call the within-sample variation *chance* or *random* variation, and part of the goal of the analysis of variance is to determine if the difference between the five sample means are what one would expect due to random variation alone or if indeed there is also a contribution from the systematic variation attributed to the aggregate types. The procedure essentially, then, separates the total variability into the following important two components:

1. Variability between aggregates, measuring systematic and random variation.
2. Variability within aggregates, measuring only random variation.

There remains then the task of determining if component 1 is significantly larger than component 2.

Many pointed questions appear at this stage concerning the preceding problem. For example, how many samples must be tested in each aggregate? This is a question that continually haunts the practitioner. In addition, what if the within-sample variation is so large that it is difficult for a statistical procedure to detect the systematic differences? Can we systematically control extraneous sources of variation and thus remove them from the portion we call random variation? We shall attempt to answer these and other questions in the following sections.

11.2 One-Way Classification

Random samples of size n are selected from each of k populations. The k different populations are often classified according to different *treatments* or *groups*. Today the term *treatment* is used very generally to refer to the various classifications, whether they be aggregates, analysts, different fertilizers, or different regions of the country. It will be assumed that the k populations are independent and normally distributed with means $\mu_1, \mu_2, \ldots, \mu_k$ and common variance σ^2. We wish to derive appropriate methods for testing the hypothesis

$$H_0: \quad \mu_1 = \mu_2 = \cdots = \mu_k$$

H_1: at least two of the means are not equal.

Let y_{ij} denote the jth observation from the ith treatment and arrange the data as in Table 11.2. Here, T_i. is the total of all observations in the sample from the ith treatment, $\bar{y}_i$. is the mean of all observations in the sample from the ith treatment, $T_{..}$ is the total of all nk observations, and $\bar{y}_{..}$ is the mean of all nk observations. Each observation may be written in the form

$$y_{ij} = \mu_i + \epsilon_{ij},$$

where ϵ_{ij} measures the deviation of the jth observation of the ith sample from the corresponding treatment mean. The ϵ_{ij} term represents random error and plays the same role as the error terms in the regression models.

Table 11.2 k Random Samples

	Treatment					
	1	2	$\cdots$	i	$\cdots$	k
	y_{11}	y_{21}	$\cdots$	y_{i1}	$\cdots$	y_{k1}
	y_{12}	y_{22}	$\cdots$	y_{i2}	$\cdots$	y_{k2}
	$\cdot$	$\cdot$		$\cdot$		$\cdot$
	$\cdot$	$\cdot$		$\cdot$		$\cdot$
	$\cdot$	$\cdot$		$\cdot$		$\cdot$
	y_{1n}	y_{2n}	$\cdots$	y_{in}	$\cdots$	y_{kn}
Total	$T_1.$	$T_2.$	$\cdots$	$T_i.$	$\cdots$	$T_k.$ $T_{..}$
Mean	$\bar{y}_1.$	$\bar{y}_2.$	$\cdots$	$\bar{y}_i.$	$\cdots$	$\bar{y}_k.$ $\bar{y}_{..}$

An alternative and preferred form of this equation is obtained by substituting $\mu_i = \mu + \alpha_i$, where μ is defined to be the mean of all the μ_i's; that is,

$$\mu = \frac{\sum_{i=1}^{k} \mu_i}{k}.$$

Hence, we may write

$$y_{ij} = \mu + \alpha_i + \epsilon_{ij},$$

subject to the constraint that $\sum_{i=1}^{k} \alpha_i = 0$. It is customary to refer to α_i as the *effect* of the ith treatment.

The null hypothesis that the k population means are equal against the alternative that at least two of the means are unequal may now be replaced by the equivalent hypothesis

$$H_0: \quad \alpha_1 = \alpha_2 = \cdots = \alpha_k = 0$$

$$H_1: \quad \text{at least one of the } \alpha_i\text{'s is not equal to zero.}$$

Our test will be based on a comparison of two independent estimates of the common population variance σ^2. These estimates will be obtained by splitting the total variability of our data into two components.

THEOREM 11.1 *Sum of Squares Identity*

$$\sum_{i=1}^{k} \sum_{j=1}^{n} (y_{ij} - \bar{y}_{..})^2 = n \sum_{i=1}^{k} (\bar{y}_{i.} - \bar{y}_{..})^2 + \sum_{i=1}^{k} \sum_{j=1}^{n} (y_{ij} - \bar{y}_{i.})^2.$$

Proof.

$$\sum_{i=1}^{k}\sum_{j=1}^{n}(y_{ij}-\bar{y}_{..})^2 = \sum_{i=1}^{k}\sum_{j=1}^{n}[(\bar{y}_{i\cdot}-\bar{y}_{..})+(y_{ij}-\bar{y}_{i\cdot})]^2$$

$$= \sum_{i=1}^{k}\sum_{j=1}^{n}[(\bar{y}_{i\cdot}-\bar{y}_{..})^2 + 2(\bar{y}_{i\cdot}-\bar{y}_{..})(y_{ij}-\bar{y}_{i\cdot}) + (y_{ij}-\bar{y}_{i\cdot})^2]$$

$$= \sum_{i=1}^{k}\sum_{j=1}^{n}(\bar{y}_{i\cdot}-\bar{y}_{..})^2$$
$$+ 2\sum_{i=1}^{k}\sum_{j=1}^{n}(\bar{y}_{i\cdot}-\bar{y}_{..})(y_{ij}-\bar{y}_{i\cdot})$$
$$+ \sum_{i=1}^{k}\sum_{j=1}^{n}(y_{ij}-\bar{y}_{i\cdot})^2.$$

The middle term is zero since

$$\sum_{j=1}^{n}(y_{ij}-\bar{y}_{i\cdot}) = \sum_{j=1}^{n}y_{ij} - n\bar{y}_{i\cdot} = \sum_{j=1}^{n}y_{ij} - n\frac{\sum_{j=1}^{n}y_{ij}}{n} = 0.$$

The first sum does not have j as a subscript and therefore may be written as

$$\sum_{i=1}^{k}\sum_{j=1}^{n}(\bar{y}_{i\cdot}-\bar{y}_{..})^2 = n\sum_{i=1}^{k}(\bar{y}_{i\cdot}-\bar{y}_{..})^2.$$

Hence

$$\sum_{i=1}^{k}\sum_{j=1}^{n}(y_{ij}-\bar{y}_{..})^2 = n\sum_{i=1}^{k}(\bar{y}_{i\cdot}-\bar{y}_{..})^2 + \sum_{i=1}^{k}\sum_{j=1}^{n}(y_{ij}-\bar{y}_{i\cdot})^2.$$

It will be convenient in what follows to identify the terms of the sum of squares identity by the following notation:

$$\text{SST} = \sum_{i=1}^{k}\sum_{j=1}^{n}(y_{ij}-\bar{y}_{..})^2 = \text{total sum of squares}$$

$$\text{SSA} = n\sum_{i=1}^{k}(\bar{y}_{i\cdot}-\bar{y}_{..})^2 = \text{treatment sum of squares}$$

$$\text{SSE} = \sum_{i=1}^{k}\sum_{j=1}^{n}(y_{ij}-\bar{y}_{i\cdot})^2 = \text{error sum of squares}.$$

The sum of squares identity can then be represented symbolically by

$$SST = SSA + SSE.$$

As implied earlier, we need to compare the appropriate measure of the between-treatment variation with the within-treatment variation in order to detect significant differences in the observations due to the treatment effects. Suppose we look at the expected value of the treatment sum of squares.

THEOREM 11.2

$$E(SSA) = (k-1)\sigma^2 + n \sum_{i=1}^{k} \alpha_i^2.$$

Proof. Looking upon SSA as a random variable whose values would undoubtedly vary if the experiment were repeated several times, we now write

$$SSA = n \sum_{i=1}^{k} (\bar{Y}_{i\cdot} - \bar{Y}_{\cdot\cdot})^2.$$

From the model

$$Y_{ij} = \mu + \alpha_i + E_{ij}$$

we obtain

$$\bar{Y}_{i\cdot} = \mu + \alpha_i + \bar{E}_{i\cdot}.$$
$$\bar{Y}_{\cdot\cdot} = \mu + \bar{E}_{\cdot\cdot}$$

since $\sum_{i=1}^{k} \alpha_i = 0$. Hence

$$SSA = n \sum_{i=1}^{k} (\alpha_i + \bar{E}_{i\cdot} - \bar{E}_{\cdot\cdot})^2$$

and

$$E(SSA) = n \sum_{i=1}^{k} \alpha_i^2 + n \sum_{i=1}^{k} E(\bar{E}_{i\cdot}^2) - nkE(\bar{E}_{\cdot\cdot}^2) + 2n \sum_{i=1}^{k} \alpha_i E(\bar{E}_{i\cdot}).$$

Recalling that the E_{ij}'s are independent variables with mean zero and variance σ^2, it follows that

$$E(\bar{E}_{i\cdot}^2) = \frac{\sigma^2}{n}, \quad E(\bar{E}_{\cdot\cdot}^2) = \frac{\sigma^2}{nk}, \quad E(\bar{E}_{i\cdot}) = 0.$$

Therefore

$$E(\text{SSA}) = n \sum_{i=1}^{k} \alpha_i^2 + k\sigma^2 - \sigma^2$$

$$= (k - 1)\sigma^2 + n \sum_{i=1}^{k} \alpha_i^2.$$

One estimate of σ^2, based on $k - 1$ degrees of freedom, is given by the treatment mean square

$$s_1^2 = \frac{\text{SSA}}{k - 1}.$$

If H_0 is true and each α_i in Theorem 11.2 is set equal to zero, we see that

$$E\left(\frac{\text{SSA}}{k - 1}\right) = \sigma^2$$

and s_1^2 is an unbiased estimate of σ^2. However, if H_1 is true, we have

$$E\left(\frac{\text{SSA}}{k - 1}\right) = \sigma^2 + \frac{n \sum_{i=1}^{k} \alpha_i^2}{k - 1}$$

and s_1^2 estimates σ^2 plus an additional term, which measures variation due to the systematic effects.

A second and independent estimate of σ^2, based on $k(n - 1)$ degrees of freedom, is the familiar formula

$$s^2 = \frac{\text{SSE}}{k(n - 1)}.$$

The estimate s^2 is unbiased regardless of the truth or falsity of the null hypothesis (see Exercise 1). It is important to note that the sum of squares identity has not only partitioned the total variability of the data, but also the total number of degrees of freedom. That is,

$$nk - 1 = k - 1 + k(n - 1).$$

When H_0 is true, the ratio

$$f = \frac{s_1^2}{s^2}$$

is a value of the random variable F having the F distribution with $k-1$ and $k(n-1)$ degrees of freedom. Since s_1^2 overestimates σ^2 when H_0 is false, we have a one-tailed test with the critical region entirely in the right tail of the distribution. The null hypothesis H_0 is rejected at the α level of significance when

$$f > f_\alpha[k-1, k(n-1)].$$

In practice one usually computes SST and SSA first and then, making use of the sum of squares identity, obtains

$$\text{SSE} = \text{SST} - \text{SSA}.$$

The previously defined formulas for SST and SSA are not in the best computational form. Equivalent and preferred formulas are given by

$$\text{SST} = \sum_{i=1}^{k} \sum_{j=1}^{n} y_{ij}^2 - \frac{T_{..}^2}{nk}$$

and

$$\text{SSA} = \frac{\sum_{i=1}^{k} T_{i.}^2}{n} - \frac{T_{..}^2}{nk}.$$

The computations in an analysis-of-variance problem are usually summarized in tabular form as shown in Table 11.3.

Table 11.3 Analysis of Variance for the One-Way Classification

Source of variation	Sum of squares	Degrees of freedom	Mean square	Computed f
Treatments	SSA	$k-1$	$s_1^2 = \dfrac{\text{SSA}}{k-1}$	$\dfrac{s_1^2}{s^2}$
Error	SSE	$k(n-1)$	$s^2 = \dfrac{\text{SSE}}{k(n-1)}$	
Total	SST	$nk-1$		

Example 11.1 Test the hypothesis $\mu_1 = \mu_2 = \cdots = \mu_5$ at the 0.05 level of significance for the data of Table 11.1 on absorption of moisture by various types of cement aggregates.

Solution.

1. H_0: $\mu_1 = \mu_2 = \cdots = \mu_5$.
2. H_1: at least two of the means are not equal.

3. $\alpha = 0.05$.
4. Critical region: $F > 2.76$ with $\nu_1 = 4$ and $\nu_2 = 25$ degrees of freedom.
5. Computations:

$$\text{SST} = 551^2 + 457^2 + \cdots + 679^2 - \frac{16,854^2}{30}$$

$$= 9,677,954 - 9,468,577$$

$$= 209,377$$

$$\text{SSA} = \frac{3,320^2 + 3,416^2 + \cdots + 3,664^2}{6} - 9,468,577$$

$$= 85,356$$

$$\text{SSE} = 209,377 - 85,356 = 124,021.$$

These results and the remaining computations are exhibited in Table 11.4.

Table 11.4 Analysis of Variance for the Data of Table 11.1

Source of variation	Sum of squares	Degrees of freedom	Mean square	Computed f
Aggregates	85,356	4	21,339	4.30
Error	124,021	25	4,961	
Total	209,377	29		

6. Conclusion: Reject H_0 and conclude that the aggregates do not have the same mean absorption.

In experimental work one often loses some of the desired observations. For example, an experiment might be conducted to determine if college students obtain different grades on the average for classes meeting at different times of the day. Because of dropouts during the semester it is entirely possible to conclude the experiment with unequal numbers of students in the various sections. The previous analysis for equal sample size will still be valid by slightly modifying the sum of squares formulas. We now assume the k random samples to be of size $n_1, n_2, \ldots, n_k$, respectively, with $N = \sum_{i=1}^{k} n_i$. The computational formulas for SST and SSA are given by

$$\text{SST} = \sum_{i=1}^{k} \sum_{j=1}^{n_i} y_{ij}^2 - \frac{T_{..}^2}{N}$$

and

$$\text{SSA} = \sum_{i=1}^{k} \frac{T_{i\cdot}^2}{n_i} - \frac{T_{\cdot\cdot}^2}{N}.$$

As before, we find SSE by subtraction. The degrees of freedom are partitioned in the same way, namely, $N - 1$ for SST, $k - 1$ for SSA, and $N - 1 - (k - 1) = N - k$ for SSE.

In concluding our discussion on the analysis of variance for the one-way classification, we state the advantages in choosing equal sample sizes over the choice of unequal sample sizes. The first advantage is that the f ratio is insensitive to slight departures from the assumption of equal variances for the k populations when the samples are of equal size. Second, the choice of equal sample size minimizes the probability of committing a type II error. Finally, the computation of SSA is simplified if the sample sizes are equal.

11.3 Tests for the Equality of Several Variances

We have already stated in Section 11.2 that the f ratio obtained from the analysis-of-variance procedure is insensitive to slight departures from the assumption of equal variances for the k populations when the samples are of equal size. This is not the case, however, if the sample sizes are unequal or if one variance is much larger than the others. Consequently, one may wish to test the hypothesis

$$H_0: \quad \sigma_1^2 = \sigma_2^2 = \cdots = \sigma_k^2$$

against the alternative

$$H_1: \quad \text{the variances are not all equal.}$$

The test most often used, called *Bartlett's test*, is based on a statistic whose sampling distribution is approximated very closely by the chi-square distribution when the k random samples are drawn from independent normal populations.

First, we compute the k sample variances $s_1^2, s_2^2, \ldots, s_k^2$ from samples of size $n_1, n_2, \ldots, n_k$, with $\sum_{i=1}^{k} n_i = N$. Second, combine the sample variances to give the pooled estimate

$$s_p^2 = \frac{\sum_{i=1}^{k} (n_i - 1)s_i^2}{N - k},$$

which is equivalent to the formula $\text{SSE}/(N - k)$ of Section 11.2. Now

$$b = 2.3026 \frac{q}{h},$$

where

$$q = (N - k) \log s_p^2 - \sum_{i=1}^{k} (n_i - 1) \log s_i^2$$

and

$$h = 1 + \frac{1}{3(k - 1)} \left(\sum_{i=1}^{k} \frac{1}{n_i - 1} - \frac{1}{N - k} \right),$$

is a value of the random variable B having approximately the chi-square distribution with $k - 1$ degrees of freedom.

The quantity q is large when the sample variances differ greatly and is equal to zero when all the sample variances are equal. Hence we reject H_0 at the α level of significance only when $b > \chi_\alpha^2$. The hazard in applying Bartlett's test is that it is very sensitive to the normality assumption. In fact, a value of $B > \chi_\alpha^2$ may result from a deviation from normality as well as from heterogeneity of variances.

Example 11.2 Use Bartlett's test to test the hypothesis of equal variances for the three populations from which the hypothetical data of Table 11.5 were selected.

Table 11.5 Hypothetical Samples

	Sample		
	A	B	C
	4	5	8
	7	1	6
	6	3	8
	6	5	9
		3	5
		4	
Total	23	21	36	40

Solution.

1. H_0: $\sigma_1^2 = \sigma_2^2 = \sigma_3^2$.
2. H_1: The variances are not all equal.
3. $\alpha = 0.05$.

4. Critical region: $B > 5.991$.
5. Computations: Referring to Table 11.5, we have $n_1 = 4$, $n_2 = 6$, $n_3 = 5$, $N = 15$, and $k = 3$. First compute

$$s_1^2 = 1.583, \quad s_2^2 = 2.300, \quad s_3^2 = 2.700,$$

and then

$$s_p^2 = \frac{(3)(1.583) + (5)(2.300) + (4)(2.700)}{12} = 2.254.$$

Now

$$q = 12 \log 2.254 - (3 \log 1.583 + 5 \log 2.300 + 4 \log 2.700)$$
$$= (12)(0.3530) - [(3)(0.1995) + (5)(0.3617) + (4)(0.4314)]$$
$$= 0.1034$$
$$h = 1 + \tfrac{1}{6}[\tfrac{1}{3} + \tfrac{1}{5} + \tfrac{1}{4} - \tfrac{1}{12}] = 1.1167.$$

Hence

$$b = \frac{(2.3026)(0.1034)}{1.1167} = 0.213.$$

6. Conclusion: Accept the hypothesis and conclude that the variances of the three populations are equal.

Although Bartlett's test is most often used in testing for homogeneity of variances, there are other methods available. A method owing to Cochran provides a computationally simple procedure but it is restricted to situations in which the sample sizes are equal. Cochran's test is particularly useful in detecting if one variance is much larger than the others. The statistic that is used is given by

$$G = \frac{\text{largest } S_i^2}{\sum\limits_{i=1}^{k} S_i^2}$$

and the hypothesis of equality of variances is rejected if $g > g_\alpha$, where the value of g_α is obtained from Table XI (see statistical Tables).

To illustrate Cochran's test let us refer again to the data of Table 11.1 on the absorption of moisture in concrete aggregates. Were we justified in assuming equal variances when we performed the analysis of variance in Example 11.1? We find that

$$s_1^2 = 12{,}134, \quad s_2^2 = 2303, \quad s_3^2 = 3594, \quad s_4^2 = 3319, \quad s_5^2 = 3455.$$

Therefore

$$g = \frac{12{,}134}{24{,}805} = 0.4892,$$

which does not exceed the tabled value $g_{0.05} = 0.5065$. Hence, we conclude
that the assumption of equal variances is reasonable.

11.4 Single-Degree-of-Freedom Comparisons

The analysis of variance in a one-way classification or the one-factor
experiment, as it is often called, merely indicates whether or not the
hypothesis of equal treatment means can be rejected. Usually an experimenter would like for his analysis to probe deeper than this. For instance,
in Example 11.1 by rejecting the null hypothesis we concluded that the
means are not all equal, but we still do not know where the differences
exist among the aggregates. The engineer might have the feeling from the
outset that aggregates 1 and 2 should have similar absorption properties
due to similar composition and that the same is true for aggregates 3 and 5
but that the two groups possibly differ a great deal. It would seem, then,
appropriate to test the hypothesis

$$H_0: \quad \mu_1 + \mu_2 - \mu_3 - \mu_5 = 0$$
$$H_1: \quad \mu_1 + \mu_2 - \mu_3 - \mu_5 \neq 0.$$

We notice that the hypothesis is a linear function of the population means
in which the coefficients sum to zero.

DEFINITION 11.1 *Any linear function of the form*

$$\omega = \sum_{i=1}^{k} c_i \mu_i, \quad \text{where} \quad \sum_{i=1}^{k} c_i = 0,$$

is called a comparison *or* contrast *in the treatment means.*

The experimenter can often make multiple comparisons by testing the
significance of contrasts in the treatment means, that is, by testing a hy-

pothesis of the type

$$H_0: \sum_{i=1}^{k} c_i \mu_i = 0$$

$$H_1: \sum_{i=1}^{k} c_i \mu_i \neq 0,$$

where $\sum_{i=1}^{k} c_i = 0$. The test is conducted by first computing a similar contrast in the sample means, namely,

$$w = \sum_{i=1}^{k} c_i \bar{y}_{i..}$$

Since $\bar{Y}_{1.}, \bar{Y}_{2.}, \ldots, \bar{Y}_{k.}$ are independent random variables having a normal distribution with means $\mu_1, \mu_2, \ldots, \mu_k$ and variances $\sigma^2/n_1, \sigma^2/n_2, \ldots, \sigma^2/n_k$, respectively, Theorem 5.11 assures us that w is a value of the normal random variable W with mean

$$\mu_W = \sum_{i=1}^{k} c_i \mu_i$$

and variance

$$\sigma_W^2 = \sigma^2 \sum_{i=1}^{k} \frac{c_i^2}{n_i}.$$

Therefore, when H_0 is true, $\mu_W = 0$ and by Example 5.5 the statistic

$$\frac{\left(\sum_{i=1}^{k} c_i \bar{Y}_{i.}\right)^2}{\sigma^2 \sum_{i=1}^{k} (c_i^2/n_i)}$$

is distributed as a chi-square random variable with 1 degree of freedom. Our hypothesis is tested at the α level of significance by computing

$$f = \frac{\left(\sum_{i=1}^{k} c_i \bar{y}_{i.}\right)^2}{s^2 \sum_{i=1}^{k} (c_i^2/n_i)} = \frac{\left(\sum_{i=1}^{k} c_i T_{i.}/n_i\right)^2}{s^2 \sum_{i=1}^{k} (c_i^2/n_i)} = \frac{SS_w}{s^2},$$

where f is a value of the random variable F having the F distribution with 1 and $N - k$ degrees of freedom and

$$SSw = \frac{\left(\sum\limits_{i=1}^{k} c_i T_{i.}/n_i\right)^2}{\sum\limits_{i=1}^{k} (c_i^2/n_i)}.$$

When the sample sizes are all equal to n,

$$SSw = \frac{\left(\sum\limits_{i=1}^{k} c_i T_{i.}\right)^2}{n \sum\limits_{i=1}^{k} . c_i^2}.$$

The quantity SSw, called the *contrast sum of squares,* indicates the portion of SSA that is explained by the contrast in question.

DEFINITION 11.2 *The two contrasts*

$$\omega_1 = \sum_{i=1}^{k} b_i \mu_i \quad and \quad \omega_2 = \sum_{i=1}^{k} c_i \mu_i$$

are said to be orthogonal *if* $\sum\limits_{i=1}^{k} b_i c_i/n_i = 0$ *or when the* n_i*'s are all equal to* n *if* $\sum\limits_{i=1}^{k} b_i c_i = 0$.

If ω_1 and ω_2 are orthogonal, then the quantities SSw_1 and SSw_2 are components of SSA each with a single degree of freedom. The treatment sum of squares with $k - 1$ degrees of freedom can be partitioned into at most $k - 1$ independent single-degree-of-freedom contrast sum of squares satisfying the identity

$$SSA = SSw_1 + SSw_2 + \cdots + SSw_{k-1}$$

if the contrasts are orthogonal to each other.

Example 11.3 Referring to Example 11.1 find the contrast sum of squares corresponding to the orthogonal contrasts

$$\omega_1 = \mu_1 + \mu_2 - \mu_3 - \mu_5$$
$$\omega_2 = \mu_1 + \mu_2 + \mu_3 + \mu_5 - 4\mu_4$$

and carry out appropriate tests of significance.

Solution. It is obvious that the two contrasts are orthogonal since $(1)(1) + (1)(1) + (-1)(1) + (0)(-4) + (-1)(1) = 0$. The second contrast indicates a comparison between aggregates 1, 2, 3, and 5, and aggregate 4. One can write down two additional contrasts orthogonal to the first two such as

$$\omega_3 = \mu_1 - \mu_2 \quad \text{(aggregate 1 versus aggregate 2)}$$
$$\omega_4 = \mu_3 - \mu_5 \quad \text{(aggregate 3 versus aggregate 5).}$$

From the data of Table 11.1, we have

$$SSw_1 = \frac{(3320 + 3416 - 3663 - 3664)^2}{6[(1)^2 + (1)^2 + (-1)^2 + (-1)^2]} = 14{,}553$$

$$SSw_2 = \frac{[3320 + 3416 + 3663 + 3664 - 4(2791)]^2}{6[(1)^2 + (1)^2 + (1)^2 + (1)^2 + (-4)^2]} = 70{,}035.$$

A more extensive analysis-of-variance table is then given by Table 11.6.

Table 11.6 Analysis of Variance Using Orthogonal Contrasts

Source of variation	Sum of squares	Degree of freedom	Mean square	Computed f
Aggregates	85,356	4	21,339	4.30
(1,2) vs. (3,5)	14,553	1	14,553	2.93
(1,2,3,5) vs. 4	70,035	1	70,035	14.12
Error	124,021	25	4,961	
Total	209,377	29		

We note that the two contrast sum of squares account for nearly all of the aggregate sum of squares. While there is a significant difference between aggregates in their absorption properties, the contrast w_1 is not significant when compared to the critical value $f_{0.05}(1,25) = 4.24$. However, the f value of 14.12 for w_2 is significant and the hypothesis

$$H_0: \quad \mu_1 + \mu_2 + \mu_3 + \mu_5 = 4\mu_4$$

is rejected.

Orthogonal contrasts are used when the experimenter is interested in partitioning the treatment variation into independent components. There are several choices available in selecting the orthogonal contrasts except for the last one. Normally the experimenter would have certain contrasts that are of interest to him. Such was the case in our example, where chemical composition suggested that aggregates (1,2) and (3,5) constitute distinct groups with different absorption properties, a postulation that was not supported by the significance test. However, the second comparison supports the conclusion that aggregate 4 seems to "stand out" from the rest. In this case, the complete partitioning of SSA was not necessary since two of the four possible independent comparisons accounted for a majority of the variation in treatments.

11.5 Multiple-Range Test

We have discussed procedures whereby multiple comparisons among the treatment means can be made after the analysis of variance has indicated that the means do differ significantly. These comparisons are made by testing the significance of particular linear combinations of the treatment means, these linear combinations representing meaningful contrasts. There are other test procedures in the form of *multiple-range tests* that are often applied for somewhat similar purposes. These tests divide the k population means into subgroups such that any two means in a subgroup do not differ significantly. Essentially then, one is testing all possible hypotheses of the type $\mu_i - \mu_j = 0$ at a single controlled significance level α. The test that we shall study in this section is called *Duncan's multiple-range test*.

Let us assume that the k random samples are all of equal size n. The range of any subset of p sample means must exceed a certain value before we consider any of the p population means to be different. This value is called the *least significant range* for the p means and is denoted by R_p, where

$$R_p = r_p \sqrt{\frac{s^2}{n}}.$$

The sample variance s^2, which is an estimate of the common variance σ^2, is obtained from the error mean square in the analysis of variance. The values of the quantity r_p, called the *least significant studentized range*, depend on the desired level of significance and the number of degrees of freedom of the error mean square. These values may be obtained from Table XII for $p = 2, 3, \ldots, 10$ means.

To illustrate the multiple-range test procedure let us consider a hypothetical example in which six treatments are compared in a one-way classification with five observations per treatment. The error mean square, obtained from the analysis-of-variance table, is $s^2 = 2.45$ with 24 degrees of freedom. First, we arrange the sample means in increasing order of magnitude:

$\bar{y}_2.$	$\bar{y}_5.$	$\bar{y}_1.$	$\bar{y}_3.$	$\bar{y}_6.$	$\bar{y}_4.$
14.50	16.75	19.84	21.12	22.90	23.20

Let $\alpha = 0.05$. Then the values of r_p are obtained from Table XII, with $\nu = 24$ degrees of freedom, for $p = 2, 3, 4, 5,$ and 6. Finally, we obtain R_p by multiplying each r_p by $\sqrt{s^2/n} = \sqrt{2.45/5} = 0.7$. The results of these computations are summarized as follows:

p	2	3	4	5	6
r_p	2.919	3.066	3.160	3.226	3.276
R_p	2.043	2.146	2.212	2.258	2.293

Comparing these least significant ranges with the differences in ordered means, we arrive at the following conclusions:

1. Since $\bar{y}_4. - \bar{y}_2. = 8.70 > R_6 = 2.293$, we conclude that $\bar{y}_4.$ and $\bar{y}_2.$ are significantly different.
2. Comparing $\bar{y}_4. - \bar{y}_5.$ and $\bar{y}_6. - \bar{y}_2.$ with R_5, we conclude that $\bar{y}_4.$ is significantly greater than $\bar{y}_5.$ and $\bar{y}_6.$ is significantly greater than $\bar{y}_2..$
3. Comparing $\bar{y}_4. - \bar{y}_1., \bar{y}_6. - \bar{y}_5.,$ and $\bar{y}_3. - \bar{y}_2.$ with R_4, we conclude that each difference is significant.
4. Comparing $\bar{y}_4. - \bar{y}_3., \bar{y}_6. - \bar{y}_1., \bar{y}_3. - \bar{y}_5.,$ and $\bar{y}_1. - \bar{y}_2.$ with R_3, we find all differences significant except for $\bar{y}_4. - \bar{y}_3..$ Therefore $\bar{y}_4., \bar{y}_3.,$ and $\bar{y}_6.$ constitute a subset of homogeneous means.
5. Comparing $\bar{y}_3. - \bar{y}_1., \bar{y}_1. - \bar{y}_5.,$ and $\bar{y}_5. - \bar{y}_2.,$ with R_2, we conclude that only $\bar{y}_3.$ and $\bar{y}_1.$ are not significantly different.

It is customary to summarize the above conclusions by drawing a line under any subset of adjacent means that are not significantly different. Thus we have

$\bar{y}_2.$	$\bar{y}_5.$	$\bar{y}_1.$	$\bar{y}_3.$	$\bar{y}_6.$	$\bar{y}_4.$
14.50	16.75	19.84	21.12	22.90	23.20

One can immediately observe from this manner of presentation that $\mu_1 = \mu_3, \mu_3 = \mu_4, \mu_3 = \mu_6,$ and $\mu_4 = \mu_6$, while all other pairs of population means are considered significantly different.

11.6 Comparing a Set of Treatments in Blocks

It often becomes necessary in analysis-of-variance problems to design the experiment in such a way that the experimental error variation due to extraneous sources can be systematically controlled. In the preceding development of the one-way analysis-of-variance problem, it was assumed that conditions remained relatively homogeneous for the *experimental units* used in the various test runs. For example, in a chemical experiment designed to determine if there is a difference in mean reaction yield among four catalysts, samples of materials to be tested are drawn from the same batches of raw materials, while other conditions such as temperature and concentration of reactants are held constant. In this case, time of the experimental runs might represent the experimental units, and if the experimenter feels that there could possibly be a slight time effect, he would *randomize* the assignment of the catalysts to the runs in order to counteract the possible trend. This type of experimental strategy, whereby the treatments (catalysts) are assigned randomly to the experimental units, is called a *completely randomized design*. As a second example of such a design, consider an experiment to compare four methods of measuring a particular physical property of a fluid substance. Suppose the sampling process is destructive; that is, once a sample of the substance has been measured by one method, it cannot be measured again by one of the other methods. If it is decided that five measurements are to be taken for each method, then 20 samples of the material are selected from a large batch *at random* and are used in the experiment to compare the four measuring devices. The experimental units are the randomly selected samples. Any variation from sample to sample will appear in the error variation, as measured by s^2 in the analysis.

If the variation due to heterogeneity in experimental units is so large that the sensitivity of detecting treatment differences is reduced due to an inflated value of s^2, a better plan might be to "block off" variation due to these units and thus reduce the extraneous variation to that accounted for by smaller or more homogeneous blocks. The simplest design calling for this strategy is a *randomized block design*. For example, suppose in the previous catalyst illustration it is known a priori that there definitely is a significant day-to-day effect on the yield and that we can measure yield for four catalysts on a given day. Rather than assigning the four catalysts to the 20 test runs completely at random, we choose, say, 5 days and run each of the four catalysts on each day, randomly assigning the catalysts to the runs within days. In this way the day-to-day variation is removed in the analysis and consequently the experimental error, which still includes any time trend *within days*, more accurately represents chance variation. Each day is referred to as a *block*.

The classical example, using a randomized block design, is an agricultural experiment in which different fertilizers are being compared for their ability to increase the yield of a particular crop. Rather than assigning fertilizers at random to many plots over a large area of variable soil composition, the fertilizers should be assigned to smaller blocks comprised of homogeneous plots. The variation between these blocks, which is most likely significant compared to the uniformity of the plots within a block, is then removed from the experimental error in the analysis of variance.

It should be obvious to the reader that the most straightforward of the randomized block designs is one in which we randomly assign each treatment once to every block. Such an experimental layout is called a *randomized complete block design*, each block constituting a single *replication* of the treatments.

11.7 Randomized Complete Block Designs

A typical layout for the randomized complete block design using three treatments in four blocks is as follows:

Block 1	*Block 2*	*Block 3*	*Block 4*
t_2	t_1	t_1	t_2
t_1	t_3	t_2	t_1
t_3	t_2	t_3	t_3

The t's denote the assignment to blocks of each of the three treatments. Of course the true allocation to units within blocks is random.

Let us generalize and consider the case of k treatments assigned to b blocks. The observation y_{ij} denotes the response of the ith treatment applied to the jth block. It will be assumed that the y_{ij}, $i = 1, 2, \ldots, k$ and $j = 1, 2, \ldots, b$, are values of independent random variables having normal distributions with mean μ_{ij} and common variance σ^2. Let us define

$$\bar{y}_{i\cdot} = i\text{th treatment mean}$$

$$\bar{y}_{\cdot\cdot} = \text{overall mean}$$

$$\bar{y}_{\cdot j} = j\text{th block mean}$$

$$T_{i\cdot} = i\text{th treatment total}$$

$$T_{\cdot j} = j\text{th block total}$$

$$T_{\cdot\cdot} = \text{overall total.}$$

The average of the population means for the ith treatment, $\mu_{i\cdot}$, is defined by

$$\mu_{i\cdot} = \frac{\sum\limits_{j=1}^{b} \mu_{ij}}{b}.$$

Similarly, the average of the population means for the jth block, $\mu_{\cdot j}$, is defined by

$$\mu_{\cdot j} = \frac{\sum\limits_{i=1}^{k} \mu_{ij}}{k},$$

and the average of the bk population means, μ, is defined by

$$\mu = \frac{\sum\limits_{i=1}^{k} \sum\limits_{j=1}^{b} \mu_{ij}}{bk}.$$

To determine if part of the variation in our observations is due to differences among the treatments, we consider the test

$$H_0': \quad \mu_{1\cdot} = \mu_{2\cdot} = \cdots = \mu_{k\cdot} = \mu$$
$$H_1': \quad \text{the } \mu_{i\cdot}\text{'s are not all equal.}$$

Likewise, to determine if part of the variation is due to differences among the blocks, we consider the test

$$H_0'': \quad \mu_{\cdot 1} = \mu_{\cdot 2} = \cdots = \mu_{\cdot b} = \mu$$
$$H_1'': \quad \text{the } \mu_{\cdot j}\text{'s are not all equal.}$$

Each observation may be written in the form

$$y_{ij} = \mu_{ij} + \epsilon_{ij},$$

where ϵ_{ij} measures the deviation of the observed value y_{ij} from the population mean μ_{ij}. The preferred form of this equation is obtained by substituting

$$\mu_{ij} = \mu + \alpha_i + \beta_j,$$

where α_i, as before, is the effect of the ith treatment and β_j is the effect of the jth block. It is assumed that the treatment and block effects are additive. Hence, we may write

$$y_{ij} = \mu + \alpha_i + \beta_j + \epsilon_{ij}.$$

Notice that the model resembles that of the one-way classification, the essential difference being the introduction of the block effect β_j. The basic concept is much like that of the one-way classification except that we must account in the analysis for the additional effect due to blocks since we are now systematically controlling variation in *two directions*. If we now impose the restrictions that

$$\sum_{i=1}^{k} \alpha_i = 0 \quad \text{and} \quad \sum_{j=1}^{b} \beta_j = 0,$$

then

$$\mu_{i\cdot} = \frac{\sum_{j=1}^{b} (\mu + \alpha_i + \beta_j)}{b} = \mu + \alpha_i$$

and

$$\mu_{\cdot j} = \frac{\sum_{i=1}^{k} (\mu + \alpha_i + \beta_j)}{k} = \mu + \beta_j.$$

The null hypothesis that the k treatment means $\mu_{i\cdot}$ are equal, and therefore equal to μ, is now equivalent to testing the hypothesis

$$H'_0: \quad \alpha_1 = \alpha_2 = \cdots = \alpha_k = 0$$
$$H'_1: \quad \text{at least one of the } \alpha_i\text{'s is not equal to zero.}$$

Likewise, the null hypothesis that the b block means $\mu_{\cdot j}$ are equal is equivalent to testing the hypothesis

$$H''_0: \quad \beta_1 = \beta_2 = \cdots = \beta_b = 0$$
$$H''_1: \quad \text{at least one of the } \beta_j\text{'s is not equal to zero.}$$

Each of these tests will be based on a comparison of independent estimates of the common population variance σ^2. These estimates will be obtained by splitting the total sum of squares of our data into three components by means of the following identity.

THEOREM 11.3 *Sum of Squares Identity*

$$\sum_{i=1}^{k} \sum_{j=1}^{b} (y_{ij} - \bar{y}_{..})^2 = b \sum_{i=1}^{k} (\bar{y}_{i.} - \bar{y}_{..})^2 + k \sum_{j=1}^{b} (\bar{y}_{.j} - \bar{y}_{..})^2$$

$$+ \sum_{i=1}^{k} \sum_{j=1}^{b} (y_{ij} - \bar{y}_{i.} - \bar{y}_{.j} + \bar{y}_{..})^2.$$

Proof. $\displaystyle \sum_{i=1}^{k} \sum_{j=1}^{b} (y_{ij} - \bar{y}_{..})^2 = \sum_{i=1}^{k} \sum_{j=1}^{b} [(\bar{y}_{i.} - \bar{y}_{..}) + (\bar{y}_{.j} - \bar{y}_{..})$

$$+ (y_{ij} - \bar{y}_{i.} - \bar{y}_{.j} + \bar{y}_{..})]^2$$

$$= \sum_{i=1}^{k} \sum_{j=1}^{b} (\bar{y}_{i.} - \bar{y}_{..})^2 + \sum_{i=1}^{k} \sum_{j=1}^{b} (\bar{y}_{.j} - \bar{y}_{..})^2$$

$$+ \sum_{i=1}^{k} \sum_{j=1}^{b} (y_{ij} - \bar{y}_{i.} - \bar{y}_{.j} + \bar{y}_{..})^2$$

$$+ 2 \sum_{i=1}^{k} \sum_{j=1}^{b} (\bar{y}_{i.} - \bar{y}_{..})(\bar{y}_{.j} - \bar{y}_{..})$$

$$+ 2 \sum_{i=1}^{k} \sum_{j=1}^{b} (\bar{y}_{i.} - \bar{y}_{..})(y_{ij} - \bar{y}_{i.} - \bar{y}_{.j} + \bar{y}_{..})$$

$$+ 2 \sum_{i=1}^{k} \sum_{j=1}^{b} (\bar{y}_{.j} - \bar{y}_{..})(y_{ij} - \bar{y}_{i.} - \bar{y}_{.j} + \bar{y}_{..}).$$

The cross-product terms are all equal to zero. Hence

$$\sum_{i=1}^{k} \sum_{j=1}^{b} (y_{ij} - \bar{y}_{..})^2 = b \sum_{i=1}^{k} (\bar{y}_{i.} - \bar{y}_{..})^2 + k \sum_{j=1}^{b} (\bar{y}_{.j} - \bar{y}_{..})^2$$

$$+ \sum_{i=1}^{k} \sum_{j=1}^{b} (y_{ij} - \bar{y}_{i.} - \bar{y}_{.j} + \bar{y}_{..})^2.$$

The sum of squares identity may be represented symbolically by

$$\text{SST} = \text{SSA} + \text{SSB} + \text{SSE},$$

where

$$\text{SST} = \sum_{i=1}^{k} \sum_{j=1}^{b} (y_{ij} - \bar{y}_{..})^2 = \text{total sum of squares}$$

$$\text{SSA} = b \sum_{i=1}^{k} (\bar{y}_{i.} - \bar{y}_{..})^2 = \text{treatment sum of squares}$$

$$\text{SSB} = k \sum_{j=1}^{b} (\bar{y}_{.j} - \bar{y}_{..})^2 = \text{block sum of squares}$$

$$\text{SSE} = \sum_{i=1}^{k} \sum_{j=1}^{b} (y_{ij} - \bar{y}_{i.} - \bar{y}_{.j} + \bar{y}_{..})^2 = \text{error sum of squares.}$$

Following the procedure outlined in Theorem 11.2, where we interpret the sum of squares as functions of the independent random variables $Y_{11}, Y_{12}, \ldots, Y_{kb}$, we can show that the expected values of the treatment, block, and error sum of squares are given by

$$E(\text{SSA}) = (k - 1)\sigma^2 + b \sum_{i=1}^{k} \alpha_i^2$$

$$E(\text{SSB}) = (b - 1)\sigma^2 + k \sum_{j=1}^{b} \beta_j^2$$

$$E(\text{SSE}) = (b - 1)(k - 1)\sigma^2.$$

One estimate of σ^2, based on $k - 1$ degrees of freedom, is given by

$$s_1^2 = \frac{\text{SSA}}{k - 1}.$$

If the treatment effects $\alpha_1 = \alpha_2 = \cdots = \alpha_k = 0$, s_1^2 is an unbiased estimate of σ^2. However, if the treatment effects are not all zero, we have

$$E\left(\frac{\text{SSA}}{k - 1}\right) = \sigma^2 + \frac{b \sum_{i=1}^{k} \alpha_i^2}{k - 1}$$

and s_1^2 overestimates σ^2. A second estimate of σ^2, based on $b - 1$ degrees of freedom, is given by

$$s_2^2 = \frac{\text{SSB}}{b - 1}.$$

The estimate s_2^2 is an unbiased estimate of σ^2 when the block effects $\beta_1 = \beta_2 = \cdots = \beta_b = 0$. If the block effects are not all zero, then

$$E\left(\frac{\text{SSB}}{b-1}\right) = \sigma^2 + \frac{k \sum_{j=1}^{b} \beta_j^2}{b-1}$$

and s_2^2 will overestimate σ^2. A third estimate of σ^2, based on $(k-1)(b-1)$ degrees of freedom and independent of s_1^2 and s_2^2, is given by

$$s^2 = \frac{\text{SSE}}{(k-1)(b-1)},$$

which is unbiased regardless of the truth or falsity of either null hypothesis.

To test the null hypothesis that the treatment effects are all equal to zero, we compute the ratio

$$f_1 = \frac{s_1^2}{s^2},$$

which is a value of the random variable F_1 having the F distribution with $k-1$ and $(k-1)(b-1)$ degrees of freedom when the null hypothesis is true. The null hypothesis is rejected at the α level of significance when $f_1 > f_\alpha[k-1, (k-1)(b-1)]$.

Similarly, to test the null hypothesis that the block effects are all equal to zero, we compute the ratio

$$f_2 = \frac{s_2^2}{s^2},$$

which is a value of the random variable F_2 having the F distribution with $b-1$ and $(k-1)(b-1)$ degrees of freedom when the null hypothesis is true. In this case the null hypothesis is rejected at the α level of significance when $f_2 > f_\alpha[b-1, (k-1)(b-1)]$.

In practice we first compute SST, SSA, and SSB, and then obtain SSE by subtraction by the formula

$$\text{SSE} = \text{SST} - \text{SSA} - \text{SSB}.$$

The degrees of freedom associated with SSE are also usually obtained by subtraction. It is not difficult to verify the identity

$$(k - 1)(b - 1) = (kb - 1) - (k - 1) - (b - 1).$$

Preferred computational formulas for the sums of squares are given as follows:

$$\text{SST} = \sum_{i=1}^{k} \sum_{j=1}^{b} y_{ij}^2 - \frac{T_{..}^2}{bk}$$

$$\text{SSA} = \frac{\sum_{i=1}^{k} T_{i.}^2}{b} - \frac{T_{..}^2}{bk}$$

$$\text{SSB} = \frac{\sum_{j=1}^{b} T_{.j}^2}{k} - \frac{T_{..}^2}{bk}.$$

The computations in an analysis-of-variance problem for a randomized complete block design may be summarized as shown in Table 11.7.

Table 11.7 Analysis of Variance for the Randomized Complete Block Design

Source of variation	Sum of squares	Degrees of freedom	Mean square	Computed f
Treatments	SSA	$k - 1$	$s_1^2 = \dfrac{\text{SSA}}{k - 1}$	$f_1 = \dfrac{s_1^2}{s^2}$
Blocks	SSB	$b - 1$	$s_2^2 = \dfrac{\text{SSB}}{b - 1}$	$f_2 = \dfrac{s_2^2}{s^2}$
Error	SSE	$(k - 1)(b - 1)$	$s^2 = \dfrac{\text{SSE}}{(b - 1)(k - 1)}$	
Total	SST	$bk - 1$		

Example 11.4 Four different machines are to be considered in the assembling of a particular product. It is decided that six different operators are to be used in an experiment to compare the machines. The operation of the machines requires a certain amount of physical dexterity and it is known that there is a difference among the operators in the speed with which they operate the machines. The basic measurements on which the machines are to be compared is *time in seconds* to completion. The data are given in Table 11.8. Test the hypothesis H_0', at the 0.05 level of significance, that the machines perform at the same mean rate of speed. Also test the hypothesis H_0'' that the operators perform at the same mean rate of speed.

Table 11.8 Time in Seconds to Assemble Product

Machine	Operator						Total
	1	*2*	*3*	*4*	*5*	*6*	
1	42.5	39.3	39.6	39.9	42.9	43.6	247.8
2	39.8	40.1	40.5	42.3	42.5	43.1	248.3
3	40.2	40.5	41.3	43.4	44.9	45.1	255.4
4	41.3	42.2	43.5	44.2	45.9	42.3	259.4
Total	163.8	162.1	164.9	169.8	176.2	174.1	1010.9

Solution.

1. (a) H_0': $\alpha_1 = \alpha_2 = \alpha_3 = \alpha_4 = 0$ (machine effects are zero).
   (b) H_0'': $\beta_1 = \beta_2 = \cdots = \beta_6 = 0$ (operator effects are zero).
2. (a) H_1': At least one of the α_i's is not equal to zero.
   (b) H_1'': At least one of the β_j's is not equal to zero.
3. $\alpha = 0.05$.
4. Critical regions: (a) $F_1 > 3.29$. (b) $F_2 > 2.90$.
5. Computations:

$$\text{SST} = 42.5^2 + 39.8^2 + \cdots + 42.3^2 - \frac{1010.9^2}{24} = 81.86$$

$$\text{SSA} = \frac{247.8^2 + 248.3^2 + 255.4^2 + 259.4^2}{6} - \frac{1010.9^2}{24} = 15.93$$

$$\text{SSB} = \frac{163.8^2 + 162.1^2 + \cdots + 174.1^2}{4} - \frac{1010.9^2}{24} = 42.09$$

$$\text{SSE} = 81.86 - 15.93 - 42.09 = 23.84.$$

These results and the remaining computations are exhibited in Table 11.9.

Table 11.9 Analysis of Variance for the Data
of Table 11.8

Source of variation	Sum of squares	Degrees of freedom	Mean square	Computed *f*
Machines	15.93	3	5.31	3.34
Operators	42.09	5	8.42	5.30
Error	23.84	15	1.59	
Total	81.86	23		

6. Conclusion:
   (a) Reject H'_0 and conclude that the machines do not perform at the same mean rate of speed.
   (b) Reject H''_0, as expected, and conclude that the operators do not perform at the same mean rate of speed.

11.8 Additional Remarks Concerning Randomized Complete Block Designs

In Chapter 7 we presented a procedure for comparing means when the observations were *paired*. The procedure involved "subtracting out" the effect due to the homogeneous pair and thus working with differences. This is a special case of a randomized complete block design with $k = 2$ treatments. The n homogeneous units to which the treatments were assigned take on the role of blocks.

If there is heterogeneity in the experimental units, the experimenter should not be mislead into believing that it is always advantageous to reduce the experimental error through the use of small homogeneous blocks. Indeed there may be instances where it would not be desirable to block. The purpose in reducing the error variance is to increase the *sensitivity* of the test for detecting differences in the treatment means. This is reflected in the power of the test procedure. (The power of the analysis-of-variance test procedure is discussed more extensively in Section 11.11.) The power for detecting certain differences among the treatment means increases with a decrease in the error variance. However, the power is also affected by the degrees of freedom with which this variance is estimated, and blocking reduces the degrees of freedom that are available from $k(b - 1)$ for the one-way classification to $(k - 1)(b - 1)$. So one could lose power by blocking if there is not a significant reduction in the error variance.

Another important assumption that is implicit in writing the model for a randomized complete block design is that the treatment and block effects were additive. This is equivalent to stating that $\mu_{ij} - \mu_{ij'} = \mu_{i'j} - \mu_{i'j'}$ or $\mu_{ij} - \mu_{i'j} = \mu_{ij'} - \mu_{i'j'}$ for every value of i, i', j, and j'. That is, the difference between the population means for blocks j and j' is the same for every treatment and the difference between the population means for treatments i and i' is the same for every block. Referring to Example 11.4, this implies that if operator 3 is 0.5 seconds faster on the average than operator 2 when machine 1 is used, then operator 3 will still be 0.5 seconds faster on the average than operator 2 when machine 2, 3, or 4 is used. Likewise, if operator 1 is 1.2 seconds faster on the average using machine 2 than on machine 4, then operator 2, 3, . . . , 6 will also be 1.2 seconds faster on the average using machine 2 than on machine 4.

In many experiments the assumption of additivity does not hold and the analysis of Section 11.7 leads to erroneous conclusions. Suppose, for instance, that operator 3 is 0.5 seconds faster on the average than operator 2 when machine 1 is used but is 0.2 seconds slower on the average than operator 2 when machine 2 is used. The operators and machines are now said to *interact*.

An inspection of Table 11.8 suggests the presence of interaction. This apparent interaction may be real or it may be due to experimental error. The analysis of Example 11.4 was based on the assumption that the apparent interaction was due entirely to experimental error. If the total variability of our data was in part due to an interaction effect, this source of variation remained a part of the error sum of squares, causing the error mean square to overestimate σ^2, and thereby increased the probability of committing a type II error. We have, in effect, assumed an incorrect model. If we let $(\alpha\beta)_{ij}$ denote the interaction effect of the ith treatment and the jth block, we can write a more appropriate model in the form

$$y_{ij} = \mu + \alpha_i + \beta_j + (\alpha\beta)_{ij} + \epsilon_{ij},$$

on which we impose the additional restrictions $\sum_{i=1}^{k} (\alpha\beta)_{ij} = \sum_{j=1}^{b} (\alpha\beta)_{ij} = 0$. One can now very easily verify that

$$E\left[\frac{SSE}{(b-1)(k-1)}\right] = \sigma^2 + \frac{\sum_{i=1}^{k}\sum_{j=1}^{b} (\alpha\beta)_{ij}^2}{(b-1)(k-1)}.$$

Thus the error mean square is seen to be a biased estimate of σ^2 when existing interaction has been ignored. It would seem necessary at this point to arrive at a procedure for the detection of interaction for cases where there is suspicion that it exists. Such a procedure requires the availability of an unbiased and independent estimate of σ^2. Unfortunately the randomized block design does not lend itself to such a test unless the experimental setup is altered. This is discussed extensively in Chapter 12.

11.9 Random Effects Models

Throughout this chapter we have dealt with analysis-of-variance procedures in which the primary goal was to study the effect on some response of certain fixed or predetermined treatments. Experiments in which the treatments or treatment levels are preselected by the experimenter as opposed to being chosen randomly are called *fixed effects* experiments or *model I* experiments. For the fixed effects model, inferences were made only on those particular treatments used in the experiment.

It is often important that the experimenter be able to draw inferences about a population of treatments by means of an experiment in which the treatments used are chosen randomly from the population. For example, a biologist may be interested in whether or not there is a significant variance in some physiological characteristic due to animal type. The animal types actually used in the experiment are then chosen randomly and represent the treatment effects. A chemist may be interested in studying the effect of analytical laboratories on the chemical analysis of some substance. He is not concerned with particular laboratories but rather with a large population of laboratories. He might then select a group of laboratories at random and allocate samples to each for analysis. The statistical inference would then involve (1) testing whether or not the laboratories contribute a nonzero variance to the analytical results, and (2) estimating the variance due to laboratories and the variance within laboratories.

The one-way random effects model, often referred to as *model II*, is written like the fixed effects model but with the terms taking on different meanings. The response

$$y_{ij} = \mu + \alpha_i + \epsilon_{ij}$$

is now a value of the random variable

$$Y_{ij} = \mu + A_i + E_{ij}$$

with $i = 1, 2, \ldots, k$ and $j = 1, 2, \ldots, n$, where the A_i's are normally and independently distributed with mean zero and variance σ_α^2 and are independent of the E_{ij}'s. As for the fixed effects model, the E_{ij}'s are also normally and independently distributed with mean zero and variance σ^2.

Note that for a model II experiment the random variable $\sum_{i=1}^{k} A_i$ assumes the value $\sum_{i=1}^{k} \alpha_i \neq 0$.

THEOREM 11.4 *For the random effects one-way analysis-of-variance model*

$$E(\text{SSA}) = (k - 1)\sigma^2 + n(k - 1)\sigma_\alpha^2$$

and

$$E(\text{SSE}) = k(n - 1)\sigma^2.$$

Proof. From the model

$$Y_{ij} = \mu + A_i + E_{ij}$$

we obtain

$$\bar{Y}_{i.} = \mu + A_i + \bar{E}_{i.}$$
$$\bar{Y}_{..} = \mu + \bar{A}_. + \bar{E}_{..}$$

Hence

$$\text{SSA} = n \sum_{i=1}^{k} (\bar{Y}_{i.} - \bar{Y}_{..})^2$$

$$= n \sum_{i=1}^{k} [(A_i - \bar{A}_.) + (\bar{E}_{i.} - \bar{E}_{..})]^2$$

and

$$E(\text{SSA}) = n \sum_{i=1}^{k} E(A_i^2) - nkE(\bar{A}^2) + n \sum_{i=1}^{k} E(\bar{E}_{i.}^2) - nkE(\bar{E}_{..}^2)$$

$$= nk\sigma_\alpha^2 - n\sigma_\alpha^2 + k\sigma^2 - \sigma^2$$

$$= (k-1)\sigma^2 + n(k-1)\sigma_\alpha^2.$$

Following the same steps as above we also find

$$\text{SSE} = \sum_{i=1}^{k} \sum_{j=1}^{n} (Y_{ij} - \bar{Y}_{i.})^2$$

$$= \sum_{i=1}^{k} \sum_{j=1}^{n} (E_{ij} - \bar{E}_{i.})^2$$

and therefore

$$E(\text{SSE}) = \sum_{i=1}^{k} \sum_{j=1}^{n} E(E_{ij}^2) - n \sum_{i=1}^{k} E(\bar{E}_{i.}^2)$$

$$= nk\sigma^2 - k\sigma^2$$

$$= n(k-1)\sigma^2.$$

Table 11.10 shows the expected mean squares for both model I and model II. The computations for model II are carried out in exactly the

Table 11.10 Expected Mean Squares for the One-Way Classification

Source of variation	Degrees of freedom	Mean squares	Expected mean squares	
			Model I	Model II
Treatments	$k-1$	s_1^2	$\sigma^2 + \dfrac{n \sum_{i=1}^{k} \alpha_i^2}{k-1}$	$\sigma^2 + n\sigma_\alpha^2$
Error	$k(n-1)$	s^2	σ^2	σ^2
Total	$nk-1$			

same way as for model I. That is, the sum of squares, degrees-of-freedom, and mean square columns in an analysis-of-variance table are the same for both models.

In the random effects model we are interested in testing the hypothesis

$$H_0: \quad \sigma_\alpha^2 = 0$$
$$H_1: \quad \sigma_\alpha^2 \neq 0.$$

It is obvious from Table 11.10 that s_1^2 and s^2 are both estimates of σ^2 when H_0 is true and that the ratio

$$f = \frac{s_1^2}{s^2}$$

is a value of the random variable F having the F distribution with $k - 1$ and $k(n - 1)$ degrees of freedom. The null hypothesis is rejected at the α level of significance when

$$f > f_\alpha[k - 1, k(n - 1)].$$

Table 11.10 can also be used to estimate the *variance components* σ^2 and σ_α^2. Since s_1^2 estimates $\sigma^2 + n\sigma_\alpha^2$ and s^2 estimates σ^2; then

$$\hat{\sigma}^2 = s^2$$

$$\hat{\sigma}_\alpha^2 = \frac{s_1^2 - s_2^2}{n}.$$

Example 11.5 The following data are coded observations on the yield of a chemical process using five batches of raw material selected randomly:

			Batch			
	1	*2*	*3*	*4*	*5*	
	9.7	10.4	15.9	8.6	9.7	
	5.6	9.6	14.4	11.1	12.8	
	8.4	7.3	8.3	10.7	8.7	
	7.9	6.8	12.8	7.6	13.4	
	8.2	8.8	7.9	6.4	8.3	
	7.7	9.2	11.6	5.9	11.7	
	8.1	7.6	9.8	8.1	10.7	
Total	55.6	59.7	80.7	58.4	75.3	329.7

Show that the batch variance component is significantly greater than zero and obtain its estimate.

Solution. The total, batch, and error sum of squares are given by

$$\text{SST} = 9.7^2 + 5.6^2 + \cdots + 10.7^2 - \frac{329.7^2}{35}$$

$$= 194.64$$

$$\text{SSA} = \frac{55.6^2 + 59.7^2 + \cdots + 75.3^2}{7} - \frac{329.7^2}{35}$$

$$= 72.60$$

$$\text{SSE} = 194.64 - 72.60 = 122.04.$$

These results, with the remaining computations, are given in Table 11.11.

Table 11.11 Analysis of Variance for Example 11.5

Source of variation	Sum of squares	Degrees of freedom	Mean square	Computed f
Batches	72.60	4	18.15	4.46
Error	122.04	30	4.07	
Total	194.64	34		

The f ratio is significant at the $\alpha = 0.05$ level, indicating that the hypothesis of a zero batch component is rejected. An estimate of the batch variance component is given by

$$\hat{\sigma}_\alpha^2 = \frac{18.15 - 4.07}{7} = 2.01.$$

In a randomized complete block experiment in which the blocks represent days it is conceivable that the experimenter would like his results to apply not only to the actual days used in the analysis but to every day in the year. He would then select the days on which he runs his experiment as well as the treatments at random and use the random effects model

$$Y_{ij} = \mu + A_i + B_j + E_{ij},$$

$i = 1, 2, \ldots, k$ and $j = 1, 2, \ldots, b$, with the A_i, B_j, and E_{ij} being independent random variables with means zero and variances σ_α^2, σ_β^2, and σ^2, respectively. The expected mean squares for model II are obtained using the same procedure as for the one-way classification and are presented along with those for model I in Table 11.12.

Table 11.12 Expected Mean Squares for the Randomized
Complete Block Design

Source of variation	Degrees of freedom	Mean square	Expected mean squares	
			Model I	Model II
Treatments	$k - 1$	s_1^2	$\sigma^2 + \dfrac{b \sum\limits_{i=1}^{k} \alpha_i^2}{k - 1}$	$\sigma^2 + b\sigma_\alpha^2$
Blocks	$b - 1$	s_2^2	$\sigma^2 + \dfrac{k \sum\limits_{j=1}^{b} \beta_j^2}{b - 1}$	$\sigma^2 + k\sigma_\beta^2$
Error	$(b - 1)(k - 1)$	s^2	σ^2	σ^2
Total	$bk - 1$			

Again the computations for the individual sum of squares and degrees of freedom are identical to those of the fixed effects model. The hypothesis

$$H_0: \quad \sigma_\alpha^2 = 0$$
$$H_1: \quad \sigma_\alpha^2 \neq 0$$

is carried out by computing

$$f = \frac{s_1^2}{s^2}$$

and rejecting H_0 when $f > f_\alpha[k - 1, (b - 1)(k - 1)]$. Similarly we can test

$$H_0: \quad \sigma_\beta^2 = 0$$
$$H_1: \quad \sigma_\beta^2 \neq 0$$

by comparing

$$f = \frac{s_2^2}{s^2}$$

with the critical point $f_\alpha[b - 1, (b - 1)(k - 1)]$.

The unbiased estimates of the variance components are given by

$$\hat{\sigma}^2 = s^2$$
$$\hat{\sigma}_\alpha^2 = \frac{s_1^2 - s^2}{b}$$
$$\hat{\sigma}_\beta^2 = \frac{s_2^2 - s^2}{k}.$$

11.10 Regression Approach to Analysis of Variance

So far we have treated the regression models and the analysis-of-variance models as two separate and unrelated topics. Although this has become the accepted approach in dealing with these procedures on an elementary level, one can treat an analysis-of-variance model as a special case of a multiple linear regression model. In this section we shall show the relationship between the two models and indicate how the analysis of variance techniques can be developed through a regression approach.

Suppose we consider two models, the multiple linear regression model

$$y_i = \beta_0 + \beta_1 x_{1i} + \beta_2 x_{2i} + \cdots + \beta_k x_{ki} + \epsilon_i$$

and the one-way classification analysis-of-variance model

$$y_{ij} = \mu + \alpha_i + \epsilon_{ij}.$$

Traditionally the two are presented as methods for handling different practical problems, the regression model being a means of arriving at a procedure for predicting some response as a function of one or more quantitative independent variables, and the analysis-of-variance model for arriving at significance tests on multiple population means. However, any mathematical model that is linear in the parameters, such as the analysis-of-variance model, can be considered a special case of the multiple linear regression model. We can use conventional matrix notation to describe how each observation is expressed as a function of the parameters for the two models. For the regression model

$$\mathbf{y} = \mathbf{X}\boldsymbol{\beta} + \boldsymbol{\epsilon},$$

or more explicitly

$$
\begin{bmatrix} y_1 \\ y_2 \\ \cdot \\ \cdot \\ \cdot \\ y_n \end{bmatrix}
=
\begin{bmatrix}
1 & x_{11} & x_{21} & \cdots & x_{k1} \\
1 & x_{12} & x_{22} & \cdots & x_{k2} \\
\cdot & \cdot & \cdot & & \cdot \\
\cdot & \cdot & \cdot & & \cdot \\
\cdot & \cdot & \cdot & & \cdot \\
1 & x_{1n} & x_{2n} & \cdots & x_{kn}
\end{bmatrix}
\begin{bmatrix} \beta_0 \\ \beta_1 \\ \cdot \\ \cdot \\ \cdot \\ \beta_k \end{bmatrix}
+
\begin{bmatrix} \epsilon_1 \\ \epsilon_2 \\ \cdot \\ \cdot \\ \cdot \\ \epsilon_n \end{bmatrix},
$$

where $\mathbf{y}$ vector on the left of the equality sign is the array of responses in the experiment. The $\mathbf{X}$ matrix has already been described and used in Section 9.2. The $\boldsymbol{\beta}$ vector is the vector of parameters appearing in the model, and the $\boldsymbol{\epsilon}$ vector completes the model by the addition of the random error. The reader will recall that the least squares estimates b_0,

$b_1, \ldots, b_k$ of the parameters $\beta_0, \beta_1, \ldots, \beta_k$ are obtained by solving the equation

$$\mathbf{Ab} = \mathbf{g},$$

where $\mathbf{A} = \mathbf{X}'\mathbf{X}$ is a nonsingular matrix and $\mathbf{g} = \mathbf{X}'\mathbf{y}$ is a vector whose elements are sums of products of elements in the columns of $\mathbf{X}$ and the elements in the vector $\mathbf{y}$. Thus the estimates are given by

$$\mathbf{b} = \mathbf{A}^{-1}\mathbf{g}.$$

Consider now the analysis-of-variance model in matrix form:

$$
\begin{bmatrix} y_{11} \\ y_{12} \\ \cdot \\ \cdot \\ \cdot \\ y_{1n} \\ y_{21} \\ y_{22} \\ \cdot \\ \cdot \\ \cdot \\ y_{2n} \\ \cdot \\ \cdot \\ y_{k1} \\ y_{k2} \\ \cdot \\ \cdot \\ \cdot \\ y_{kn} \end{bmatrix}
=
\begin{bmatrix}
1 & 1 & 0 & \cdots & 0 \\
1 & 1 & 0 & \cdots & 0 \\
\cdot & \cdot & \cdot & & \cdot \\
\cdot & \cdot & \cdot & & \cdot \\
\cdot & \cdot & \cdot & & \cdot \\
1 & 1 & 0 & \cdots & 0 \\
1 & 0 & 1 & \cdots & 0 \\
1 & 0 & 1 & \cdots & 0 \\
\cdot & \cdot & \cdot & & \cdot \\
\cdot & \cdot & \cdot & & \cdot \\
1 & 0 & 1 & \cdots & 0 \\
\cdot & \cdot & \cdot & & \cdot \\
\cdot & \cdot & \cdot & & \cdot \\
1 & 0 & 0 & \cdots & 1 \\
1 & 0 & 0 & \cdots & 1 \\
\cdot & \cdot & \cdot & & \cdot \\
\cdot & \cdot & \cdot & & \cdot \\
\cdot & \cdot & \cdot & & \cdot \\
1 & 0 & 0 & \cdots & 1
\end{bmatrix}
\begin{bmatrix} \mu \\ \alpha_1 \\ \alpha_2 \\ \cdot \\ \cdot \\ \cdot \\ \alpha_k \end{bmatrix}
+
\begin{bmatrix} \epsilon_{11} \\ \epsilon_{12} \\ \cdot \\ \cdot \\ \cdot \\ \epsilon_{1n} \\ \epsilon_{21} \\ \epsilon_{22} \\ \cdot \\ \cdot \\ \cdot \\ \epsilon_{2n} \\ \cdot \\ \cdot \\ \epsilon_{k1} \\ \epsilon_{k2} \\ \cdot \\ \cdot \\ \cdot \\ \epsilon_{kn} \end{bmatrix}.
$$

Again each observation is expressed as a function of the parameters. Here the very important $\mathbf{X}$ matrix, the matrix of experimental conditions, consists of ones and zeros. Similar formulations can be written for the randomized complete block model.

Let us apply the least squares approach to the one-way analysis-of-variance model. The normal equations are given by

$$
\begin{bmatrix}
nk & n & n & \cdots & n \\
n & n & 0 & \cdots & 0 \\
n & 0 & n & \cdots & 0 \\
\cdot & \cdot & \cdot & & \cdot \\
\cdot & \cdot & \cdot & & \cdot \\
\cdot & \cdot & \cdot & & \cdot \\
n & 0 & 0 & \cdots & n
\end{bmatrix}
\begin{bmatrix}
\hat{\mu} \\
\hat{\alpha}_1 \\
\hat{\alpha}_2 \\
\cdot \\
\cdot \\
\cdot \\
\hat{\alpha}_k
\end{bmatrix}
=
\begin{bmatrix}
T_{..} \\
T_{1.} \\
T_{2.} \\
\cdot \\
\cdot \\
\cdot \\
T_{k.}
\end{bmatrix}
$$

At this stage it is simple to illustrate why a distinction is made in the presentation of the two models. The last k columns of the **A** matrix for the analysis-of-variance model add to the first column and thus the matrix is *singular*, implying that there is no unique solution to the estimating equations. This initially seems like a sad and serious drawback as far as the model is concerned. In fact we say that the parameters in the model are not *estimable*. The reader will recall that significance tests were performed on the population means, $\mu_1 = \mu + \alpha_1$, $\mu_2 = \mu + \alpha_2$, ..., $\mu_k = \mu + \alpha_k$, and, in formulating the test procedure, the linear constraint $\sum_{i=1}^{k} \alpha_i = 0$ was applied. Thus the α_i's take on the role of deviations (plus or minus) of the treatment or population means from the overall mean μ. Testing equality of population means then becomes equivalent to testing that the α_i's $(i = 1, 2, \ldots, k)$ are all zero.

With the constraint that the α_i's sum to zero, the estimating equations can be solved to yield

$$
\hat{\mu} = \frac{T_{..}}{nk} = \bar{y}_{..}
$$

$$
\hat{\alpha}_i = \frac{T_{i.}}{n} - \frac{T_{..}}{nk} = \bar{y}_{i.} - \bar{y}_{..}, \qquad i = 1, 2, \ldots, k.
$$

While these estimates are not unique, since they are dependent on the constraint that was applied to the α_i's, they do give us a basis for using the general regression procedure outlined in Section 9.5 to determine if the deletion of the α_i's from the model significantly increases the error sum of squares, thereby providing us, in the regression context, with a test of hypothesis of no significant treatment effects.

If we approach the hypothesis testing problem for the one-way analysis-of-variance model following the multiple regression procedures in Chapter 9, we might begin by computing the regression sum of squares for the parameters $\alpha_1, \alpha_2, \ldots, \alpha_k$. These parameters take on the same role as the coefficients β_1, β_2, $\ldots$, β_k in the multiple linear regression model. We would then compute the regression sum of squares

$$R(\alpha_1,\alpha_2,\ldots,\alpha_k) = \text{SSR}$$

$$= b_0 g_0 + b_1 g_1 + \cdots + b_k g_k - \frac{\left(\sum_{i=1}^{k}\sum_{j=1}^{n} y_{ij}\right)^2}{nk}$$

$$= b_1 g_1 + \cdots + b_k g_k$$

$$= \hat{\alpha}_1 g_1 + \cdots + \hat{\alpha}_k g_k.$$

The right side of the estimating equations give $g_1 = T_{1\cdot}$, $g_2 = T_{2\cdot}$, $\ldots$, $g_k = T_{k\cdot}$. Hence

$$R(\alpha_1,\alpha_2,\ldots,\alpha_k) = \sum_{i=1}^{k}\left(\frac{T_{i\cdot}}{n} - \frac{T_{\cdot\cdot}}{nk}\right) T_{i\cdot}$$

$$= \sum_{i=1}^{k} \frac{T_{i\cdot}^2}{n} - \frac{T_{\cdot\cdot}^2}{nk}$$

$$= \text{SSA},$$

with $k - 1$ degrees of freedom rather than k. One degree of freedom is lost on account of the single linear restraint imposed on the treatment effects. The error sum of squares with $(nk - 1) - (k - 1) = k(n - 1)$ degrees of freedom is given by

$$\text{SSE} = \text{SST} - R(\alpha_1,\alpha_2,\ldots,\alpha_k)$$

$$= \text{SST} - \text{SSA},$$

which is identical to the expression developed earlier in this chapter.

The hypothesis that the regression on the α_i's is insignificant, that is, $\alpha_i = 0$ for all i's is tested by forming the ratio

$$f = \frac{R(\alpha_1,\alpha_2,\ldots,\alpha_k)/(k - 1)}{\text{SSE}/k(n - 1)} = \frac{\text{SSA}/(k - 1)}{s^2}.$$

A value of $f > f_\alpha[(k - 1), k(n - 1)]$ implies that regression is significantly increased and consequently the error sum of squares is significantly decreased by including the treatment effects in the model.

The regression approach to analysis-of-variance-type models can be extended to the randomized complete block design and to the factorial models of Chapter 12.

11.11 Power of Analysis-of-Variance Tests

As we indicated earlier, the research worker is often plagued by the problem of not knowing how large a sample to choose. In conducting a one-way fixed effects analysis of variance with n observations per treatment, the main objective is to test the hypothesis of equality of treatment means, namely,

$$H_0: \quad \alpha_1 = \alpha_2 = \cdots = \alpha_k = 0$$

$H_1:$ at least one of the α_i's are not equal to zero.

Quite often, however, the experimental error variance, σ^2, is so large that the test procedure will be insensitive to actual differences among the k treatment means. In Section 11.2 the expected values of the mean squares for the one-way model were given by

$$E(S_1^2) = E\left(\frac{\text{SSA}}{k-1}\right) = \sigma^2 + \frac{n \sum_{i=1}^{k} \alpha_i^2}{(k-1)}$$

$$E(S^2) = E\left(\frac{\text{SSE}}{k(n-1)}\right) = \sigma^2.$$

Thus for a given deviation from the null hypothesis H_0, as measured by $n \sum_{i=1}^{k} \alpha_i^2/(k-1)$, large values of σ^2 decrease the chance of obtaining a value $f = s_1^2/s^2$ that is in the critical region for the test. The sensitivity of the test describes the ability of the procedure to detect differences in the population means and is measured by the power of the test (see Section 7.5), which is merely $1 - \beta$, where β is the probability of accepting a false hypothesis. We can interpret the power for our analysis-of-variance tests, then, as the probability that the F statistic is in the critical region when, in fact, the null hypothesis is false and the treatment means do differ. For the one-way analysis-of-variance test, the power, $1 - \beta$, is given by

$$1 - \beta = Pr\left[\frac{S_1^2}{S^2} > f_\alpha(\nu_1, \nu_2) \mid H_1 \text{ is true}\right]$$

$$= Pr\left[\frac{S_1^2}{S^2} > f_\alpha(\nu_1, \nu_2) \mid \frac{n \sum_{i=1}^{k} \alpha_i^2}{k-1}\right].$$

The term $f_\alpha(\nu_1,\nu_2)$ is, of course, the upper tail critical point of the F distribution with ν_1 and ν_2 degrees of freedom. For given values of $\sum_{i=1}^{k} \alpha_i^2/(k-1)$ and σ^2, the power can be increased by using a larger sample size n. The problem becomes one of designing the experiment with a value of n so that the power requirements are met. For example, we might require that for specific values of $\sum_{i=1}^{k} \alpha_i^2 \neq 0$ and σ^2 the hypothesis be rejected with probability 0.9. When the power of the test is low, it severely limits the scope of the inferences that can be drawn from the experimental data.

Fixed Effects Case. In the analysis of variance the power depends on the distribution of the F ratio under the alternative hypothesis that the treatment means differ. Therefore, in the case of the one-way fixed effects model, we require the distribution of S_1^2/S^2 when, in fact, $\sum_{i=1}^{k} \alpha_i^2 \neq 0$. Of course when the hypothesis is true, $\alpha_i = 0$ for $i = 1, 2, \ldots, k$, and the statistic follows the F distribution with $k-1$ and $N-k$ degrees of freedom. If $\sum_{i=1}^{k} \alpha_i^2 \neq 0$, the ratio follows a *noncentral F distribution*.

The basic random variable of the noncentral F is denoted by F'. Let $f'_\alpha(\nu_1,\nu_2,\lambda)$ be a value of F' with parameters ν_1, ν_2, and λ. The parameters ν_1 and ν_2 of the distribution are the degrees of freedom associated with S_1^2 and S^2, respectively, and λ is called the *noncentrality parameter*. When $\lambda = 0$, the noncentral F simply reduces to the ordinary F distribution with ν_1 and ν_2 degrees of freedom.

For the fixed effects, one-way analysis of variance with sample sizes $n_1, n_2, \ldots, n_k$, we define

$$\lambda = \frac{\sum_{i=1}^{k} n_i\alpha_i^2}{2\sigma^2}.$$

If we have tables of the noncentral F at our disposal, the power for detecting a particular alternative is obtained by evaluating the following probability:

$$1 - \beta = Pr\left[\frac{S_1^2}{S} > f_\alpha(k-1, N-k) \mid \lambda = \frac{\sum_{i=1}^{k} n_i\alpha_i^2}{2\sigma^2}\right]$$
$$= Pr[F' > f_\alpha(k-1, N-k)].$$

Although the noncentral F is normally defined in terms of λ, it is more convenient, for purposes of tabulation, to work with

$$\phi^2 = \frac{2\lambda}{\nu_1 + 1}.$$

Table XV (see Statistical Tables) gives graphs of the power of the analysis of variance as a function of ϕ for various values of ν_1, ν_2, and the significance level α. These *power charts* can be used not only for the fixed effects models discussed in this chapter but also for the multifactor models of Chapter 12. It remains now to give a procedure whereby the noncentrality parameter λ, and thus ϕ, can be found for these fixed effects cases.

The noncentrality parameter λ can be written in terms of the *expected values of the numerator mean square* of the F ratio in the analysis of variance. We have

$$\lambda = \frac{\nu_1[E(S_i^2)]}{2\sigma^2} - \frac{\nu_1}{2}$$

and thus

$$\phi^2 = \frac{[E(S_i^2) - \sigma^2]}{\sigma^2}\left(\frac{\nu_1}{\nu_1 + 1}\right).$$

Expressions for λ and ϕ^2 for the one-way classification and the randomized complete block design are given in Table 11.13.

Table 11.13 Noncentrality Parameter λ and ϕ^2
for Fixed Effects Models

	One-way classification	Randomized complete block
λ	$\dfrac{\sum\limits_{i=1}^{k} n_i \alpha_i^2}{2\sigma^2}$	$\dfrac{b\sum\limits_{i=1}^{k} \alpha_i^2}{2\sigma^2}$
ϕ^2	$\dfrac{\sum\limits_{i=1}^{k} n_i \alpha_i^2}{k\sigma^2}$	$\dfrac{b\sum\limits_{i=1}^{k} \alpha_i^2}{k\sigma^2}$

Note from Table XV that for given values of ν_1 and ν_2, the power of the test increases with increasing values of ϕ. The value of λ depends, of course, on σ^2 and in a practical problem one may often need to substitute the error mean square as an estimate in determining ϕ^2.

Example 11.6 In a randomized block experiment four treatments are to be compared in six blocks, resulting in 15 degrees of freedom for error. Are six blocks sufficient if the power of our test for detecting differences among the treatment means, at the 0.05 level of significance, is to be at least 0.8 when the true means are $\mu_1. = 5.0$, $\mu_2. = 7.0$, $\mu_3. = 4.0$, and $\mu_4. = 4.0$? An estimate of σ^2 to be used in the computation of the power is given by $\hat{s}^2 = 2.0$.

Solution. Recall that the treatment means are given by $\mu_i. = \mu + \alpha_i$. If we invoke the restriction that $\sum_{i=1}^{4} \alpha_i = 0$, we have

$$\mu = \frac{\sum_{i=1}^{4} \mu_i.}{4} = 5.0,$$

and then $\alpha_1 = 0$, $\alpha_2 = 2.0$, $\alpha_3 = -1.0$, and $\alpha_4 = -1.0$. Therefore

$$\phi^2 = \frac{b \sum_{i=1}^{k} \alpha_i^2}{k\sigma^2} = \frac{(6)(6)}{(4)(2)} = 4.5,$$

from which we obtain $\phi = 2.121$. Using Table XV, the power is found to be approximately 0.89 and thus the power requirements are met. This means that if the value of $\sum_{i=1}^{4} \alpha_i^2 = 6$ and $\sigma^2 = 2.0$, the use of six blocks will result in rejecting the hypothesis of equal treatment means with probability 0.89.

Random Effects Case. In the fixed effects case, the computation of power requires the use of the noncentral F distribution. Such is not the case in the random effects model. In fact, the power is computed very simply by the use of the standard F tables. Consider, for example, the one-way random effects model, n observations per treatment, with the hypothesis

$$H_0: \quad \sigma_\alpha^2 = 0$$
$$H_1: \quad \sigma_\alpha^2 \neq 0.$$

When H_1 is true the ratio

$$f = \frac{\mathrm{SSA}[(k-1)(\sigma^2 + n\sigma_\alpha^2)]}{\mathrm{SSE}/k(n-1)\sigma^2} = \frac{s_1^2}{s^2(1 + n\sigma_\alpha^2/\sigma^2)}$$

is a value of the random variable F having the F distribution with $k - 1$ and $k(n - 1)$ degrees of freedom. The problem becomes one, then, of determining the probability of rejecting H_0 under the condition that the true treatment variance component $\sigma_\alpha^2 \neq 0$. We have then

$$
\begin{aligned}
1 - \beta &= Pr\left\{\frac{S_1^2}{S^2} > f_\alpha[(k - 1), k(n - 1)] \mid H_1 \text{ is true}\right\} \\
&= Pr\left\{\frac{S_1^2}{S^2(1 + n\sigma_\alpha^2/\sigma^2)} > \frac{f_\alpha[(k - 1), k(n - 1)]}{1 + n\sigma_\alpha^2/\sigma^2}\right\} \\
&= Pr\left\{F > \frac{f_\alpha[(k - 1), k(n - 1)]}{1 + n\sigma_\alpha^2/\sigma^2}\right\}.
\end{aligned}
$$

Note that as n increases the value $f_\alpha[(k - 1), k(n - 1)]/(1 + n\sigma_\alpha^2/\sigma^2)$ approaches zero, resulting in an increase in the power of the test. An illustration of the power for this kind of situation is given in Figure 11.1.

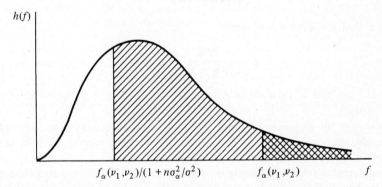

Figure 11.1 Power for the random effects one-way analysis of variance.

The crosshatched area is the significance level α, while the entire shaded area is the power of the test.

Example 11.7 Suppose in a one-way classification it is of interest to test for the significance of the variance component σ_α^2. Four treatments are to be used in the experiment with five observations per treatment. What will be the probability of rejecting the hypothesis $\sigma_\alpha^2 = 0$, when in fact the treatment variance component is $(3/4)\sigma^2$?

Solution. Using an $\alpha = 0.05$ significance level, we have

$$1 - \beta = Pr\left\{F > \frac{f_{0.05}(3,16)}{1 + (5)(3)/4}\right\}$$

$$= Pr\left[F > \frac{f_{0.05}(3,16)}{4.75}\right]$$

$$= Pr\left(F > \frac{3.24}{4.75}\right)$$

$$= Pr(F > 0.864).$$

Using Theorem 5.19 and then Table A-7c of *Introduction to Statistical Analysis* by Dixon and Massey, we see that

$$1 - \beta \simeq 0.46.$$

Therefore only about 46% of the time will the test procedure detect a variance component that is $(3/4)\sigma^2$.

EXERCISES

1. Show that the error mean square

$$s^2 = \frac{\text{SSE}}{k(n - 1)}$$

for the analysis of variance in a one-way classification is an unbiased estimate of σ^2.

2. Show that the computing formula for SSA, in the analysis of variance of the one-way classification, is equivalent to the corresponding term in the identity of Theorem 11.1.

3. Six different machines are being considered for use in manufacturing rubber seals. The machines are being compared with respect to tensile strength of the product. A random sample of four seals from each machine is used to determine whether or not the mean tensile strength varies from machine to machine. The following are the tensile strength measurements in pounds per square inch $\times 10^{-2}$:

		Machine			
1	*2*	*3*	*4*	*5*	*6*
17.5	16.4	20.3	14.6	17.5	18.3
16.9	19.2	15.7	16.7	19.2	16.2
15.8	17.7	17.8	20.8	16.5	17.5
18.6	15.4	18.9	18.9	20.5	20.1

Perform the analysis of variance at the 0.05 level of significance and indicate whether or not the treatment means differ significantly.

4. Three sections of the same elementary mathematics course are taught by three teachers. The final grades were recorded as follows:

	Teacher	
A	B	C
73	88	68
89	78	79
82	48	56
43	91	91
80	51	71
73	85	71
66	74	87
60	77	41
45	31	59
93	78	68
36	62	53
77	76	79
	96	15
	80	
	56	

Is there a significant difference in the average grades given by the three teachers? Use a 0.05 level of significance.

5. Test for homogeneity of variances in Exercise 4.

6. Four laboratories are being used to perform chemical analyses. Samples of the same material are sent to the laboratories for analysis as part of the study to determine whether or not they give, on the average, the same results. The analytical results for the four laboratories are as follows:

	Laboratory		
A	B	C	D
58.7	62.7	55.9	60.7
61.4	64.5	56.1	60.3
60.9	63.1	57.3	60.9
59.1	59.2	55.2	61.4
58.2	60.3	58.1	62.3

(a) Use Bartlett's test to determine if the within-laboratory variances are significantly different at the $\alpha = 0.05$ level of significance.

(b) If the hypothesis in (a) is accepted, perform the analysis of variance and give conclusions concerning the laboratories.

(c) Extend the analysis of variance to make significance tests on the following contrasts: (1) B versus A, C, D; (2) C versus A and D; and (3) A versus D.

7. An investigation was conducted to determine the source of reduction in yield of a certain chemical product. It was known that the loss in yield occurred in the mother liquor, that is, the material removed at the filtration stage. It was felt that different blends of the original material may result in different yield reductions at the mother liquor stage. The following are results of the percent reduction for three batches at each of four preselected blends:

		Blend		
1	*2*	*3*	*4*
25.6	25.2	20.8	31.6
24.3	28.6	26.7	29.8
27.9	24.7	22.2	34.3

(a) Perform the analysis of variance at the $\alpha = 0.05$ level of significance.

(b) Use Duncan's multiple-range test to determine which blends differ.

8. Four different kinds of fertilizer f_1, f_2, f_3, and f_4 are used to study the yield of beans. The soil is divided into three blocks each containing four homogeneous plots. The yields in pounds per acre and the corresponding treatments are as follows:

Block 1	*Block 2*	*Block 3*
$f_1 = 42.7$	$f_3 = 50.9$	$f_4 = 51.1$
$f_3 = 48.5$	$f_1 = 50.0$	$f_2 = 46.3$
$f_4 = 32.8$	$f_2 = 38.0$	$f_1 = 51.9$
$f_2 = 39.3$	$f_4 = 40.2$	$f_3 = 53.5$

(a) Conduct an analysis of variance using the randomized complete block model.

(b) Use single-degree-of-freedom contrasts to make the following comparisons among the fertilizers: (1) (f_1, f_3) versus (f_2, f_4); (2) f_1 versus f_3.

9. Show that the computing formula for SSB, in the analysis of variance of the randomized complete block design, is equivalent to the corresponding term in the identity of Theorem 11.3.

10. An experiment is conducted in which four treatments are to be compared in five blocks. The following data are generated:

Treatment	Block				
	1	2	3	4	5
1	12.8	10.6	11.7	10.7	11.0
2	11.7	14.2	11.8	9.9	13.8
3	11.5	14.7	13.6	10.7	15.9
4	12.6	16.5	15.4	9.6	17.1

Perform the analysis of variance, separating out the treatment, block, and error sums of squares. Use a 0.05 level of significance to test the hypothesis that there is no difference between the treatment means.

11. The following data show the effect of four operators, chosen randomly, on the output of a particular machine:

Operator			
1	2	3	4
175.4	168.5	170.1	175.2
171.7	162.7	173.4	175.7
173.0	165.0	175.7	180.1
170.5	164.1	170.7	183.7

Perform a model II analysis of variance and compute an estimate of the operator variance component and the experimental error variance component.

12. Assuming a random effects model for Exercise 10, compute estimates of the treatment and block variance components.

13. (a) Using a regression approach for the randomized complete block design, obtain the normal equations $\mathbf{Ab} = \mathbf{g}$ in matrix form.
    (b) Show that $R(\beta_1, \beta_2, \ldots, \beta_b \mid \alpha_1, \alpha_2, \ldots, \alpha_k) = \text{SSB}$.

14. In Exercise 10, if one uses an $\alpha = 0.05$ level test, how many blocks are needed in order that we accept the hypothesis of equality of treatment means with probability 0.1 when, in fact,

$$\frac{\sum_{i=1}^{4} \alpha_i^2}{\sigma^2} = 2.0?$$

15. In Exercise 11, if we are interested in testing for the significance of the operator variance component, do we have large enough samples to ensure with a probability as large as 0.95 a significant variance component if the true σ_α^2 is $1.5\sigma^2$? If not, how many runs are necessary for each operator? Use a 0.05 level of significance.

12 FACTORIAL EXPERIMENTS

12.1 Two-Factor Experiments

Consider a situation in which it is of interest to study the effect of *two* *factors* A and B on some response. For example, in a chemical experiment we would like to simultaneously vary the reaction pressure and reaction time and study the effect of each on the yield. In a biological experiment, it is of interest to study the effect of drying time and temperature on the amount of solids (percent by weight) left in samples of yeast. The term *factor* is used in a general sense to denote any feature of the experiment such as temperature, time, or pressure that may be varied from trial to trial. We define the *levels* of a factor to be the actual values used in the experiment.

In each of these cases it is important not only to determine if the two factors have an influence on the response but also if there is a significant interaction between the two factors. As far as terminology is concerned, the experiment described here is a two-way classification or a two-factor experiment with a completely randomized design. That is, no restrictions such as blocking are made on the experimental units. In the case of the yeast example, the various treatment combinations of temperature and drying time would be assigned randomly to the samples of yeast.

12.2 Interaction in Two-Factor Experiments

In the randomized block model discussed previously it was assumed that one observation on each treatment is taken in each block. If the model assumption is correct, namely, if blocks and treatments are the only real effects and interaction does not exist, the expected value of the error mean square is the experimental error variance σ^2. Suppose, however, there is interaction occurring between treatments and blocks as indicated by the model

$$y_{ij} = \mu + \alpha_i + \beta_j + (\alpha\beta)_{ij} + \epsilon_{ij}$$

of Section 11.8. The expected value of the error mean square was then given as

$$E\left[\frac{\text{SSE}}{(b-1)(k-1)}\right] = \sigma^2 + \frac{\sum\limits_{i=1}^{k}\sum\limits_{j=1}^{b}(\alpha\beta)_{ij}^2}{(b-1)(k-1)}.$$

The treatment and blocks effects do not appear in the expected error mean square but the interaction effects do. Thus if there is interaction in the model, the error mean square reflects variation due to experimental error *plus* an interaction contribution and, for this experimental plan, there is no way of separating them.

From an experimenter's point of view it should seem necessary to arrive at a significance test on the existence of interaction by separating true error variation from that due to interaction. The effects of factors A and B, often called the *main effects*, take on a different meaning in the presence of interaction. In the previous biological example the effect that drying time has on the amount of solids left in the yeast might very well depend on the temperature that the samples are exposed to. In general there could very well be experimental situations in which factor A has a positive effect on the response at one level of factor B, while at a different level of factor B the effect of A is negative. We use the term *positive effect* here to indicate that the yield or response increases as the levels of a given factor increase according to some defined order. In the same sense a *negative effect* corresponds to a decrease in yield for increasing levels of the factor. Consider, for example, the following hypothetical data taken on two factors each at three levels:

A	b_1	b_2	b_3	*Total*
		B		
a_1	4.4	8.8	5.2	18.4
a_2	7.5	8.5	2.4	18.4
a_3	9.7	7.9	0.8	18.4
Total	21.6	25.2	8.4	55.2

Clearly, the effect of A is positive at b_1 and negative at b_3. These differences in the levels of A at different levels of B are of interest to the experimenter but an ordinary significance test on factor A would yield a value of zero for SSA since the totals for each level of A are all of the same magnitude. We say then that the presence of interaction is *masking* the effect of factor A. Therefore, if we consider the average influence of A, over all three levels of B, there is no effect. However, this is most likely not what is pertinent to the experimenter.

We suggest that the experimenter first attempt to detect interaction with a test of significance. Then if interaction is not significant, he would proceed to make tests on the effects of the main factors. If the data indicate the presence of interaction, this might well dictate the need to observe the influence of each factor at fixed levels of the other.

Interaction and experimental error are separated in the two-factor experiment only if multiple observations are taken at the various treatment combinations. To ease the computations involved, there should be the same number, n, of observations at each combination. These should be true replications, not just repeated measurements. For example, in the yeast illustration, if we take $n = 2$ observations at each combination of temperature and drying time, there should be two separate samples and not merely repeated measurements on the same sample. This is important because now the measure of experimental error comes from variation between readings *within* treatment combinations and thus indicates true or pure experimental error.

12.3 Two-Factor Analysis of Variance

To present general formulas for the analysis of variance of a two-factor experiment using repeated observations, we shall consider the case of n replications of the treatment combinations determined by a levels of factor A and b levels of factor B. The observations may be classified by means of a rectangular array in which the rows represent the levels of factor A and the columns represent the levels of factor B. Each treatment combination defines a cell in our array. Thus we have ab cells, each cell containing n observations. Denoting the kth observation taken at the ith level of factor A and the jth level of factor B by y_{ijk}, the abn observations are shown in Table 12.1.

The observations in the (ij)th cell constitute a random sample of size n from a population that is assumed to be normally distributed with mean μ_{ij} and variance σ^2. All ab populations are assumed to have the same variance σ^2. Let us define the following useful symbols, some of which are used in Table 12.1:

$T_{ij.}$ = sum of the observations in the (ij)th cell

$T_{i..}$ = sum of the observations for the ith level of factor A

$T_{.j.}$ = sum of the observations for the jth level of factor B

$T_{...}$ = sum of all abn observations

$\bar{y}_{ij.}$ = mean of the observations in the (ij)th cell

$\bar{y}_{i..}$ = mean of the observations for the ith level of factor A

$\bar{y}_{.j.}$ = mean of the observations for the jth level of factor B

$\bar{y}_{...}$ = mean of all abn observations.

Each observation in Table 12.1 may be written in the form

$$y_{ijk} = \mu_{ij} + \epsilon_{ijk},$$

Table 12.1 Two-Factor Experiment with n Replications

A	B				Total	Mean
	1	2	$\cdots$	b		
1	y_{111}	y_{121}	$\cdots$	y_{1b1}	$T_{1..}$	$\bar{y}_{1..}$
	y_{112}	y_{122}	$\cdots$	y_{1b2}		
	.	.		.		
	.	.		.		
	.	.		.		
	y_{11n}	y_{12n}	$\cdots$	y_{1bn}		
2	y_{211}	y_{221}	$\cdots$	y_{2b1}	$T_{2..}$	$\bar{y}_{2..}$
	y_{212}	y_{222}	$\cdots$	y_{2b2}		
	.	.		.		
	.	.		.		
	y_{21n}	y_{22n}	$\cdots$	y_{2bn}		
.	.	.		.	.	.
.	.	.		.	.	.
.	.	.		.	.	.
a	y_{a11}	y_{a21}	$\cdots$	y_{ab1}	$T_{a..}$	$\bar{y}_{a..}$
	y_{a12}	y_{a22}	$\cdots$	y_{ab2}		
	.	.		.		
	.	.		.		
	.	.		.		
	y_{a1n}	y_{a2n}	$\cdots$	y_{abn}		
Total	$T_{.1.}$	$T_{.2.}$	$\cdots$	$T_{.b.}$	$T_{...}$	
Mean	$\bar{y}_{.1.}$	$\bar{y}_{.2.}$	$\cdots$	$\bar{y}_{.b.}$		$\bar{y}_{...}$

where ϵ_{ijk} measures the deviations of the observed y_{ijk} values in the (ij)th cell from the population mean μ_{ij}. If we let $(\alpha\beta)_{ij}$ denote the interaction effect of the ith level of factor A and the jth level of factor B, α_i the effect of the ith level of factor A, β_j the effect of the jth level of factor B, and μ the overall mean, we can write

$$\mu_{ij} = \mu + \alpha_i + \beta_j + (\alpha\beta)_{ij},$$

and then

$$y_{ijk} = \mu + \alpha_i + \beta_j + (\alpha\beta)_{ij} + \epsilon_{ijk},$$

on which we impose the restrictions

$$\sum_{i=1}^{a} \alpha_i = 0, \quad \sum_{j=1}^{b} \beta_j = 0, \quad \sum_{i=1}^{a} (\alpha\beta)_{ij} = 0, \quad \sum_{j=1}^{b} (\alpha\beta)_{ij} = 0.$$

The three hypotheses to be tested are as follows:

1. H_0': $\alpha_1 = \alpha_2 = \cdots = \alpha_a = 0$
    H_1': at least one of the α_i's is not equal to zero

2. H_0'': $\beta_1 = \beta_2 = \cdots = \beta_b = 0$
    H_1'': at least one of the β_j's is not equal to zero

3. H_0''': $(\alpha\beta)_{11} = (\alpha\beta)_{12} = \cdots = (\alpha\beta)_{ab} = 0$
    H_1''': at least one of the $(\alpha\beta)_{ij}$'s is not equal to zero.

Each of these tests will be based on a comparison of independent estimates of σ^2 provided by the splitting of the total sum of squares of our data into four components by means of the following identity.

THEOREM 12.1 *Sum of Squares Identity*

$$\sum_{i=1}^{a} \sum_{j=1}^{b} \sum_{k=1}^{n} (y_{ijk} - \bar{y}_{...})^2 = bn \sum_{i=1}^{a} (\bar{y}_{i..} - \bar{y}_{...})^2$$

$$+ an \sum_{j=1}^{b} (\bar{y}_{.j.} - \bar{y}_{...})^2 + n \sum_{i=1}^{a} \sum_{j=1}^{b} (\bar{y}_{ij.} - \bar{y}_{i..} - \bar{y}_{.j.} + \bar{y}_{...})^2$$

$$+ \sum_{i=1}^{a} \sum_{j=1}^{b} \sum_{k=1}^{n} (y_{ijk} - \bar{y}_{ij.})^2.$$

Proof.

$$\sum_{i=1}^{a} \sum_{j=1}^{b} \sum_{k=1}^{n} (y_{ijk} - \bar{y}_{...})^2 = \sum_{i=1}^{a} \sum_{j=1}^{b} \sum_{k=1}^{n} [(\bar{y}_{i..} - \bar{y}_{...}) + (\bar{y}_{.j.} - \bar{y}_{...})$$

$$+ (\bar{y}_{ij.} - \bar{y}_{i..} - \bar{y}_{.j.} + \bar{y}_{...}) + (y_{ijk} - \bar{y}_{ij.})]^2$$

$$= \sum_{i=1}^{a} \sum_{j=1}^{b} \sum_{k=1}^{n} (\bar{y}_{i..} - \bar{y}_{...})^2$$

$$+ \sum_{i=1}^{a} \sum_{j=1}^{b} \sum_{k=1}^{n} (\bar{y}_{.j.} - \bar{y}_{...})^2$$

$$+ \sum_{i=1}^{a} \sum_{j=1}^{b} \sum_{k=1}^{n} (\bar{y}_{ij.} - \bar{y}_{i..} - \bar{y}_{.j.} + \bar{y}_{...})^2$$

$$+ \sum_{i=1}^{a} \sum_{j=1}^{b} \sum_{k=1}^{n} (y_{ijk} - \bar{y}_{ij.})^2$$

$$+ \text{ 6 cross-product terms.}$$

The cross-product terms are all equal to zero. Hence

$$\sum_{i=1}^{a} \sum_{j=1}^{b} \sum_{k=1}^{n} (y_{ijk} - \bar{y}_{...})^2 = bn \sum_{i=1}^{a} (\bar{y}_{i..} - \bar{y}_{...})^2 + an \sum_{j=1}^{b} (\bar{y}_{.j.} - \bar{y}_{...})^2$$

$$+ n \sum_{i=1}^{a} \sum_{j=1}^{b} (\bar{y}_{ij.} - \bar{y}_{i..} - \bar{y}_{.j.} + \bar{y}_{...})^2$$

$$+ \sum_{i=1}^{a} \sum_{j=1}^{b} \sum_{k=1}^{n} (y_{ijk} - \bar{y}_{ij.})^2.$$

Symbolically, we write the sum of squares identity as

$$\text{SST} = \text{SSA} + \text{SSB} + \text{SS}(AB) + \text{SSE},$$

where SSA and SSB are called the sum of squares for the main effects A and B, respectively, SS(AB) is called the interaction sum of squares for A and B, and SSE is the error sum of squares. The degrees of freedom are partitioned according to the identity

$$abn - 1 = (a - 1) + (b - 1) + (a - 1)(b - 1) + ab(n - 1).$$

Dividing each of the sum of squares on the right side of the sum of squares identity by their corresponding number of degrees of freedom, we obtain the four independent estimates

$$s_1^2 = \frac{\text{SSA}}{a - 1}, \quad s_2^2 = \frac{\text{SSB}}{b - 1}, \quad s_3^2 = \frac{\text{SS(AB)}}{(a - 1)(b - 1)}, \quad s^2 = \frac{\text{SSE}}{ab(n - 1)}$$

of σ^2. If we interpret the sum of squares as functions of the independent random variables Y_{111}, Y_{112}, ..., Y_{abn}, it is not difficult to verify that

$$E(S_1^2) = E\left[\frac{\text{SSA}}{(a-1)}\right] = \sigma^2 + \frac{nb\sum_{i=1}^{a}\alpha_i^2}{a-1}$$

$$E(S_2^2) = E\left[\frac{\text{SSB}}{(b-1)}\right] = \sigma^2 + \frac{na\sum_{j=1}^{b}\beta_j^2}{b-1}$$

$$E(S_3^2) = E\left[\frac{\text{SS(AB)}}{(a-1)(b-1)}\right] = \sigma^2 + \frac{n\sum_{i=1}^{a}\sum_{j=1}^{b}(\alpha\beta)_{ij}^2}{(a-1)(b-1)}$$

$$E(S^2) = E\left[\frac{\text{SSE}}{ab(n-1)}\right] = \sigma^2,$$

from which we immediately conclude that all four estimates of σ^2 are unbiased when H_0', H_0'', and H_0''' are true.

To test the hypothesis H_0', that the effects of factors A are all equal to zero, we compute the ratio

$$f_1 = \frac{s_1^2}{s^2},$$

which is a value of the random variable F_1 having the F distribution with $a-1$ and $ab(n-1)$ degrees of freedom when H_0' is true. The null hypothesis is rejected at the α level of significance when $f_1 > f_\alpha[a-1, ab(n-1)]$. Similarly, to test the hypothesis H_0'', that the effects of factor B are all equal to zero, we compute the ratio

$$f_2 = \frac{s_2^2}{s^2},$$

which is a value of the random variable F_2 having the F distribution with $b-1$ and $ab(n-1)$ degrees of freedom when H_0'' is true. This hypothesis is rejected at the α level of significance when $f_2 > f_\alpha[b-1, ab(n-1)]$. Finally, to test the hypothesis H_0''', that the interaction effects are all equal to zero, we compute the ratio

$$f_3 = \frac{s_3^2}{s^2},$$

which is a value of the random variable F_3 having the F distribution with $(a-1)(b-1)$ and $ab(n-1)$ degrees of freedom when H_0''' is true. We conclude that interaction is present when $f_3 > f_\alpha[(a-1)(b-1), ab(n-1)]$.

As indicated in Section 12.2, it is advisable to conduct the test for interaction before attempting to draw inferences on the main effects. If interaction is not significant, the experimenter can proceed to test the main effects. However, a significant interaction could very well imply that the data should be analyzed in a somewhat different manner—perhaps observing the effect of factor A at fixed levels of factor B, and so forth.

The computations in an analysis-of-variance problem, for a two-factor experiment with n replications, are usually summarized as in Table 12.2.

Table 12.2 Analysis of Variance for the Two-Factor Experiment with n Replications

Source of variation	Sum of squares	Degrees of freedom	Mean square	Computed f
Main effects				
A	SSA	$a-1$	$s_1^2 = \dfrac{\text{SSA}}{a-1}$	$f_1 = \dfrac{s_1^2}{s^2}$
B	SSB	$b-1$	$s_2^2 = \dfrac{\text{SSB}}{b-1}$	$f_2 = \dfrac{s_2^2}{s^2}$
Interaction				
AB	SS(AB)	$(a-1)(b-1)$	$s_3^2 = \dfrac{\text{SS}(AB)}{(a-1)(b-1)}$	$f_3 = \dfrac{s_3^2}{s^2}$
Error	SSE	$ab(n-1)$	$s^2 = \dfrac{\text{SSE}}{ab(n-1)}$	
Total	SST	$abn-1$		

The sums of squares are usually obtained by constructing the following table of totals

			B		
A	1	2	$\cdots$	b	Total
1	$T_{11.}$	$T_{12.}$	$\cdots$	$T_{1b.}$	$T_{1..}$
2	$T_{21.}$	$T_{22.}$	$\cdots$	$T_{2b.}$	$T_{2..}$
.	.	.		.	.
.	.	.		.	.
.	.	.		.	.
a	$T_{a1.}$	$T_{a2.}$	$\cdots$	$T_{ab.}$	$T_{a..}$
Total	$T_{.1.}$	$T_{.2.}$	$\cdots$	$T_{.b.}$	$T_{...}$

and using the following computational formulas:

$$\text{SST} = \sum_{i=1}^{a} \sum_{j=1}^{b} \sum_{k=1}^{n} y_{ijk}^2 - \frac{T_{...}^2}{abn}$$

$$\text{SSA} = \frac{\sum_{i=1}^{a} T_{i..}^2}{bn} - \frac{T_{...}^2}{abn}$$

$$\text{SSB} = \frac{\sum_{j=1}^{b} T_{.j.}^2}{an} - \frac{T_{...}^2}{abn}$$

$$\text{SS(AB)} = \frac{\sum_{i=1}^{a} \sum_{j=1}^{b} T_{ij.}^2}{n} - \frac{\sum_{i=1}^{a} T_{i..}^2}{bn} - \frac{\sum_{j=1}^{b} T_{.j.}^2}{an} + \frac{T_{...}^2}{abn}.$$

As before, SSE is obtained by subtraction using the formula

$$\text{SSE} = \text{SST} - \text{SSA} - \text{SSB} - \text{SS(AB)}.$$

Example 12.1 In an experiment conducted to determine which of three different missile systems is preferable, the propellant burning rate for 24 static firings was measured. Four different propellant types were used. The experiment yielded duplicate observations of burning rates at each combination of the treatments. The data, after coding, are given in Table 12.3.

Table 12.3 Propellant Burning Rates

Missile System	Propellant type			
	b_1	b_2	b_3	b_4
a_1	34.0	30.1	29.8	29.0
	32.7	32.8	26.7	28.9
a_2	32.0	30.2	28.7	27.6
	33.2	29.8	28.1	27.8
a_3	28.4	27.3	29.7	28.8
	29.3	28.9	27.3	29.1

Use a 0.05 level of significance to test the following hypotheses: (a) H_0': There is no difference in the mean propellant burning rates when different missile systems are used. (b) H_0'': There is no difference in the mean propellant burning rates of the four propellant types. (c) H_0''': There is no interaction between the different missile systems and the different propellant types.

Solution.

1. (a) H_0': $\alpha_1 = \alpha_2 = \alpha_3 = 0$
   (b) H_0'': $\beta_1 = \beta_2 = \beta_3 = \beta_4 = 0$.
   (c) H_0''': $(\alpha\beta)_{11} = (\alpha\beta)_{12} = \cdots = (\alpha\beta)_{34} = 0$.
2. (a) H_1':   at least one of the α_i's is not equal to zero.
   (b) H_1'':   at least one of the β_j's is not equal to zero.
   (c) H_1''':   at least one of the $(\alpha\beta)_{ij}$'s is not equal to zero.
3. $\alpha = 0.05$.
4. Critical regions: (a) $F_1 > 3.89$, (b) $F_2 > 3.49$, (c) $F_3 > 3.00$.
5. Computations: From Table 12.3 we first construct the following table of totals:

	b_1	b_2	b_3	b_4	*Total*
a_1	66.7	62.9	56.5	57.9	244.0
a_2	65.2	60.0	56.8	55.4	237.4
a_3	57.7	56.2	57.0	57.9	228.8
Total	189.6	179.1	170.3	171.2	710.2

Now

$$\text{SST} = 34.0^2 + 32.7^2 + \cdots + 29.1^2 - \frac{710.2^2}{24}$$

$$= 21{,}107.68 - 21{,}016.00 = 91.68$$

$$\text{SSA} = \frac{244.0^2 + 237.4^2 + 228.8^2}{8} - \frac{710.2^2}{24}$$

$$= 21{,}030.52 - 21{,}016.00 = 14.52$$

$$\text{SSB} = \frac{189.6^2 + 179.1^2 + 170.3^2 + 171.2^2}{6} - \frac{710.2^2}{24}$$

$$= 21{,}056.08 - 21{,}016.00 = 40.08$$

$$\text{SS(AB)} = \frac{66.7^2 + 65.2^2 + \cdots + 57.9^2}{2} - 21{,}030.52$$

$$- 21{,}056.08 + 21{,}016.00$$

$$= 22.17$$

$$\text{SSE} = 91.68 - 14.52 - 40.08 - 22.17 = 14.91.$$

These results, with the remaining computations, are given in Table 12.4.

Table 12.4 Analysis of Variance for the Data of Table 12.3

Source of variation	Sum of squares	Degrees of freedom	Mean square	Computed f
Missile system	14.52	2	7.26	5.85
Propellant type	40.08	3	13.36	10.77
Interaction	22.17	6	3.70	2.98
Error	14.91	12	1.24	
Total	91.68	23		

6. Conclusions:
   (a) Accept H_0''' and conclude that there is no interaction between the different missile systems and the different propellant types.
   (b) Reject H_0' and conclude that different missile systems result in different mean propellant burning rates.
   (c) Reject H_0'' and conclude that the mean propellant burning rates are not the same for the four propellant types.

Example 12.2 Referring to Example 12.1, choose two orthogonal contrasts to partition the sum of squares for the missile systems into single-degree-of-freedom components to be used in comparing systems 1 and 2 with 3 and system 1 versus system 2.

Solution. The contrast for comparing systems 1 and 2 with 3 is given by

$$\omega_1 = \mu_1. + \mu_2. - 2\mu_3..$$

A second contrast, orthogonal to ω_1, for comparing system 1 with system 2, is given by $\omega_2 = \mu_1. - \mu_2..$ The single-degree-of-freedom sum of squares are as follows:

$$SS\omega_1 = \frac{[244.0 + 237.4 - 2(228.8)]^2}{8[(1)^2 + (1)^2 + (-2)^2]} = 11.80$$

$$SS\omega_2 = \frac{(244.0 - 237.4)^2}{8[(1)^2 + (-1)^2]} = 2.72.$$

Notice that $SS\omega_1 + SS\omega_2 = SSA$, as expected. The computed f values corresponding to ω_1 and ω_2 are, respectively,

$$f_1 = \frac{11.80}{1.24} = 9.5$$

and

$$f_2 = \frac{2.72}{1.24} = 2.2.$$

Compared to the critical value $f_{0.05}(1,12) = 4.75$, we find f_1 to be significant. Thus the first contrast indicates that the hypothesis

$$H_0: \quad \frac{\mu_{1\cdot} + \mu_{2\cdot}}{2} = \mu_{3\cdot}.$$

is rejected. Since $f_2 < 4.75$, the mean burning rates of the first and second systems are not significantly different.

If the hypothesis of no interaction in Example 12.1 is true, as indicated by the f ratio in the analysis-of-variance table, we are able to make the *general* comparisons of Example 12.2 regarding our missile systems rather than separate comparisons for each propellant. Likewise, we might make general comparisons among the propellants rather than separate comparisons for each missile system. For example, we could compare propellants 1 and 2 with 3 and 4 and also propellant 1 versus propellant 2. The resulting f ratios, each with 1 and 12 degrees of freedom, turn out to be 24.86 and 7.41, respectively, and both are significant at the 0.05 level of significance. The indication is then that propellant 1 gives the highest mean burning rate. A prudent experimenter might be somewhat cautious in making overall conclusions in a problem such as this one where the f ratio for interaction is barely below the 0.05 critical value. This is far from overwhelming evidence that interaction between the factors does not exist. In fact a quick inspection of the data, namely, the cell totals, points out possible evidence of interaction. For example, the overall evidence, 189.6 versus 171.2, certainly indicates that propellant 1 is superior, in terms of a higher burning rate, to propellant 4. However, if we restrict ourselves to system 3, where we have a total of 57.7 for propellant 1 as opposed to 57.9 for propellant 4, there appears to be little or no difference between propellants 1 and 4. In fact there appears to be a stabilization of burning rates for the different propellants if we operate with system 3. There is certainly overall evidence, 244.0 versus 228.8, that indicates that system 1 gives a higher burning rate than system 3, but if we restrict ourselves to propellant 4, this conclusion does not appear to hold.

12.4 Three-Factor Experiments

In this section we consider an experiment with three factors A, B, and C at a, b, and c levels, respectively, in a completely randomized experimental design. Assume again that we have n observations for each of the abc treatment combinations. We shall proceed to outline significance tests for the three main effects and interactions involved. It is hoped that the reader can then use the description given here to generalize the analysis to $k > 3$ factors.

The model for the three-factor experiment is given by

$$y_{ijkl} = \mu + \alpha_i + \beta_j + \gamma_k + (\alpha\beta)_{ij} + (\alpha\gamma)_{ik} + (\beta\gamma)_{jk} + (\alpha\beta\gamma)_{ijk} + \epsilon_{ijkl},$$

$i = 1, 2, \ldots, a; j = 1, 2, \ldots, b; k = 1, 2, \ldots, c;$ and $l = 1, 2, \ldots, n$— where α_i, β_j, and γ_k are the main effects; $(\alpha\beta)_{ij}$, $(\alpha\gamma)_{ik}$, and $(\beta\gamma)_{jl}$ are the two-factor interaction effects that have the same interpretation as in the two-factor experiment. The term $(\alpha\beta\gamma)_{ijk}$ is called the three-factor interaction effect, namely, a term that represents a nonadditivity of the $(\alpha\beta)_{ij}$ over the different levels of the factor C. As before, the sum of all main effects is zero and the sum over any subscript of the two- and three-factor interaction effects is zero. In many experimental situations these higher-order interactions are insignificant and their mean squares reflect only random variation, but we shall outline the analysis in its most general detail.

Again, in order that valid significance tests can be made, we must assume that the errors are values of independent and normally distributed random variables, each with zero mean and common variance σ^2.

The general philosophy concerning the analysis is the same as that discussed for the one- and two-factor experiments. The sum of squares is partitioned into eight terms, each representing a source of variation from which we obtain independent estimates of σ^2 when all the main effects and interaction effects are zero. If the effects of any given factor or interaction are not all zero, then the mean square will estimate the error variance plus a component due to the systematic effect in question.

Let us proceed directly to the computational procedure for obtaining the sums of squares in the three-factor analysis of variances. We require the following notation:

$T_{\ldots}$ = sum of all $abcn$ observations

$T_{i\ldots}$ = sum of the observations for the ith level of factor A

$T_{.j..}$ = sum of the observations for the jth level of factor B

$T_{..k.}$ = sum of the observations for the kth level of factor C

$T_{ij..}$ = sum of the observations for the ith level of A and the jth level of B

$T_{i \cdot k \cdot}$ = sum of the observations for the ith level of A and the kth level of C

$T_{\cdot jk \cdot}$ = sum of the observations for the jth level of B and kth level of C

$T_{ijk \cdot}$ = sum of the observations for the (ijk)th treatment combination.

In practice it is advantageous to construct the following two-way tables of totals and subtotals:

A	B				$Total$
	1	2	$\cdots$	b	
1	$T_{11k \cdot}$	$T_{12k \cdot}$	$\cdots$	$T_{1bk \cdot}$	$T_{1 \cdot k \cdot}$
2	$T_{21k \cdot}$	$T_{22k \cdot}$	$\cdots$	$T_{2bk \cdot}$	$T_{2 \cdot k \cdot}$
$\cdot$	$\cdot$	$\cdot$		$\cdot$	$\cdot$
$\cdot$	$\cdot$	$\cdot$		$\cdot$	$\cdot$
$\cdot$	$\cdot$	$\cdot$		$\cdot$	$\cdot$
a	$T_{a1k \cdot}$	$T_{a2k \cdot}$	$\cdots$	$T_{abk \cdot}$	$T_{a \cdot k \cdot}$
Total	$T_{\cdot 1k \cdot}$	$T_{\cdot 2k \cdot}$	$\cdots$	$T_{\cdot bk \cdot}$	$T_{\cdot \cdot k \cdot}$

$k = 1, 2 \ldots, c$

A	B				$Total$
	1	2	$\cdots$	b	
1	$T_{11 \cdot \cdot}$	$T_{12 \cdot \cdot}$	$\cdots$	$T_{1b \cdot \cdot}$	$T_{1 \cdot \cdot \cdot}$
2	$T_{21 \cdot \cdot}$	$T_{22 \cdot \cdot}$	$\cdots$	$T_{2b \cdot \cdot}$	$T_{2 \cdot \cdot \cdot}$
$\cdot$	$\cdot$	$\cdot$		$\cdot$	$\cdot$
$\cdot$	$\cdot$	$\cdot$		$\cdot$	$\cdot$
$\cdot$	$\cdot$	$\cdot$		$\cdot$	$\cdot$
a	$T_{a1 \cdot \cdot}$	$T_{a2 \cdot \cdot}$	$\cdots$	$T_{ab \cdot \cdot}$	$T_{a \cdot \cdot \cdot}$
Total	$T_{\cdot 1 \cdot \cdot}$	$T_{\cdot 2 \cdot \cdot}$	$\cdots$	$T_{\cdot b \cdot \cdot}$	$T_{\cdot \cdot \cdot \cdot}$

A	C				$Total$
	1	2	$\cdots$	c	
1	$T_{1 \cdot 1 \cdot}$	$T_{1 \cdot 2 \cdot}$	$\cdots$	$T_{1 \cdot c \cdot}$	$T_{1 \cdot \cdot \cdot}$
2	$T_{2 \cdot 1 \cdot}$	$T_{2 \cdot 2 \cdot}$	$\cdots$	$T_{2 \cdot c \cdot}$	$T_{2 \cdot \cdot \cdot}$
$\cdot$	$\cdot$	$\cdot$		$\cdot$	$\cdot$
$\cdot$	$\cdot$	$\cdot$		$\cdot$	$\cdot$
$\cdot$	$\cdot$	$\cdot$		$\cdot$	$\cdot$
a	$T_{a \cdot 1 \cdot}$	$T_{a \cdot 2 \cdot}$	$\cdots$	$T_{a \cdot c \cdot}$	$T_{a \cdot \cdot \cdot}$
Total	$T_{\cdot \cdot 1 \cdot}$	$T_{\cdot \cdot 2 \cdot}$	$\cdots$	$T_{\cdot \cdot c \cdot}$	$T_{\cdot \cdot \cdot \cdot}$

B	C				
	1	*2*	$\cdots$	c	*Total*
1	$T_{\cdot 11 \cdot}$	$T_{\cdot 12 \cdot}$	$\cdots$	$T_{\cdot 1c \cdot}$	$T_{\cdot 1 \cdot \cdot}$
2	$T_{\cdot 21 \cdot}$	$T_{\cdot 22 \cdot}$	$\cdots$	$T_{\cdot 2c \cdot}$	$T_{\cdot 2 \cdot \cdot}$
.	.	.		.	.
.	.	.		.	.
.	.	.		.	.
b	$T_{\cdot b1 \cdot}$	$T_{\cdot b2 \cdot}$	$\cdots$	$T_{\cdot bc \cdot}$	$T_{\cdot b \cdot \cdot}$
Total	$T_{\cdot \cdot 1 \cdot}$	$T_{\cdot \cdot 2 \cdot}$	$\cdots$	$T_{\cdot \cdot c \cdot}$	$T_{\cdot \cdot \cdot \cdot}$

The sums of squares are computed by substituting the appropriate totals into the following computational formulas:

$$\text{SST} = \sum_{i=1}^{a} \sum_{j=1}^{b} \sum_{k=1}^{c} \sum_{l=1}^{n} y_{ijkl}^2 - \frac{T_{\cdots}^2}{abcn}$$

$$\text{SSA} = \frac{\sum_{i=1}^{a} T_{i\cdots}^2}{bcn} - \frac{T_{\cdots}^2}{abcn}$$

$$\text{SSB} = \frac{\sum_{j=1}^{b} T_{\cdot j \cdot \cdot}^2}{acn} - \frac{T_{\cdots}^2}{abcn}$$

$$\text{SSC} = \frac{\sum_{k=1}^{c} T_{\cdot \cdot k \cdot}^2}{abn} - \frac{T_{\cdots}^2}{abcn}$$

$$\text{SS}(AB) = \frac{\sum_{i=1}^{a} \sum_{j=1}^{b} T_{ij\cdot\cdot}^2}{cn} - \frac{\sum_{i=1}^{a} T_{i\cdots}^2}{bcn} - \frac{\sum_{j=1}^{b} T_{\cdot j \cdot \cdot}^2}{acn} + \frac{T_{\cdots}^2}{abcn}$$

$$\text{SS}(AC) = \frac{\sum_{i=1}^{a} \sum_{k=1}^{c} T_{i\cdot k \cdot}^2}{bn} - \frac{\sum_{i=1}^{a} T_{i\cdots}^2}{bcn} - \frac{\sum_{k=1}^{c} T_{\cdot \cdot k \cdot}^2}{abn} + \frac{T_{\cdots}^2}{abcn}$$

$$\text{SS}(BC) = \frac{\sum_{j=1}^{b} \sum_{k=1}^{c} T_{\cdot jk \cdot}^2}{an} - \frac{\sum_{j=1}^{b} T_{\cdot j \cdot \cdot}^2}{acn} - \frac{\sum_{k=1}^{c} T_{\cdot \cdot k \cdot}^2}{abn} + \frac{T_{\cdots}^2}{abcn}$$

$$\text{SS}(ABC) = \frac{\sum_{i=1}^{a} \sum_{j=1}^{b} \sum_{k=1}^{c} T_{ijk\cdot}^2}{n} - \frac{\sum_{i=1}^{a} \sum_{j=1}^{b} T_{ij\cdot\cdot}^2}{cn} - \frac{\sum_{i=1}^{a} \sum_{k=1}^{c} T_{i\cdot k \cdot}^2}{bn}$$

$$- \frac{\sum_{j=1}^{b} \sum_{k=1}^{c} T_{\cdot jk \cdot}^2}{an} + \frac{\sum_{i=1}^{a} T_{i\cdots}^2}{bcn} + \frac{\sum_{j=1}^{b} T_{\cdot j \cdot \cdot}^2}{acn} + \frac{\sum_{k=1}^{c} T_{\cdot \cdot k \cdot}^2}{abn} - \frac{T_{\cdots}^2}{abcn},$$

and SSE, as usual, is obtained by subtraction. The computations in an analysis-of-variance problem, for a three-factor experiment with n replications, are summarized in Table 12.5.

Table 12.5 Analysis of Variance for the Three-Factor Experiment with n Replications

Source of variation	Sum of squares	Degrees of freedom	Mean square	Computed f
Main effects				
A	SSA	$a - 1$	s_1^2	$f_1 = \dfrac{s_1^2}{s^2}$
B	SSB	$b - 1$	s_2^2	$f_2 = \dfrac{s_2^2}{s^2}$
C	SSC	$c - 1$	s_3^2	$f_3 = \dfrac{s_3^2}{s^2}$
Two-factor interactions				
AB	SS(AB)	$(a - 1)(b - 1)$	s_4^2	$f_4 = \dfrac{s_4^2}{s^2}$
AC	SS(AC)	$(a - 1)(c - 1)$	s_5^2	$f_5 = \dfrac{s_5^2}{s^2}$
BC	SS(BC)	$(b - 1)(c - 1)$	s_6^2	$f_6 = \dfrac{s_6^2}{s^2}$
Three-factor interactions				
ABC	SS(ABC)	$(a - 1)(b - 1)(c - 1)$	s_7^2	$f_7 = \dfrac{s_7^2}{s^2}$
Error	SSE	$abc(n - 1)$	s^2	
Total	SST	$abcn - 1$		

For the three-factor experiment with a single replicate we may use the analysis of Table 12.5 by setting $n = 1$ and using the ABC interaction for our error sum of squares. In this case we are assuming that the ABC interaction is zero and that $SS(ABC)$ represents variation due only to experimental error and thereby provides an estimate of the error variance.

Example 12.3 In the production of a particular material three variables are of interest: A the operator effect (three operators), B the catalyst used in the experiment (three catalysts), and C the washing time of the product following the cooling process (15 minutes and 20 minutes). Three runs were made at each combination of factors. It was felt that all interactions among the factors should be studied. The coded yields are as follows:

	Washing time, C					
	15 minutes			20 minutes		
	B			B		
A	1	2	3	1	2	3
1	10.7	10.3	11.2	10.9	10.5	12.2
	10.8	10.2	11.6	12.1	11.1	11.7
	11.3	10.5	12.0	11.5	10.3	11.0
2	11.4	10.2	10.7	9.8	12.6	10.8
	11.8	10.9	10.5	11.3	7.5	10.2
	11.5	10.5	10.2	10.9	9.9	11.5
3	13.6	12.0	11.1	10.7	10.2	11.9
	14.1	11.6	11.0	11.7	11.5	11.6
	14.5	11.5	11.5	12.7	10.9	12.2

Perform an analysis of variance to test for significant effects.

Solution. First we construct the following two-way tables:

C (15 minutes)	B			Total
	1	2	3	
A				
1	32.8	31.0	34.8	98.6
2	34.7	31.6	31.4	97.7
3	42.2	35.1	33.6	110.9
Total	109.7	97.7	99.8	307.2

C (20 minutes)	B			Total
	1	2	3	
A				
1	34.5	31.9	34.9	101.3
2	32.0	30.0	32.5	94.5
3	35.1	32.6	35.7	103.4
Total	101.6	94.5	103.1	299.2

		B		
A	1	2	3	Total
1	67.3	62.9	69.7	199.9
2	66.7	61.6	63.9	192.2
3	77.3	67.7	69.3	214.3
Total	211.3	192.2	202.9	606.4

	C		
A	1	2	Total
1	98.6	101.3	199.9
2	97.7	94.5	192.2
3	110.9	103.4	214.3
Total	307.2	299.2	606.4

	C		
B	1	2	Total
1	109.7	101.6	211.3
2	97.7	94.5	192.2
3	99.8	103.1	202.9
Total	307.2	299.2	606.4

Now

$$\text{SST} = 10.7^2 + 10.8^2 + \cdots + 12.2^2 - \frac{606.4^2}{54}$$

$$= 6{,}872.84 - 6{,}809.65 = 63.19$$

$$\text{SSA} = \frac{199.9^2 + 192.2^2 + 214.3^2}{18} - \frac{606.4^2}{54}$$

$$= 6{,}823.63 - 6{,}809.65 = 13.98$$

$$\text{SSB} = \frac{211.3^2 + 192.2^2 + 202.9^2}{18} - \frac{606.4^2}{54}$$

$$= 6{,}819.83 - 6{,}809.65 = 10.18$$

$$\text{SSC} = \frac{307.2^2 + 299.2^2}{27} - \frac{606.4^2}{54}$$

$$= 6{,}810.83 - 6{,}809.65 = 1.18$$

$$\text{SS(AB)} = \frac{67.3^2 + 66.7^2 + \cdots + 69.3^2}{6}$$
$$- 6{,}823.63 - 6{,}819.83 + 6{,}809.65$$

$$= 4.78$$

$$\text{SS(AC)} = \frac{98.6^2 + 97.7^2 + \cdots + 103.4^2}{9}$$
$$- 6{,}823.63 - 6{,}810.83 + 6{,}809.65$$

$$= 2.92$$

$$SS(BC) = \frac{109.7^2 + 97.7^2 + \cdots + 103.1^2}{9}$$
$$- 6{,}819.83 - 6{,}810.83 + 6{,}809.65$$

$$= 3.64$$

$$SS(ABC) = \frac{32.8^2 + 34.7^2 + \cdots + 35.7^2}{3}$$
$$- 6{,}838.59 - 6{,}827.73 - 6{,}824.65$$
$$+ 6{,}823.63 + 6{,}819.83 + 6{,}810.83 - 6{,}809.65$$

$$= 4.89$$

$$SSE = 63.19 - 13.98 - 10.18 - 1.18 - 4.78 - 2.92 - 3.64 - 4.89$$
$$= 21.62.$$

These results, with the remaining computations, are given in Table 12.6.

Table 12.6 Analysis of Variance for Example 12.3

Source of variation	Sum of squares	Degrees of freedom	Mean square	Computed f
Main effects				
A	13.98	2	6.99	11.65
B	10.18	2	5.09	8.48
C	1.18	1	1.18	1.97
Two-factor interactions				
AB	4.78	4	1.20	2.00
AC	2.92	2	1.46	2.43
BC	3.64	2	1.82	3.03
Three-factor interactions				
ABC	4.89	4	1.22	2.03
Error	21.62	36	0.60	
Total	63.19	53		

None of the interactions show a significant effect at the $\alpha = 0.05$ level. The operator and catalyst effects are significant, while the washing time has no significant effect on the yield for the range used in the experiment.

12.5 Discussion of Specific Multifactor Models

We have described the three-factor model and its analysis in the most general form by including all possible interactions in the model. Of course, there are many situations in which it is known a priori that the model should not contain certain interactions. We can then take advantage of this knowledge by combining or pooling the sums of squares corresponding to negligible interactions with the error sum of squares to form a new estimator for σ^2 with a larger number of degrees of freedom. For example, in a metallurgy experiment designed to study the effect on film thickness of three important processing variables, suppose it is known that factor A, acid concentration, does not interact with factors B and C. The sums of squares SSA, SSB, SSC, and SS(BC) are computed using the methods described in Section 12.4. The mean squares for the remaining effects will now all independently estimate the error variance σ^2. Therefore we form our new error mean square by pooling SS(AB), SS(AC), SS(ABC), and SSE, along with the corresponding degrees of freedom. The resulting denominator for the significance tests is then the error mean square given by

$$s^2 = \frac{\text{SS(AB)} + \text{SS(AC)} + \text{SS(ABC)} + \text{SSE}}{(a-1)(b-1) + (a-1)(c-1) + (a-1)(b-1)(c-1) + abc(n-1)}.$$

Computationally, of course, one obtains the pooled sum of squares and the pooled degrees of freedom by subtraction once SST and the sums of squares for the existing effects are computed. The analysis-of-variance table would then take the form of Table 12.7.

Table 12.7 Analysis of Variance with Factor A Noninteracting

Source of variation	Sum of squares	Degrees of freedom	Mean square	Computed f
Main effects				
A	SSA	$a-1$	s_1^2	$f_1 = \dfrac{s_1^2}{s^2}$
B	SSB	$b-1$	s_2^2	$f_2 = \dfrac{s_2^2}{s^2}$
C	SSC	$c-1$	s_3^2	$f_3 = \dfrac{s_3^2}{s^2}$
Interaction effect				
BC	SS(BC)	$(b-1)(c-1)$	s_4^2	$f_4 = \dfrac{s_4^2}{s^2}$
Error	SSE	Substraction	s^2	
Total	SST	$abcn-1$		

In our analysis of the two-factor experiment in Section 12.3 a completely randomized design was used. By interpreting the levels of factor A in Table 12.7 as different blocks, we then have the analysis-of-variance procedure for a two-factor experiment in a randomized block design. For example, if we interpret the operators in Example 12.3 as blocks and assume no interaction between blocks and the other two factors, the analysis of variance takes the form of Table 12.8 rather than that given

Table 12.8 Analysis of Variance for a Two-Factor Experiment in a Randomized Block Design

Source of variation	Sum of squares	Degrees of freedom	Mean square	Computed f
Blocks	13.98	2	6.99	9.45
Main effects				
B	10.18	2	5.09	6.88
C	1.18	1	1.18	1.59
Interaction				
BC	3.64	2	1.82	2.46
Error	34.21	46	0.74	
Total	63.19	53		

in Table 12.6. The reader can easily verify that the error mean square is also given by

$$s^2 = \frac{4.78 + 2.92 + 4.89 + 21.62}{4 + 2 + 4 + 36} = 0.74,$$

which demonstrates the pooling of the sums of squares for the nonexisting interaction effects.

12.6 Model II Factorial Experiments

In a two-factor experiment with random effects we have the model

$$Y_{ijk} = \mu + A_i + B_j + (AB)_{ij} + E_{ijk},$$

$i = 1, 2, \ldots, a; j = 1, 2, \ldots, b;$ and $k = 1, 2, \ldots, n$, where the A_i, B_j, $(AB)_{ij}$, and E_{ijk} are independent random variables with zero means and variances σ_α^2, σ_β^2, $\sigma_{\alpha\beta}^2$, and σ^2, respectively. The sum of squares for the model II experiments are computed in exactly the same way as for the model I experiments. We are now interested in testing hypotheses of the form

$$H_0': \sigma_\alpha^2 = 0 \qquad H_0'': \sigma_\beta^2 = 0 \qquad H_0''': \sigma_{\alpha\beta}^2 = 0$$
$$H_1': \sigma_\alpha^2 \neq 0 \qquad H_1'': \sigma_\beta^2 \neq 0 \qquad H_1''': \sigma_{\alpha\beta}^2 \neq 0,$$

where the denominator in the f ratio is not necessarily the error mean square. The appropriate denominator can be determined by examining the expected values of the various mean squares. These are shown in Table 12.9.

Table 12.9 Expected Mean Squares for a
Model II Two-Factor Experiment

Source of variation	Degrees of freedom	Mean square	Expected mean square
A	$a - 1$	s_1^2	$\sigma^2 + n\sigma_{\alpha\beta}^2 + bn\sigma_\alpha^2$
B	$b - 1$	s_2^2	$\sigma^2 + n\sigma_{\alpha\beta}^2 + an\sigma_\beta^2$
AB	$(a - 1)(b - 1)$	s_3^2	$\sigma^2 + n\sigma_{\alpha\beta}^2$
Error	$ab(n - 1)$	s^2	σ^2
Total	$abn - 1$		

From Table 12.9 we see that H_0' and H_0'' are tested by using s_3^2 in the denominator of the f ratio, while H_0''' is tested using s^2 in the denominator. The unbiased estimates of the variance components are given by

$$\hat{\sigma}^2 = s^2$$

$$\hat{\sigma}_{\alpha\beta}^2 = \frac{s_3^2 - s^2}{n}$$

$$\hat{\sigma}_\alpha^2 = \frac{s_1^2 - s_3^2}{bn}$$

$$\hat{\sigma}_\beta^2 = \frac{s_2^2 - s_3^2}{an}.$$

The expected mean squares for the three-factor experiment with random effects in a completely randomized design are shown in Table 12.10.

Table 12.10 Expected Mean Squares for a
Model II Three-Factor Experiment

Source of variation	Degrees of freedom	Mean square	Expected mean square
A	$a - 1$	s_1^2	$\sigma^2 + n\sigma_{\alpha\beta\gamma}^2 + cn\sigma_{\alpha\beta}^2 + bn\sigma_{\alpha\gamma}^2 + bcn\sigma_\alpha^2$
B	$b - 1$	s_2^2	$\sigma^2 + n\sigma_{\alpha\beta\gamma}^2 + cn\sigma_{\alpha\beta}^2 + an\sigma_{\beta\gamma}^2 + acn\sigma_\beta^2$
C	$c - 1$	s_3^2	$\sigma^2 + n\sigma_{\alpha\beta\gamma}^2 + bn\sigma_{\alpha\gamma}^2 + an\sigma_{\beta\gamma}^2 + abn\sigma_\gamma^2$
AB	$(a - 1)(b - 1)$	s_4^2	$\sigma^2 + n\sigma_{\alpha\beta\gamma}^2 + cn\sigma_{\alpha\beta}^2$
AC	$(a - 1)(c - 1)$	s_5^2	$\sigma^2 + n\sigma_{\alpha\beta\gamma}^2 + bn\sigma_{\alpha\gamma}^2$
BC	$(b - 1)(c - 1)$	s_6^2	$\sigma^2 + n\sigma_{\alpha\beta\gamma}^2 + an\sigma_{\beta\gamma}^2$
ABC	$(a - 1)(b - 1)(c - 1)$	s_7^2	$\sigma^2 + n\sigma_{\alpha\beta\gamma}^2$
Error	$abc(n - 1)$	s^2	σ^2
Total	$abcn - 1$		

It is evident from the expected mean squares of Table 12.10 that one can form appropriate f ratios for testing all two-factor and three-factor interaction variance components. However, to test a hypothesis of the form

$$H_0: \quad \sigma_\alpha^2 = 0$$
$$H_1: \quad \sigma_\alpha^2 \neq 0,$$

there appears to be no appropriate f ratio unless we have found one or more of the two-factor interaction variance components not significant. Suppose, for example, that we have compared s_5^2 with s_7^2 and found $\sigma_{\alpha\gamma}^2$ to be negligible. We could then argue that the term $\sigma_{\alpha\gamma}^2$ should be dropped from all the expected mean squares of Table 12.10; then the ratio s_1^2/s_4^2 provides a test for the significance of the variance component σ_α^2. Therefore, if we are to test hypotheses concerning the variance component of the main effects it is necessary first to investigate the significance of the two-factor interaction components. An approximate test derived by Satterthwaite may be used when certain two-factor interaction variance components are found to be significant and hence must remain a part of the expected mean square.

Example 12.4 In a study to determine which are the important sources of variation in an industrial process, three measurements are taken on yield for three operators chosen randomly and four batches of raw materials chosen randomly. It was decided that a significance test should be made at the 0.05 level of significance to determine if the variance components due to batches, operators, and interaction are significant. In addition, estimates of variance components are to be computed. The data are as follows, with the response being percent by weight:

Operators	Batches			
	1	*2*	*3*	*4*
	66.9	68.3	69.0	69.3
1	68.1	67.4	69.8	70.9
	67.2	67.7	67.5	71.4
	66.3	68.1	69.7	69.4
2	65.4	66.9	68.8	69.6
	65.8	67.6	69.2	70.0
	65.6	66.0	67.1	67.9
3	66.3	66.9	66.2	68.4
	65.2	67.3	67.4	68.7

Solution. The sums of squares are found in the usual way with the results given by

$$SSA \text{ (operators)} = 21.45$$
$$SSB \text{ (batches)} = 52.21$$
$$SS(AB) \text{ (interaction)} = 5.04$$
$$SSE \text{ (error)} = 8.27.$$

All other computations are carried out and exhibited in Table 12.11. Since

Table 12.11 Analysis of Variance for Example 12.4

Source of variation	Sum of squares	Degrees of freedom	Mean square	Computed f
Operators	21.45	2	10.45	12.44
Batches	52.21	3	17.40	20.71
Interaction	5.04	6	0.84	2.47
Error	8.27	24	0.34	
Total	86.97	35		

$f_{0.05}(2,24) = 3.40$, $f_{0.05}(3,24) = 3.01$, and $f_{0.05}(6,24) = 2.51$ we find the operator and batch variance components to be significant, while the interaction variance is not significant at the $\alpha = 0.05$ level. Estimates of the main effect variance components are given by

$$\hat{\sigma}_\alpha^2 = \frac{10.45 - 0.84}{12} = 0.82$$

$$\hat{\sigma}_\beta^2 = \frac{17.40 - 0.84}{9} = 1.84.$$

12.7 Choice of Sample Size

Our study of factorial experiments throughout this chapter has been restricted to the use of a completely randomized design with the exception of Section 12.5, where we demonstrated the analysis of a two-factor experiment in a randomized block design. The completely randomized design is very easy to lay out and the analysis is simple to perform; however, it should be used only when the number of treatment combinations is small and the experimental material is homogeneous. Although the randomized block design is ideal for dividing a large group of heterogeneous units into subgroups of homogeneous units, it is generally difficult to obtain uniform blocks with enough units to which a large number of treatment combinations may be assigned. This[4] disadvantage may be overcome by choosing a design from the catalog of *incomplete block designs*. These designs allow one to investigate differences among t treatments arranged in b blocks containing k experimental units, where $k < t$.

Once the experimenter has selected a completely randomized design, he must decide if the number of replications is sufficient to yield tests in the analysis of variance with high power. If not, he must add additional replications, which in turn may force him into a randomized complete block design. Had he started with a randomized block design it would still be necessary to determine if the number of blocks is sufficient to yield powerful tests. Basically, then, we are back to the question of sample size.

The power of a fixed effects test or the probability of rejecting H_0 when the alternative H_1 is true, for a given sample size, is found from Table XV by computing the noncentrality parameter λ and the function ϕ discussed in Section 11.11. Expressions for λ and ϕ^2 for the two-factor and three-factor fixed effects experiments are given in Table 12.12.

Table 12.12 Noncentrality Parameter λ and ϕ^2 for Two-Factor and Three-Factor Models

	Two-factor experiments		*Three-factor experiments*		
	A	B	A	B	C
λ	$\dfrac{bn\sum_{i=1}^{a}\alpha_i^2}{2\sigma^2}$	$\dfrac{an\sum_{j=1}^{b}\beta_j^2}{2\sigma^2}$	$\dfrac{bcn\sum_{i=1}^{a}\alpha_i^2}{2\sigma^2}$	$\dfrac{acn\sum_{j=1}^{b}\beta_j^2}{2\sigma^2}$	$\dfrac{abn\sum_{k=1}^{c}\gamma_k^2}{2\sigma^2}$
ϕ^2	$\dfrac{bn\sum_{i=1}^{a}\alpha_i^2}{a\sigma^2}$	$\dfrac{an\sum_{j=1}^{b}\beta_j^2}{b\sigma^2}$	$\dfrac{bcn\sum_{i=1}^{a}\alpha_i^2}{a\sigma^2}$	$\dfrac{acn\sum_{j=1}^{b}\beta_j^2}{b\sigma^2}$	$\dfrac{abn\sum_{k=1}^{c}\gamma_k^2}{c\sigma^2}$

The results of Section 11.11 for the random effects model can be extended very easily to the two- and three-factor models. Once again the general procedure is based on the values of the expected mean squares. For example, if we are testing $\sigma_\alpha^2 = 0$ in a two-factor experiment by computing the ratio s_1^2/s_3^2 (see Table 12.9), then

$$f = \frac{s_1^2/(\sigma^2 + n\sigma_{\alpha\beta}^2 + bn\sigma_\alpha^2)}{s_3^2/(\sigma^2 + n\sigma_{\alpha\beta}^2)}$$

is a value of the random variable F having the F distribution with $a - 1$

and $(a - 1)(b - 1)$ degrees of freedom, and the power of the test is given by

$$
\begin{aligned}
1 - \beta &= Pr\left\{\frac{S_1^2}{S_3^2} > f_\alpha[(a - 1), (a - 1)(b - 1)] \mid \sigma_\alpha^2 \neq 0\right\} \\
&= Pr\left\{F > \frac{f_\alpha[(a - 1), (a - 1)(b - 1)](\sigma^2 + n\sigma_{\alpha\beta}^2)}{\sigma^2 + n\sigma_{\alpha\beta}^2 + bn\sigma_\alpha^2}\right\}.
\end{aligned}
$$

EXERCISES

1. An experiment was conducted to study the effect of temperature and type of oven on the life of a particular component being tested. Four types of ovens and three temperature levels were used in the experiment. Twenty-four pieces were assigned randomly, 2 to each combination of treatments, and the following results recorded:

	Oven			
Temperature (degrees)	O_1	O_2	O_3	O_4
500	227	214	225	260
	221	259	236	229
550	187	181	232	246
	208	179	198	273
600	174	198	178	206
	202	194	213	219

(a) Test the hypothesis of no interaction between type of oven and temperature at the $\alpha = 0.05$ level of significance.

(b) If the significance test in (a) is accepted, test the hypotheses that the ovens and temperatures have no effect on the life of the component.

2. A study was made to determine if humidity conditions have an effect on the force required to pull apart pieces of glued plastic. Three different types of plastic were tested using four different levels of humidity. The results, in pounds, are given as follows:

	Humidity			
Plastic type	30%	50%	70%	90%
A	39.0	33.1	33.8	33.0
	42.8	37.8	30.7	32.9
B	36.9	27.2	29.7	28.5
	41.0	26.8	29.1	27.9
C	27.4	29.2	26.7	30.9
	30.3	29.9	32.0	31.5

(a) Assuming a model I experiment, perform an analysis of variance and test the hypothesis of no interaction between humidity and plastic type at the 0.05 level of significance.

(b) Perform an analysis of variance using only plastics A and B, and once again test for the presence of interaction.

(c) Use a single-degree-of-freedom comparison to evaluate the force required at 30% humidity versus (50%, 70%, 90%).

(d) Repeat part (c) using only plastic C.

In all tests use s^2 from the overall analysis of variance of part (a) as the denominator in the f ratio.

3. The following data are taken in a study involving three factors A, B, and C, all fixed effects:

	C_1			C_2			C_3		
	B_1	B_2	B_3	B_1	B_2	B_3	B_1	B_2	B_3
A_1	15.0	14.8	15.9	16.8	14.2	13.2	15.8	15.5	19.2
	18.5	13.6	14.8	15.4	12.9	11.6	14.3	13.7	13.5
	22.1	12.2	13.6	14.3	13.0	10.1	13.0	12.6	11.1
A_2	11.3	17.2	16.1	18.9	15.4	12.4	12.7	17.3	7.8
	14.6	15.5	14.7	17.3	17.0	13.6	14.2	15.8	11.5
	18.2	14.2	13.4	16.1	18.6	15.2	15.9	14.6	12.2

(a) Perform tests of significance on all interactions at the $\alpha = 0.05$ level.

(b) Perform tests of significance on the main effects at the $\alpha = 0.05$ level.

(c) Give an explanation of how a significant interaction has masked the effect of factor C.

4. Consider an experimental situation involving factors A, B, and C, where we assume a three-way fixed effects model of the form

$$y_{ijkl} = \mu + \alpha_i + \beta_j + \gamma_k + (\beta\gamma)_{jk} + \epsilon_{ijkl}.$$

All other interactions are considered to be nonexistent or negligible. The data were recorded as follows:

	B_1			B_2		
	C_1	C_2	C_3	C_1	C_2	C_3
A_1	4.0	3.4	3.9	4.4	3.1	3.1
	4.9	4.1	4.3	3.4	3.5	3.7
A_2	3.6	2.8	3.1	2.7	2.9	3.7
	3.9	3.2	3.5	3.0	3.2	4.2
A_3	4.8	3.3	3.6	3.6	2.9	2.9
	3.7	3.8	4.2	3.8	3.3	3.5
A_4	3.6	3.2	3.2	2.2	2.9	3.6
	3.9	2.8	3.4	3.5	3.2	4.3

  (a) Perform a test of significance on the BC interaction at the $\alpha = 0.05$ level.

  (b) Perform tests of significance on the main effects A, B, and C using a pooled error mean square at the $\alpha = 0.05$ level.

5. To estimate the various components of variability in a filtration process, the percent of material lost in the mother liquor was measured for 12 experimental conditions, three runs on each condition. Three filters and four operators were selected at random to use in the experiment resulting in the following measurements:

Filter	Operator 1	2	3	4
1	16.2	15.9	15.6	14.9
	16.8	15.1	15.9	15.2
	17.1	14.5	16.1	14.9
2	16.6	16.0	16.1	15.4
	16.9	16.3	16.0	14.6
	16.8	16.5	17.2	15.9
3	16.7	16.5	16.4	16.1
	16.9	16.9	17.4	15.4
	17.1	16.8	16.9	15.6

  (a) Test the hypothesis of no interaction variance component between filters and operators at the $\alpha = 0.05$ level of significance.

  (b) Test the hypotheses that the operators and the filters have no effect on the variability of the filtration process.

  (c) Estimate the components of variance due to filters, operators, and experimental error.

6. Consider the following analysis of variance for a model II experiment:

Source of variation	Degrees of freedom	Mean square
A	3	140
B	1	480
C	2	325
AB	3	15
AC	6	24
BC	2	18
ABC	6	2
Error	24	5
Total	47	

Test for significant variance components among all main effects and interaction effects at the 0.01 level of significance (a) using a pooled estimate of error when appropriate (b) not pooling sums of squares of insignificant effects.

7. Are two observations for each treatment combination in Exercise 4 sufficient if the power of our test for detecting differences among the levels of factor C at the 0.05 level of significance is to be at least 0.8 when $\gamma_1 = -0.2$, $\gamma_2 = 0.4$, and $\gamma_3 = -0.2$. Use the same pooled estimate of σ^2 that was used in the analysis of variance.

8. Using the estimates of the variance components in Exercise 5, evaluate the power when we test the variance component due to filters to be zero.

13 2^k FACTORIAL EXPERIMENTS

13.1 Introduction

In almost any experimental study in which statistical procedures are applied to a collection of scientific data, the methods involve performing certain operations or computations on the sample information, followed by the drawing of inferences about the population or populations studied. Often there are characteristics of the experiment that are subject to the control of the experimenter, quantities such as sample size, number of levels of the factors, treatment combinations to be used, and so forth. These *experimental parameters* can often have a great effect on the precision with which hypotheses are tested or estimation is accomplished.

We have already been exposed to certain experimental design concepts. The sampling plan for the simple t test on the mean of a normal population and also the analysis of variance involve randomly allocating prechosen treatments to experimental units. The randomized block design, where treatments are assigned to units within relatively homogeneous blocks, involves restricted randomization.

In this chapter we given special attention to experimental designs in which the experimental plan calls for the study of the effect on a response of k factors, each at two levels. These are commonly known as 2^k factorial experiments. We often denote the levels as "high" and "low," even though this notation may be arbitrary in the case of qualitative variables.

The complete factorial design requires that each level of every factor occur with each level of every other factor, giving a total of 2^k treatment combinations. We shall denote the higher levels of the factors $A, B, C, \ldots$ by the letters $a, b, c, \ldots$ and the lower levels of each factor by the notation (1). In the presence of other letters we omit the symbol (1). For example, the treatment combination in a 2^4 experiment that contains the high levels of factors B and C and the low levels of factors A and D is written simply as bc. The treatment combination that consists of the low level of all factors in the experiment is denoted by the symbol (1). In the case of a 2^3 experiment, the eight possible treatment combinations are (1), a, b, c, ab, ac, bc, and abc.

The factorial experiment allows the effect of each and every factor to be estimated and tested independently through the usual analysis of variance. In addition, the interaction effects are easily assessed. The disadvantage, of course, with the factorial experiment is the excessive amount of experimentation that is required. For example, if it is desired to study the effect of eight variables, $2^8 = 256$ treatment combinations are required. In many instances we can obtain considerable information by using only a fraction of the experimental runs. This type of design is called a *fractional factorial design* and will be considered in Sections 13.6 and 13.7.

13.2 Analysis of Variance

Consider initially a 2^2 factorial plan in which there are n experimental observations per treatment combination. Extending our previous notation, we now interpret the symbols (1), a, b, and ab to be the total yields for each of the four treatment combinations. Table 13.1 gives a two-way table of these total yields.

Table 13.1 2^2 **Factorial Experiment**

	\multicolumn{2}{c}{B}	$Mean$
A	(1)	b	$\dfrac{b + (1)}{2n}$
	a	ab	$\dfrac{ab + a}{2n}$
Mean	$\dfrac{a + (1)}{2n}$	$\dfrac{ab + b}{2n}$	

Clearly, there will be exactly one single-degree-of-freedom contrast for the means of each factor A and B, which we shall write as

$$w_A = \frac{ab + a - b - (1)}{2n}$$

$$w_B = \frac{ab + b - a - (1)}{2n}.$$

The contrast w_A is seen to be the difference between the mean response at the low and high levels of factor A. In fact, we call w_A the *main effect* of A. Similarly, w_B is the main effect of factor B. Apparent interaction in the data is observed by inspecting the difference between $ab - b$ and $a - (1)$ or between $ab - a$ and $b - (1)$ in Table 13.1. Hence a third contrast in the treatment totals, orthogonal to these main effect contrasts, is the *interaction effect* given by

$$w_{AB} = \frac{ab - a - b + (1)}{2n}.$$

We take advantage of the fact that in the 2^2 factorial, or for that matter in the general 2^k factorial experiment, each main effect and interaction effect has associated with it a single degree of freedom. Therefore we can write $2^k - 1$ orthogonal single-degree-of-freedom contrasts in the treatment combinations, each representing variation due to some main or interaction effect. Thus, under the usual independence and normality assumptions in the experimental model, we can make tests to determine if the contrast reflects systematic variation or merely chance or random variation. The sums of squares for each contrast is found by following the procedures given in Section 11.4. Writing $T_1.. = b + (1)$, $T_2.. = ab + a$, $c_1 = -1$, and $c_2 = 1$, where $T_1..$ and $T_2..$ are the totals of $2n$ observations, we have

$$\text{SSA} = \text{SS}w_A = \frac{\left(\sum\limits_{i=1}^{2} c_i T_i..\right)^2}{2n \sum\limits_{i=1}^{2} c_i^2}$$

$$= \frac{[ab + a - b - (1)]^2}{2^2 n},$$

with 1 degree of freedom. Similarly, we find

$$\text{SSB} = \frac{[ab + b - (1) - a]^2}{2^2 n}$$

$$\text{SS(AB)} = \frac{[ab + (1) - a - b]^2}{2^2 n},$$

each with 1 degree of freedom, while the error sum of squares, with $2^2(n-1)$ degrees of freedom, is obtained by subtraction from the formula

$$SSE = SST - SSA - SSB - SS(AB).$$

In computing the sums of squares for the main effects A and B and the interaction effect AB, it is convenient to present the total yields of the treatment combinations along with the appropriate algebraic signs for each contrast as in Table 13.2.

Table 13.2 Signs for Contrasts in a 2^2
Factorial Experiment

Treatment combination	Factorial effect		
	A	B	AB
(1)	−	−	+
a	+	−	−
b	−	+	−
ab	+	+	+

The main effects are obtained as simple comparisons between the low and high levels. Therefore, we assign a positive sign to the treatment combination that is at the high level of a given factor and a negative sign to the treatment combination at the lower level. The positive and negative signs for the interaction effect are obtained by multiplying the corresponding signs of the contrasts of the interacting factors.

Let us now consider an experiment using three factors A, B, and C with levels (1), a; (1), b; and (1), c, respectively. This is a 2^3 factorial experiment giving the eight treatment combinations (1), a, b, c, ab, ac, bc, and abc. The treatment combinations and the appropriate algebraic signs for each contrast used in computing the sums of squares for the main effects and interaction effects are presented in Table 13.3.

An inspection of Table 13.3 reveals that for the 2^3 experiment any two contrasts among the seven are orthogonal and therefore the seven effects are assessed independently. The sum of squares for, say, the ABC interaction with 1 degree of freedom is given by

$$SS(ABC) = \frac{[abc + a + b + c - (1) - ab - ac - bc]^2}{2^3 n}.$$

For a 2^k factorial experiment the single-degree-of-freedom sums of squares for the main effects and interaction effects are obtained by squaring the

Table 13.3 Signs for Contrasts in a 2^3 Factorial Experiment

Treatment combination	Factorial effect						
	A	B	C	AB	AC	BC	ABC
(1)	−	−	−	+	+	+	−
a	+	−	−	−	−	+	+
b	−	+	−	−	+	−	+
c	−	−	+	+	−	−	+
ab	+	+	−	+	−	−	−
ac	+	−	+	−	+	−	−
bc	−	+	+	−	−	+	−
abc	+	+	+	+	+	+	+

appropriate contrasts in the treatment totals and dividing by $2^k n$, where n is the number of replications of the treatment combinations.

The orthogonality property has the same importance here as it did in the material on comparisons discussed in Chapter 12. Orthogonality of contrasts implies that the estimated effects and thus the sums of squares are independent. This independence is easily illustrated in a 2^3 factorial experiment if the yields, with factor A at its high level, are increased by an amount x in Table 13.3. Only the A contrast leads to a larger sum of squares since the x effect cancels out in the formation of the six remaining contrasts as a result of the two positive and two negative signs associated with treatment combinations in which A is at the high level.

13.3 Yates' Technique for Computing Contrasts

It is laborious to write out the table of positive and negative signs for large experiments. A systematic tabular technique for deriving the factorial effects has been developed by Yates. The treatment combinations and the observations must be written down in *standard form*. For one factor the standard form is (1), a. For two factors we add b and ab, derived by multiplying the first two treatment combinations by the additional letter b. For three factors we add c, ac, bc, and abc, derived by multiplying the first four treatment combinations by the additional letter c, and so on. In the case of three factors the standard order is then

$$(1), \quad a, \quad b, \quad ab, \quad c, \quad ac, \quad bc, \quad abc.$$

Yates' method is carried out in the following steps:

1. Place the treatment combinations and the corresponding total yields in a column in standard order.

2. Obtain the top half of a column marked (1) by adding the first two yields, then the next two, and so forth. The bottom half is obtained by subtracting the first from the second of each of these same pairs.
3. Repeat the operation using the results in column (1) to obtain column (2). This operation is continued until we have k columns for a 2^k experiment.
4. The first value of the kth column will be the grand total of the yields in the experiment. Each remaining number will be a contrast in the treatment totals. Finally, the sum of squares for the main effects and interaction effects are obtained by squaring the entries in column (k) and dividing by $2^k n$, where n is the number of replications and 2^k is the sum of the squares of the coefficients of the individual contrasts.

As an illustration we outline the procedure in Table 13.4 for the 2^3 factorial experiment.

Example 13.1 In a metallurgy experiment it is desired to test the effect of four factors and their interactions on the concentration (percent by weight) of a particular phosphorus compound in casting material. The variables are (A), percent phosphorus in the refinement; (B), percent remelted material; (C), fluxing time; and (D), holding time. The four factors are varied in a 2^4 factorial experiment with two castings taken and the content measured at each treatment combination. Using Yates' technique, perform the analysis of variance of the following data:

Treatment combination	Weight % of phosphorus compound		
	Replication 1	Replication 2	Total
(1)	30.3	28.6	58.9
a	28.5	31.4	59.9
b	24.5	25.6	50.1
ab	25.9	27.2	53.1
c	24.8	23.4	48.2
ac	26.9	23.8	50.7
bc	24.8	27.8	52.6
abc	22.2	24.9	47.1
d	31.7	33.5	65.2
ad	24.6	26.2	50.8
bd	27.6	30.6	58.2
abd	26.3	27.8	54.1
cd	29.9	27.7	57.6
acd	26.8	24.2	51.0
bcd	26.4	24.9	51.3
abcd	26.9	29.3	56.2
Total	428.1	436.9	865.0

Table 13.4 Yates' Technique for a 2^3 Factorial

Treatment combination	(1)	(2)	(3)	Identification
(1)	$(1) + a$	$(1) + a + b + ab$	$(1) + a + b + ab + c + ac + bc + abc$	Total
a	$b + ab$	$c + ac + bc + abc$	$a - (1) + ab - b + ac - c + abc - bc$	A contrast
b	$c + ac$	$a - (1) + ab - b$	$b + ab - (1) - a + bc + abc - c - ac$	B contrast
ab	$bc + abc$	$ac - c + abc - bc$	$ab - b - a + (1) + abc - bc - ac + c$	AB contrast
c	$a - (1)$	$b + ab - (1) - a$	$c + ac + bc + abc - (1) - a - b - ab$	C contrast
ac	$ab - b$	$bc + abc - c - ac$	$ac - c + abc - bc - a + (1) - ab + b$	AC contrast
bc	$ac - c$	$ab - b - a + (1)$	$bc + abc - c - ac - b - ab + (1) + a$	BC contrast
abc	$abc - bc$	$abc - bc - ac + c$	$abc - bc - ac + c - ab + b + a - (1)$	ABC contrast

Solution.

Table 13.5 Yates' Technique for a 2^4 Example

Treatment combination	Treatment total	(1)	(2)	(3)	(4)	Sum of squares
(1)	58.9	118.8	222.0	420.6	865.0	
a	59.9	103.2	198.6	444.4	−19.2	11.52
b	50.1	98.9	228.3	1.0	−19.6	12.00
ab	53.1	99.7	216.1	−20.2	15.8	7.80
c	48.2	116.0	4.0	−14.8	−35.6	39.61
ac	50.7	112.3	−3.0	−4.8	9.8	3.00
bc	52.6	108.6	−18.5	−6.0	19.0	11.28
abc	47.1	107.5	−1.7	21.8	−8.8	2.42
d	65.2	1.0	−15.6	−23.4	23.8	17.70
ad	50.8	3.0	0.8	−12.2	−21.2	14.05
bd	58.2	2.5	−3.7	−7.0	10.0	3.13
abd	54.1	−5.5	−1.1	16.8	27.8	24.15
cd	57.6	−14.4	2.0	16.4	11.2	3.92
acd	51.0	−4.1	−8.0	2.6	23.8	17.70
bcd	51.3	−6.6	10.3	−10.0	−13.8	5.95
abcd	56.2	4.9	11.5	1.2	11.2	3.92

Table 13.5 gives the 16 treatment totals and outlines Yates' technique for computing the individual sums of squares. The total sum of squares is given by

$$\text{SST} = 30.3^2 + 28.5^2 + \cdots + 29.3^2 - \frac{865^2}{32} = 217.51.$$

We can now set up the analysis of variance as in Table 13.6. Note that the interactions BC, AD, ABD, and ACD are significant when compared with $f_{0.05}(1,16) = 4.49$. The tests on the main effects (which in the presence of interactions may be regarded as the effects *averaged* over the levels of the other factors) indicate significance in each case.

Very often the experimenter knows in advance that certain interactions in a 2^k factorial experiment are negligible and should not be included in the model. For example, in a 2^4 factorial experiment he may postulate a model that contains only two-factor interaction effects and then pool the sums of squares and corresponding degrees of freedom of the remaining higher-ordered interactions with the pure error. In fact this is often done in lieu of taking several replications.

Table 13.6 Analysis of Variance for the Data of Table 13.5

Source of variation	Sum of squares	Degrees of freedom	Mean square	Computed f
Main effects:				
A	11.52	1	11.52	4.68
B	12.00	1	12.00	4.90
C	39.61	1	39.61	16.10
D	17.70	1	17.70	7.20
Two-factor interactions:				
AB	7.80	1	7.80	3.17
AC	3.00	1	3.00	1.22
AD	14.05	1	14.05	5.71
BC	11.28	1	11.28	4.59
BD	3.13	1	3.13	1.27
CD	3.92	1	3.92	1.59
Three-factor interactions:				
ABC	2.42	1	2.42	0.98
ABD	24.15	1	24.15	9.82
ACD	17.70	1	17.70	7.20
BCD	5.95	1	5.95	2.42
Four-factor interaction:				
ABCD	3.92	1	3.92	1.59
Error	39.36	16	2.46	
Total	217.51	31		

13.4 Factorial Experiments in Incomplete Blocks

The 2^k factorial experiment lends itself to partitioning into *incomplete blocks*. For a k-factor experiment, it is often useful to use a design in 2^p blocks ($p < k$) when the entire 2^k treatment combinations cannot be applied under homogeneous conditions. The disadvantage with this experimental setup is that certain effects are completely sacrificed owing to the blocking, the amount of sacrifice depending on the number of blocks required. For example, suppose that the eight treatment combinations in a 2^3 factorial experiment must be run in two blocks of size 4. One possible arrangement is given by

Block 1	Block 2
(1)	a
ab	b
ac	c
bc	abc

If one assumes the usual model with the additive block effect, this effect cancels out in the formation of the contrasts on all effects except ABC. To illustrate, let x denote the contribution to the yield owing to the difference between blocks. Writing the yields as

Block 1	*Block 2*
(1)	$a + x$
ab	$b + x$
ac	$c + x$
bc	$abc + x$

we see that the ABC contrast and also the contrast comparing the two blocks are both given by

$$ABC \text{ contrast} = (abc + x) + (c + x) + (b + x) + (a + x) - (1) - ab$$
$$- ac - bc$$
$$= abc + a + b + c - (1) - ab - ac - bc + 4x.$$

Therefore we are measuring the ABC effect plus the block effect and there is no way of assessing the ABC interaction effect independent of blocks. We say then that the ABC interaction is *completely confounded with blocks*. By necessity, information on ABC has been sacrificed. On the other hand, the block effect cancels out in the formation of all other contrasts. For example, the A contrast is given by

$$A \text{ contrast} = (abc + x) + (a + x) + ab + ac - (b + x) - (c + x) - bc$$
$$- (1)$$
$$= abc + a + ab + ac - b - c - bc - (1),$$

as in the case of a completely randomized design. We say that the effects A, B, C, AB, AC, and BC are orthogonal to blocks. Generally, for a 2^k factorial experiment in 2^p blocks, the number of effects confounded with blocks is 2^{p-1}, which is equivalent to the degrees of freedom for blocks.

When two blocks are to be used with a 2^k factorial, one effect, usually a high-order interaction, is chosen as the *defining contrast*. This effect is to be confounded with blocks. The additional $2^k - 2$ effects are orthogonal with the defining contrast and thus with blocks.

Suppose we represent the defining contrast as $A^{\gamma_1}B^{\gamma_2}C^{\gamma_3} \ldots$, where γ_i is either zero or 1. This generates the expression

$$L = \gamma_1 + \gamma_2 + \cdots + \gamma_k,$$

which is evaluated for each of the 2^k treatment combinations. The L values are then reduced (modulo 2) to either zero or 1 and thereby determine to which block the treatment combinations are assigned. In other words, the treatment combinations are divided into two blocks according to whether the L values leave a remainder of zero or 1 when divided by 2.

Example 13.2 Determine the values of L (modulo 2) for a 2^3 factorial experiment when the defining contrast is ABC.

Solution. With ABC the defining contrast, we have

$$L = \gamma_1 + \gamma_2 + \gamma_3,$$

which is applied to each treatment combination as follows:

$$
\begin{aligned}
(1): \quad & L = 0 + 0 + 0 = 0 = 0 && \text{(modulo 2)} \\
a: \quad & L = 1 + 0 + 0 = 1 = 1 && \text{(modulo 2)} \\
b: \quad & L = 0 + 1 + 0 = 1 = 1 && \text{(modulo 2)} \\
ab: \quad & L = 1 + 1 + 0 = 2 = 0 && \text{(modulo 2)} \\
c: \quad & L = 0 + 0 + 1 = 1 = 1 && \text{(modulo 2)} \\
ac: \quad & L = 1 + 0 + 1 = 2 = 0 && \text{(modulo 2)} \\
bc: \quad & L = 0 + 1 + 1 = 2 = 0 && \text{(modulo 2)} \\
abc: \quad & L = 1 + 1 + 1 = 3 = 1 && \text{(modulo 2).}
\end{aligned}
$$

The blocking arrangement, in which ABC is confounded, is given as before by

Block 1	Block 2
(1)	a
ab	b
ac	c
bc	abc

The A, B, C, AB, AC, and BC effects and sums of squares are computed in the usual way, ignoring blocks.

The block containing the treatment combination (1) in Example 13.2 is called the *principal block*. This block forms an algebraic group with respect to multiplication when the exponents are reduced to the modulo 2 base. For example, the property of closure holds since $(ab)(bc) = ab^2c = ac$, $(ab)(ab) = a^2b^2 = (1)$, and so forth.

If the experimenter is required to allocate the treatment combinations to four blocks, two defining contrasts are chosen by the experimenter. A third effect known as their *generalized interaction* is automatically confounded with blocks, these three effects corresponding to the three degrees of freedom for blocks. The procedure for constructing the design is best explained through an example. Suppose for a 2^4 factorial it is decided that AB and CD are the defining contrasts. The third effect confounded, their generalized interaction, is formed by multiplying together the initial two modulo 2. Thus the effect

$$(AB)(CD) = ABCD$$

is also confounded with blocks. We construct the design by calculating the expressions

$$L_1 = \gamma_1 + \gamma_2 \qquad (AB)$$
$$L_2 = \gamma_3 + \gamma_4 \qquad (CD)$$

modulo 2 for each of the 16 treatment combinations to generate the following blocking scheme:

Block 1	Block 2	Block 3	Block 4
(1)	a	c	ac
ab	b	abc	bc
cd	acd	d	ad
abcd	bcd	abd	bd
$L_1 = 0$	$L_1 = 1$	$L_1 = 0$	$L_1 = 1$
$L_2 = 0$	$L_2 = 0$	$L_2 = 1$	$L_2 = 1$

A shortcut procedure can be used to construct the remaining blocks after the principal block has been generated. We begin by placing any treatment combination not in the principal block in a second block and build the block by multiplying (modulo 2) by the treatment combinations in the principal block. In the preceding example the second, third, and fourth blocks are generated as follows:

Block 2	Block 3	Block 4
$a(1) = a$	$c(1) = c$	$ac(1) = ac$
$a(ab) = b$	$c(ab) = abc$	$ac(ab) = bc$
$a(cd) = acd$	$c(cd) = d$	$ac(cd) = ad$
$a(abcd) = bcd$	$c(abcd) = abd$	$ac(abcd) = bd$

The analysis for the case of four blocks is quite simple. All effects that are orthogonal to blocks (those other than the defining contrasts) are computed in the usual fashion. In fact Yates' technique can be used on the entire experiment, but the sums of squares for the three confounded effects are then added together to form the sum of squares due to blocks.

The general scheme for the 2^k factorial experiment in 2^p blocks is not difficult. We select p defining contrasts such that none is the generalized interaction of any two in the group. Since there are $2^p - 1$ degrees of freedom for blocks, we have $2^p - 1 - p$ additional effects confounded with blocks. For example, in a 2^6 factorial experiment in eight blocks, we might choose ACF, $BCDE$, and $ABDF$ as the defining contrasts. Then

$$(ACF)(BCDE) = ABDEF$$
$$(ACF)(ABDF) = BCD$$
$$(BCDE)(ABDF) = ACEF$$
$$(ACF)(BCDE)(ABDF) = E$$

are the additional four effects confounded with blocks. This is not a desirable blocking scheme since one of the confounded effects is the main effect E. The design is constructed by evaluating

$$L_1 = \gamma_1 + \gamma_3 + \gamma_6$$
$$L_2 = \gamma_2 + \gamma_3 + \gamma_4 + \gamma_5$$
$$L_3 = \gamma_1 + \gamma_2 + \gamma_4 + \gamma_6$$

and assigning treatment combinations to blocks according to the following scheme:

Block 1: $L_1 = 0$, $L_2 = 0$, $L_3 = 0$

Block 2: $L_1 = 0$, $L_2 = 0$, $L_3 = 1$

Block 3: $L_1 = 0$, $L_2 = 1$, $L_3 = 0$

Block 4: $L_1 = 0$, $L_2 = 1$, $L_3 = 1$

Block 5: $L_1 = 1$, $L_2 = 0$, $L_3 = 0$

Block 6: $L_1 = 1$, $L_2 = 0$, $L_3 = 1$

Block 7: $L_1 = 1$, $L_2 = 1$, $L_3 = 0$

Block 8: $L_1 = 1$, $L_2 = 1$, $L_3 = 1$.

The shortcut procedure that was illustrated for the case of four blocks also applies here. Hence we can construct the remaining seven blocks from the principal block.

Example 13.3 It is of interest to study the effect of five factors on some response with the assumption that interactions involving three, four, and five of the factors are negligible. We shall divide the 32 treatment combinations into four blocks using the defining contrasts $BCDE$ and $ABCD$. Thus $(BCDE)(ABCD) = AE$ is also confounded with blocks. The experimental design and the observations are given in Table 13.7.

Table 13.7 Data for a 2^5 Experiment in Four Blocks

Block 1	*Block 2*	*Block 3*	*Block 4*
(1) = 30.6	a = 32.4	b = 32.6	e = 30.7
bc = 31.5	abc = 32.4	c = 31.9	bce = 31.7
bd = 32.4	abd = 32.1	d = 33.3	bde = 32.2
cd = 31.5	acd = 35.3	bcd = 33.0	cde = 31.8
abe = 32.8	be = 31.5	ae = 32.0	ab = 32.0
ace = 32.1	ce = 32.7	abce = 33.1	ac = 33.1
ade = 32.4	de = 33.4	abde = 32.9	ad = 32.2
abcde = 31.8	bcde = 32.9	acde = 35.0	abcd = 32.3

The allocation of treatment combinations to experimental units within blocks is, of course, random. By pooling the unconfounded three, four, and five factor interactions to form the error term, perform the analysis of variance for the data of Table 13.7.

Solution. The sums of squares for each of the 31 contrasts are computed by Yates' method and the block sum of squares is found to be

$$SS(blocks) = SS(ABCD) + SS(BCDE) + SS(AE)$$
$$= 7.538.$$

The analysis of variance is given in Table 13.8. None of the two-factor interactions are significant at the $\alpha = 0.05$ level when compared to

Table 13.8 Analysis of Variance for the Data of Table 13.7

Source of variation	Sum of squares	Degrees of freedom	Mean square	Computed f
Main effect:				
A	3.251	1	3.251	6.32
B	0.320	1	0.320	0.62
C	1.361	1	1.361	2.64
D	4.061	1	4.061	7.89
E	0.005	1	0.005	0.01
Two-factor interaction:				
AB	1.531	1	1.531	2.97
AC	1.125	1	1.125	2.18
AD	0.320	1	0.320	0.62
BC	1.201	1	1.201	2.33
BD	1.711	1	1.711	3.32
BE	0.020	1	0.020	0.04
CD	0.045	1	0.045	0.09
CE	0.001	1	0.001	0.002
DE	0.001	1	0.001	0.002
Blocks (ABCD, BCDE, AE)	7.538	3	2.513	4.88
Error	7.208	14	0.515	

$f_{0.05}(1,14) = 4.60$. The main effects A and D are significant and both give positive effects on the response as we go from the low to the high level. Notice that the block effects are also significant when compared to $f_{0.05}(3,14) = 3.34$. However, there is no way to determine whether the significant block effects are due to actual differences in the blocks or perhaps due to a significant interaction that has been confounded with blocks.

13.5 Partial Confounding

It is possible to confound any effect with blocks by the methods described in Section 13.4. Suppose we consider a 2^3 factorial experiment in two blocks with three complete replications. If ABC is confounded with blocks in all three replicates, we can proceed as before and determine single-degree-of-freedom sums of squares for all main effects and two-factor interaction effects. The sum of squares for blocks has 5 degrees of freedom leaving $23 - 5 - 6 = 12$ degrees of freedom for error.

Now let us confound ABC in one replicate, AC in the second, and BC in the third. The plan for this type of experiment would be as follows:

Block			Block			Block	
1	*2*		*1*	*2*		*1*	*2*
abc	ab		abc	ab		abc	ab
a	ac		ac	bc		bc	ac
b	bc		b	a		a	b
c	(1)		(1)	c		(1)	c

Replicate 1	Replicate 2	Replicate 3
ABC confounded	AC confounded	BC confounded

The effects ABC, AC, and BC are said to be *partially* confounded with blocks. These three effects can be estimated from two of the three replicates. The ratio 2/3 serves as a measure of the extent of the confounding. Yates calls this ratio the *relative information* on the confounded effects. This ratio gives the amount of information available on the partially confounded effect relative to that available on an unconfounded effect.

The analysis-of-variance layout is given in Table 13.9. The sums of squares for blocks and for the unconfounded effects A, B, C, and AB are found in the usual way. The sum of squares for AC, BC, and ABC are computed from the two replicates in which the particular effect is not confounded. One must be careful to divide by 16 instead of 24 when obtaining the sums of squares for the partially confounded effects since we are only using 16 observations. In Table 13.9 the primes are inserted with the degrees of freedom as a reminder that these effects are partially confounded and require special calculations.

Table 13.9 Analysis of Variance with
Partial Confounding

Source of variation	Degrees of freedom
Blocks	5
A	1
B	1
C	1
AB	1
AC	1'
BC	1'
ABC	1'
Error	11
Total	23

13.6 Fractional Factorial Experiments

The 2^k factorial experiment can become quite demanding, in terms of the number of experimental units required, when k is large. One of the real advantages with this experimental plan is that it allows a degree of freedom for each interaction. However, in many experimental situations, it is known that certain interactions are negligible and thus it would be a waste of experimental effort to use the complete factorial experiment. In fact, the experimenter may have an economic constraint that disallows taking observations at all of the 2^k treatment combinations. When k is large we can often make use of a *fractional factorial experiment* in which perhaps one-half, one-fourth, or even one-eighth of the total factorial plan is actually carried out.

The construction of the half-replicate design is identical to the allocation of the 2^k factorial experiment into two blocks. We begin by selecting a defining contrast that is to be completely sacrificed. We then construct the two blocks accordingly and choose either of them as the experimental plan.

Consider a 2^4 factorial experiment in which we wish to use a half-replicate. The defining contrast $ABCD$ is chosen and thus an appropriate experimental plan would be to select the principal block consisting of the following treatment combinations:

$$\{(1), ab, ac, ad, bc, bd, cd, abcd\}$$

With this plan, we have contrasts on all effects except $ABCD$. Clearly

$$A \text{ contrast} = ab + ac + ad + abcd - (1) - bc - bd - cd$$
$$AB \text{ contrast} = abcd + ab + (1) + cd - ac - ad - bc - bd,$$

with similar expressions for the contrasts of the remaining main effects and interaction effects. However, with no more than 8 of the 16 observations in our fractional design, only 7 of the 14 unconfounded contrasts are orthogonal. Consider for example the CD contrast given by

$$CD \text{ contrast} = abcd + cd + (1) + ab - ac - ad - bc - bd.$$

Observe that this is also the single-degree-of-freedom contrast for AB. The word *aliases* is given to two factorial effects that have the same contrast. Therefore AB and CD are aliases. In the 2^k factorial experiments, the alias of any factorial effect is its generalized interaction with the defining contrast. For example, if $ABCD$ is the defining contrast, then the alias of A is $A(ABCD) = BCD$. It can be seen then that the complete alias structure in a half-replicate of a 2^4 factorial experiment, using $ABCD$ as the defining contrast, is (the symbol $\equiv$ implies *aliased with*)

$$A \equiv BCD$$
$$B \equiv ACD$$
$$C \equiv ABD$$
$$D \equiv ABC$$
$$AB \equiv CD$$
$$AC \equiv BD$$
$$AD \equiv BC.$$

Without supplementary statistical evidence, there is no way of explaining which of two aliased effects are actually providing the influence on the response. In a sense they *share a degree of freedom*. Herein lies the disadvantage in fractional factorial experiments. They have their greatest use when k is quite large and there is some a priori knowledge concerning the interactions. In the example presented, the main effects can be estimated if the three factor interactions are known to be negligible. For testing purposes the only possible procedure, in the absence of either an outside measure of experimental error or a replication of the experiment, would be to pool the sums of squares associated with the two-factor interactions. This, of course, is desirable only if these interactions represent negligible effects.

The construction of the 1/4 fraction or quarter-replicate is identical to the procedure whereby one assigns 2^k treatment combinations to four blocks. This involves the sacrificing of two defining contrasts along with their generalized interaction. Any of the four resulting blocks serves as an appropriate set of experimental runs. Each effect has three aliases, which are given by the generalized interaction with the three defining contrasts. Suppose in a 1/4 fraction of a 2^6 factorial experiment we use $ACEF$ and $BDEF$ as the defining contrast, resulting in

$$(ACEF)(BDEF) = ABCD$$

also being sacrificed. Using $L_1 = 0$, $L_2 = 0$ (modulo 2), where

$$L_1 = \gamma_1 + \gamma_3 + \gamma_5 + \gamma_6$$
$$L_2 = \gamma_2 + \gamma_4 + \gamma_5 + \gamma_6,$$

we have an appropriate set of experimental runs given by

$\{(1), \; abcd, \; ef, \; abcdef, \; cde, \; cdf, \; abe, \; abf, \; acef, \; bdef, \; ac, \; bd, \; adf, \; ade, \; bcf, \; bce\}$

and the alias structure for the main effects is written

$$A \equiv CEF \equiv ABDEF \equiv BCD$$
$$B \equiv ABCEF \equiv DEF \equiv ACD$$
$$C \equiv AEF \equiv BCDEF \equiv ABD$$
$$D \equiv ACDEF \equiv BEF \equiv ABC$$
$$E \equiv ACF \equiv BDF \equiv ABCDE$$
$$F \equiv ACE \equiv BDE \equiv ABCDF,$$

each with a single degree of freedom. For the two-factor interactions,

$$AB \equiv BCEF \equiv ADEF \equiv CD$$
$$AC \equiv EF \equiv ABCDEF \equiv BD$$
$$AD \equiv CDEF \equiv ABEF \equiv BC$$
$$AE \equiv CF \equiv ABDF \equiv BCDE$$
$$AF \equiv CE \equiv ABDE \equiv BCDF$$
$$BE \equiv ABCF \equiv DF \equiv ACDE$$
$$BF \equiv ABCE \equiv DE \equiv ACDF.$$

Here, of course, there is some aliasing among the two-factor interactions. The remaining two degrees of freedom are accounted for by the following groups:

$$ADF \equiv CDE \equiv ABE \equiv BCF$$
$$ABF \equiv BCE \equiv ADE \equiv CDF.$$

It becomes evident that one should always be aware of what the alias structure is for a fractional factorial experiment before he finally recommends the experimental plan. Proper choice of defining contrasts is important since it dictates the alias structure. For example, if one would like to study main effects and all two-factor interactions in an experiment involving eight factors and it is known that interactions involving three or more factors are negligible, a very practical design would be one in which the defining contrasts are $ACEGH$ and $BDEFGH$, resulting in a third, namely,

$$(ACEGH)(BDEFGH) = ABCDF.$$

All main effects and two-factor interactions are not aliased with one another and are therefore estimable. The analysis of variance would contain the following:

Main effects	8 single degrees of freedom
Two factor interactions	28 single degrees of freedom
Error	27 pooled degrees of freedom
Total	63 degrees of freedom

For the 1/8 and higher fractional factorials, the method of constructing the design generalizes. Of course, the aliasing can become quite extensive. For example, with a 1/8 fraction, each effect has seven aliases. The design is constructed by selecting three defining contrasts as if eight blocks were being constructed. Four additional effects are sacrificed and any one of the eight blocks can be properly used as the design.

13.7 Analysis of Fractional Factorial Experiments

The difficulty in making formal significance tests using data from fractional factorial experiments lies in the determination of the proper error term. Unless there are data available from prior experiments, the error must come from a pooling of contrasts representing effects that are presumed to be negligible.

Sums of squares for individual effects are found using essentially the same procedures given for the complete factorial. One can form a contrast in the treatment combinations by constructing the usual table of positive and negative signs. For example, for a half-replicate of a 2^3 factorial experiment, with ABC the defining contrast, one possible set of treatment combinations and the appropriate algebraic signs for each contrast used in computing the sums of squares for the various effects are presented in Table 13.10.

Note that in Table 13.10 the A and BC contrasts are identical, illustrating the aliasing. Also $B \equiv AC$ and $C \equiv AB$. In this situation we have three orthogonal contrasts representing the 3 degrees of freedom

Table 13.10 Signs for Contrasts in a Half-Replicate of
a 2^3 Factorial Experiment

Treatment combination	*Factorial effect*						
	A	*B*	*C*	*AB*	*AC*	*BC*	*ABC*
a	+	−	−	−	−	+	+
b	−	+	−	−	+	−	+
c	−	−	+	+	−	−	+
abc	+	+	+	+	+	+	+

available. If two observations are obtained for each of the four treatment combinations, we would then have an estimate of the error variance with 4 degrees of freedom. Assuming the interaction effects to be negligible we could test all the main effects for significance.

The sum of squares for any main effect, say A, is given by

$$\text{SSA} = \frac{(a - b - c + abc)^2}{2^2 n}.$$

In general, the single-degree-of-freedom sum of squares for any effect in a 2^{-p} fraction of a 2^k factorial experiment ($k > p$), is obtained by squaring contrasts in the treatment totals selected and dividing by $2^{k-p}n$, where n is the number of replications of these treatment combinations.

Example 13.4 Suppose we wish to use a half-replicate to study the effects of five factors, each at two levels, on some response and it is known that whatever the effect of each factor, it will be constant for each level of the other factors. Let the defining contrast be $ABCDE$, causing main effects to be aliased with four-factor interactions. The pooling of contrasts involving interactions provides $15 - 5 = 10$ degrees of freedom for error. Perform an analysis of variance on the following data, testing all main effects for significance at the 0.05 level:

Treatment	Response	Treatment	Response
a	11.3	bcd	14.1
b	15.6	abe	14.2
c	12.7	ace	11.7
d	10.4	ade	9.4
e	9.2	bce	16.2
abc	11.0	bde	13.9
abd	8.9	cde	14.7
acd	9.6	abcde	13.2

Solution. The sums of squares for the main effects are

$$\text{SSA} = \frac{(11.3 - 15.6 - \cdots - 14.7 + 13.2)^2}{2^{5-1}} = \frac{(-17.5)^2}{16} = 19.14$$

$$\text{SSB} = \frac{(-11.3 + 15.6 - \cdots - 14.7 + 13.2)^2}{2^{5-1}} = \frac{(18.1)^2}{16} = 20.48$$

$$\text{SSC} = \frac{(-11.3 - 15.6 + \cdots + 14.7 + 13.2)^2}{2^{5-1}} = \frac{(10.3)^2}{16} = 6.63$$

$$\text{SSD} = \frac{(-11.3 - 15.6 - \cdots + 14.7 + 13.2)^2}{2^{5-1}} = \frac{(-7.7)^2}{16} = 3.71$$

$$\text{SS(E)} = \frac{(-11.3 - 15.6 - \cdots + 14.7 + 13.2)^2}{2^{5-1}} = \frac{(8.9)^2}{16} = 4.95,$$

where the factor E is enclosed in parentheses to avoid confusion with the error sum of squares. The total sum of squares is

$$\text{SST} = 11.3^2 + 15.6^2 + \cdots + 13.2^2 - \frac{196.1^2}{16} = 85.74.$$

All other calculations and tests of significance are summarized in Table 13.11. The tests indicate that factor A has a significant negative effect

Table 13.11 Analysis of Variance for the Data of a
Half-Replicate of a 2^5 Factorial Experiment

Source of variation	Sum of squares	Degrees of freedom	Mean square	Computed f
Main effect				
A	19.14	1	19.14	6.21
B	20.48	1	20.48	6.65
C	6.63	1	6.63	2.15
D	3.71	1	3.71	1.20
E	4.95	1	4.95	1.61
Error	30.83	10	3.08	
Total	85.74	15		

on the response, while factor B has a significant positive effect. Factors C, D, and E are not significant at the 0.05 level.

EXERCISES

1. The following data were obtained from a 2^3 factorial experiment replicated three times:

Treatment combination	Replicate 1	Replicate 2	Replicate 3
(1)	12	19	10
a	15	20	16
b	24	16	17
ab	23	17	27
c	17	25	21
ac	16	19	19
bc	24	23	29
abc	28	25	20

Evaluate the sums of squares for all factorial effects by the contrast method.

2. The effects of four factors on some response are to be studied. Each factor is varied at two levels in a 2^4 factorial arrangement and the following data recorded:

Treatment combination	Response
(1)	23.8
a	19.6
b	29.9
ab	25.7
c	26.5
ac	22.6
bc	32.6
abc	28.6
d	21.6
ad	17.5
bd	27.5
abd	23.7
cd	24.6
acd	20.9
bcd	31.1
abcd	26.7

Assuming all three- and four-factor interactions to be negligible, analyze the given data by Yates' technique.

3. A preliminary experiment is conducted to study the effects of four factors and their interactions on the output of a certain machining operation. Two runs are made at each of the treatment combinations in order to supply a measure of pure experimental error. Two levels of each factor are used re-

sulting in the following data:

Treatment combination	Replicate 1	Replicate 2
(1)	7.9	9.6
a	9.1	10.2
b	8.6	5.8
c	10.4	12.0
d	7.1	8.3
ab	11.1	12.3
ac	16.4	15.5
ad	7.1	8.7
bc	12.6	15.2
bd	4.7	5.8
cd	7.4	10.9
abc	21.9	21.9
abd	9.8	7.8
acd	13.8	11.2
bcd	10.2	11.1
abcd	12.8	14.3

Use Yates' method to make tests on all main effects and interactions.

4. In a 2^3 factorial experiment with three replications, show the block arrangement and indicate by means of an analysis-of-variance table the effects to be tested and their degrees of freedom, when the AB interaction is confounded with blocks.

5. The following coded data represent the strength of a certain type of bread-wrapper stock produced under 16 different conditions, the latter representing two levels of each of four process variables. An operator effect was introduced into the model since it was necessary to obtain half the experimental runs under operator 1 and half under operator 2. It was felt that operators do have an effect on the quality of the product.

Operator 1	Operator 2
(1) = 18.8	a = 14.7
ab = 16.5	b = 15.1
ac = 17.8	c = 14.7
bc = 17.3	abc = 19.0
d = 13.5	ad = 16.9
abd = 17.6	bd = 17.5
acd = 18.5	cd = 18.2
bcd = 17.6	abcd = 20.1

(a) Assuming all interactions are negligible, make significance tests for the factors A, B, C, and D.

(b) What interaction is confounded with operators?

6. Divide the treatment combinations of a 2^4 factorial experiment into four blocks by confounding ABC and ABD. What additional effect is also confounded with blocks?

7. An experiment was conducted to determine the breaking strength of a certain alloy containing five different metals A, B, C, D, and E. Two different percentages of each metal were used in forming the $2^5 = 32$ different alloys. Since only eight alloys could be tested on a given day, the experiment was conducted over a period of 4 days in which the $ABDE$ and the AE effects were confounded with days. The experimental data were recorded as follows:

Treatment combination	Breaking strength	Treatment combination	Breaking strength
(1)	21.4	e	29.5
a	32.5	ae	31.3
b	28.1	be	33.0
ab	25.7	abe	23.7
c	34.2	ce	26.1
ac	34.0	ace	25.9
bc	23.5	bce	35.2
abc	24.7	$abce$	30.4
d	32.6	de	28.5
ad	29.0	ade	36.2
bd	30.1	bde	24.7
abd	27.3	$abde$	29.0
cd	22.0	cde	31.3
acd	35.8	$acde$	34.7
bcd	26.8	$bcde$	26.8
$abcd$	36.4	$abcde$	23.7

(a) Set up the blocking scheme for the 4 days.
(b) What additional effect is confounded with days?
(c) Use Yates' technique to obtain the sums of squares for all main effects.

8. By confounding ABC in two replicates and AB in the third, show the block arrangement and the analysis-of-variance table for a 2^3 factorial experiment with three replicates. What is the relative information on the confounded effects?

9. The following experiment was run to study main effects and all interactions. Four factors are used at two levels each. The experiment is replicated and two blocks are necessary in each replication. The data are as follows:

Replicate I		Replicate II	
Block 1	Block 2	Block 3	Block 4
(1) = 17.1	a = 15.5	(1) = 18.7	a = 17.0
d = 16.8	b = 14.8	ab = 18.6	b = 17.1
ab = 16.4	c = 16.2	ac = 18.5	c = 17.2
ac = 17.2	ad = 17.2	ad = 18.7	d = 17.6
bc = 16.8	bd = 18.3	bc = 18.9	abc = 17.5
abd = 18.1	cd = 17.3	bd = 17.0	abd = 18.3
acd = 19.1	abc = 17.7	cd = 18.7	acd = 18.4
bcd = 18.4	$abcd$ = 19.2	$abcd$ = 19.8	bcd = 18.3

(a) What effect is confounded with blocks in the first replication of the experiment? In the second replication?

(b) Conduct an appropriate analysis of variance showing tests on all main effects and interaction effects.

10. Construct a design involving 12 runs in which two factors are varied at two levels each. You are further restricted in that blocks of size 2 must be used and you must be able to make significance tests on both main effects and the interaction effect.

11. List the aliases for the various effects in a 2^5 factorial experiment when the defining contrast is $ACDE$.

12. Construct a 1/4 fraction of a 2^6 factorial design using $ABCD$ and $BDEF$ as the defining contrasts. Show what effects are aliased with the six main effects.

13. Show the blocking scheme for a 2^7 factorial experiment in eight blocks of size 16 each, using $ABCD$, $CDEFG$, and BDF as defining contrasts. Indicate what interactions are completely sacrificed in the experiment.

14. Seven factors are varied at two levels in an experiment involving only 16 trials. A 1/8 fraction of a 2^7 factorial experiment is used with the defining contrasts being ACD, BEF, and CEG. The data are as follows:

Treatment combination	Response
(1)	31.6
ad	28.7
abce	33.1
cdef	33.6
acef	33.7
bcde	34.2
abdf	32.5
bf	27.8
acg	31.1
cdg	32.0
beg	32.8
adefg	35.3
efg	32.4
abdeg	35.3
bcdfg	35.6
abcfg	35.1

Perform an analysis of variance on all seven main effects, assuming interactions are negligible.

REFERENCES

Anderson, R. L., and T. A. Bancroft, *Statistical Theory in Research*, New York: McGraw-Hill Book Co., Inc., 1952.

Bennett, C. A., and N. L. Franklin, *Statistical Analysis in Chemistry and the Chemical Industry*, New York: John Wiley & Sons, Inc., 1954.

Bowker, A. H., and G. J. Lieberman, *Engineering Statistics*, Englewood Cliffs, N.J.: Prentice-Hall Inc., 1959.

Brownlee, K. A., *Statistical Theory and Methodology in Science and Engineering*, 2nd ed., New York: John Wiley & Sons, Inc., 1965.

Chew, V., *Experimental Designs in Industry*, New York: John Wiley & Sons, Inc., 1958.

Cochran, W. G., "Some Consequences When the Assumptions for the Analysis of Variance Are Not Satisfied," *Biometrics*, Vol. 3, 1947.

Cochran, W. G., "Some Methods for Strengthening the Common Chi-Square Tests," *Biometrics*, Vol. 10, 1954.

Cochran, W. G., and G. M. Cox, *Experimental Designs*, 2nd ed., New York: John Wiley & Sons, Inc., 1957.

Davies, O. L., *The Design and Analysis of Industrial Experiments*, New York: Hafner Publishing Co., 1956.

Dixon, W. J., and F. J. Massey, Jr., *Introduction to Statistical Analysis*, 3rd ed., New York: McGraw-Hill Book Co., Inc., 1969.

Draper, N., and H. Smith, *Applied Regression Analysis*, New York: John Wiley & Sons, Inc., 1966.

Fehr, H. F., L. N. H. Bunt, and G. Grossman, *An Introduction to Sets, Probability and Hypothesis Testing*, Boston: D. C. Heath and Company, 1964.

Freund, J. E., *Mathematical Statistics*, Englewood Cliffs, N.J.: Prentice-Hall, Inc., 1962.

Guenther, W. C., *Analysis of Variance*, Englewood Cliffs, N.J.: Prentice-Hall, Inc., 1964.

Guenther, W. C., *Concepts of Statistical Inference*, New York: McGraw-Hill Book Co., Inc., 1965.

Guenther, W. C., *Concepts of Probability*, New York: McGraw-Hill Book Co., Inc., 1968.

Guttman, I., and S. S. Wilks, *Introductory Engineering Statistics*, New York: John Wiley & Sons, Inc., 1965.

Hicks, C. R., *Fundamental Concepts in the Design of Experiments*, New York: Holt, Rinehart & Winston, 1964.

Hodges, J. L., Jr., and E. L. Lehmann, *Basic Concepts of Probability and Statistics*, San Francisco: Holden-Day, Inc., 1964.

Hogg, R. V., and A. T. Craig, *Introduction to Mathematical Statistics*, 3rd ed., New York: The Macmillan Company, 1970.

Johnson, N. L., and F. C. Leone, *Statistics and Experimental Design: In Engineering and the Physical Sciences*, Vols. I & II, New York: John Wiley & Sons, Inc., 1964.

Kempthorne, O., *The Design and Analysis of Experiments*, New York: John Wiley & Sons, Inc., 1952.

Larson, H. J., *Introduction to Probability Theory and Statistical Inference*, New York: John Wiley & Sons, Inc., 1969.

Li, C. C., *Introduction to Experimental Statistics*, New York: McGraw-Hill Book Co., Inc., 1964.

Li, J. C. R., *Introduction to Statistical Inference*, Ann Arbor, Mich.: J. W. Edwards, Publisher, Inc., 1957.

Meyer, P. L., *Introductory Probability and Statistical Applications*, Reading, Mass.: Addison-Wesley Publishing Co., Inc., 1965.

Miller, I., and J. E. Freund, *Probability and Statistics for Engineers*, Englewood Cliffs, N.J.: Prentice-Hall, Inc., 1965.

Myers, R. H., *Response Surface Methodology*, Boston: Allyn and Bacon, Inc., 1971.

Ostle, B., *Statistics in Research*, 2nd ed., Ames: Iowa State University Press, 1964.

Pearson, E. S., and H. O. Hartley, *Biometrika Tables for Statisticians*, Vol. I, 3rd ed., Cambridge: Cambridge University Press, 1966.

Satterthwaite, F. E., "An Approximate Distribution of Estimates of Variance Components," *Biometrics*, Vol. 2, 1946.

Scheffé, H., *The Analysis of Variance*, New York: John Wiley & Sons, Inc., 1959.

Walpole, R. E., *Introduction to Statistics*, New York: The Macmillan Company, 1968.

Wine, R. L., *Statistics for Scientists and Engineers*, Englewood Cliffs, N.J.: Prentice-Hall, Inc., 1964.

STATISTICAL TABLES

Table I Squares and Square Roots

n	n^2	$\sqrt{n}$	$\sqrt{10n}$	n	n^2	$\sqrt{n}$	$\sqrt{10n}$
1.0	1.00	1.000	3.162	5.5	30.25	2.345	7.416
1.1	1.21	1.049	3.317	5.6	31.36	2.366	7.483
1.2	1.44	1.095	3.464	5.7	32.49	2.387	7.550
1.3	1.69	1.140	3.606	5.8	33.64	2.408	7.616
1.4	1.96	1.183	3.742	5.9	34.81	2.429	7.681
1.5	2.25	1.225	3.873	6.0	36.00	2.449	7.746
1.6	2.56	1.265	4.000	6.1	37.21	2.470	7.810
1.7	2.89	1.304	4.123	6.2	38.44	2.490	7.874
1.8	3.24	1.342	4.243	6.3	39.69	2.510	7.937
1.9	3.61	1.378	4.359	6.4	40.96	2.530	8.000
2.0	4.00	1.414	4.472	6.5	42.25	2.550	8.062
2.1	4.41	1.449	4.583	6.6	43.56	2.569	8.124
2.2	4.84	1.483	4.690	6.7	44.89	2.588	8.185
2.3	5.29	1.517	4.796	6.8	46.24	2.608	8.246
2.4	5.76	1.549	4.899	6.9	47.61	2.627	8.307
2.5	6.25	1.581	5.000	7.0	49.00	2.646	8.367
2.6	6.76	1.612	5.099	7.1	50.41	2.665	8.426
2.7	7.29	1.643	5.196	7.2	51.84	2.683	8.485
2.8	7.84	1.673	5.292	7.3	53.29	2.702	8.544
2.9	8.41	1.703	5.385	7.4	54.76	2.720	8.602
3.0	9.00	1.732	5.477	7.5	56.25	2.739	8.660
3.1	9.61	1.761	5.568	7.6	57.76	2.757	8.718
3.2	10.24	1.789	5.657	7.7	59.29	2.775	8.775
3.3	10.89	1.817	5.745	7.8	60.84	2.793	8.832
3.4	11.56	1.844	5.831	7.9	62.41	2.811	8.888
3.5	12.25	1.871	5.916	8.0	64.00	2.828	8.944
3.6	12.96	1.897	6.000	8.1	65.61	2.846	9.000
3.7	13.69	1.924	6.083	8.2	67.24	2.864	9.055
3.8	14.44	1.949	6.164	8.3	68.89	2.881	9.110
3.9	15.21	1.975	6.245	8.4	70.56	2.898	9.165
4.0	16.00	2.000	6.325	8.5	72.25	2.915	9.220
4.1	16.81	2.025	6.403	8.6	73.96	2.933	9.274
4.2	17.64	2.049	6.481	8.7	75.69	2.950	9.327
4.3	18.49	2.074	6.557	8.8	77.44	2.966	9.381
4.4	19.36	2.098	6.633	8.9	79.21	2.983	9.434
4.5	20.25	2.121	6.708	9.0	81.00	3.000	9.487
4.6	21.16	2.145	6.782	9.1	82.81	3.017	9.539
4.7	22.09	2.168	6.856	9.2	84.64	3.033	9.592
4.8	23.04	2.191	6.928	9.3	86.49	3.050	9.644
4.9	24.01	2.214	7.000	9.4	88.36	3.066	9.695
5.0	25.00	2.236	7.071	9.5	90.25	3.082	9.747
5.1	26.01	2.258	7.141	9.6	92.16	3.098	9.798
5.2	27.04	2.280	7.211	9.7	94.09	3.114	9.849
5.3	28.09	2.302	7.280	9.8	96.04	3.130	9.899
5.4	29.16	2.324	7.348	9.9	98.01	3.146	9.950

Table II Binomial Probability Sums $\sum_{x=0}^{r} b(x; n, p)$

		\multicolumn{10}{c}{p}									
n	r	0.10	0.20	0.25	0.30	0.40	0.50	0.60	0.70	0.80	0.90
5	0	0.5905	0.3277	0.2373	0.1681	0.0778	0.0312	0.0102	0.0024	0.0003	0.0000
	1	0.9185	0.7373	0.6328	0.5282	0.3370	0.1875	0.0870	0.0308	0.0067	0.0005
	2	0.9914	0.9421	0.8965	0.8369	0.6826	0.5000	0.3174	0.1631	0.0579	0.0086
	3	0.9995	0.9933	0.9844	0.9692	0.9130	0.8125	0.6630	0.4718	0.2627	0.0815
	4	1.0000	0.9997	0.9990	0.9976	0.9898	0.9688	0.9222	0.8319	0.6723	0.4095
	5	1.0000	1.0000	1.0000	1.0000	1.0000	1.0000	1.0000	1.0000	1.0000	1.0000
10	0	0.3487	0.1074	0.0563	0.0282	0.0060	0.0010	0.0001	0.0000	0.0000	0.0000
	1	0.7361	0.3758	0.2440	0.1493	0.0464	0.0107	0.0017	0.0001	0.0000	0.0000
	2	0.9298	0.6778	0.5256	0.3828	0.1673	0.0547	0.0123	0.0016	0.0001	0.0000
	3	0.9872	0.8791	0.7759	0.6496	0.3823	0.1719	0.0548	0.0106	0.0009	0.0000
	4	0.9984	0.9672	0.9219	0.8497	0.6331	0.3770	0.1662	0.0474	0.0064	0.0002
	5	0.9999	0.9936	0.9803	0.9527	0.8338	0.6230	0.3669	0.1503	0.0328	0.0016
	6	1.0000	0.9991	0.9965	0.9894	0.9452	0.8281	0.6177	0.3504	0.1209	0.0128
	7	1.0000	0.9999	0.9996	0.9984	0.9877	0.9453	0.8327	0.6172	0.3222	0.0702
	8	1.0000	1.0000	1.0000	0.9999	0.9983	0.9893	0.9536	0.8507	0.6242	0.2639
	9	1.0000	1.0000	1.0000	1.0000	0.9999	0.9990	0.9940	0.9718	0.8926	0.6513
	10	1.0000	1.0000	1.0000	1.0000	1.0000	1.0000	1.0000	1.0000	1.0000	1.0000
15	0	0.2059	0.0352	0.0134	0.0047	0.0005	0.0000	0.0000	0.0000	0.0000	0.0000
	1	0.5490	0.1671	0.0802	0.0353	0.0052	0.0005	0.0000	0.0000	0.0000	0.0000
	2	0.8159	0.3980	0.2361	0.1268	0.0271	0.0037	0.0003	0.0000	0.0000	0.0000
	3	0.9444	0.6482	0.4613	0.2969	0.0905	0.0176	0.0019	0.0001	0.0000	0.0000
	4	0.9873	0.8358	0.6865	0.5155	0.2173	0.0592	0.0094	0.0007	0.0000	0.0000
	5	0.9978	0.9389	0.8516	0.7216	0.4032	0.1509	0.0338	0.0037	0.0001	0.0000
	6	0.9997	0.9819	0.9434	0.8689	0.6098	0.3036	0.0951	0.0152	0.0008	0.0000
	7	1.0000	0.9958	0.9827	0.9500	0.7869	0.5000	0.2131	0.0500	0.0042	0.0003
	8	1.0000	0.9992	0.9958	0.9848	0.9050	0.6964	0.3902	0.1311	0.0181	0.0003
	9	1.0000	0.9999	0.9992	0.9963	0.9662	0.8491	0.5968	0.2784	0.0611	0.0023
	10	1.0000	1.0000	0.9999	0.9993	0.9907	0.9408	0.7827	0.4845	0.1642	0.0127
	11	1.0000	1.0000	1.0000	0.9999	0.9981	0.9824	0.9095	0.7031	0.3518	0.0556
	12	1.0000	1.0000	1.0000	1.0000	0.9997	0.9963	0.9729	0.8732	0.6020	0.1841
	13	1.0000	1.0000	1.0000	1.0000	1.0000	0.9995	0.9948	0.9647	0.8329	0.4510
	14	1.0000	1.0000	1.0000	1.0000	1.0000	1.0000	0.9995	0.9953	0.9648	0.7941
	15	1.0000	1.0000	1.0000	1.0000	1.0000	1.0000	1.0000	1.0000	1.0000	1.0000
20	0	0.1216	0.0115	0.0032	0.0008	0.0000	0.0000	0.0000	0.0000	0.0000	0.0000
	1	0.3917	0.0692	0.0243	0.0076	0.0005	0.0000	0.0000	0.0000	0.0000	0.0000
	2	0.6769	0.2061	0.0913	0.0355	0.0036	0.0002	0.0000	0.0000	0.0000	0.0000
	3	0.8670	0.4114	0.2252	0.1071	0.0160	0.0013	0.0001	0.0000	0.0000	0.0000
	4	0.9568	0.6296	0.4148	0.2375	0.0510	0.0059	0.0003	0.0000	0.0000	0.0000
	5	0.9887	0.8042	0.6172	0.4164	0.1256	0.0207	0.0016	0.0000	0.0000	0.0000
	6	0.9976	0.9133	0.7858	0.6080	0.2500	0.0577	0.0065	0.0003	0.0000	0.0000
	7	0.9996	0.9679	0.8982	0.7723	0.4159	0.1316	0.0210	0.0013	0.0000	0.0000
	8	0.9999	0.9900	0.9591	0.8867	0.5956	0.2517	0.0565	0.0051	0.0001	0.0000
	9	1.0000	0.9974	0.9861	0.9520	0.7553	0.4119	0.1275	0.0171	0.0006	0.0000
	10	1.0000	0.9994	0.9961	0.9829	0.8725	0.5881	0.2447	0.0480	0.0026	0.0000
	11	1.0000	0.9999	0.9991	0.9949	0.9435	0.7483	0.4044	0.1133	0.0100	0.0001
	12	1.0000	1.0000	0.9998	0.9987	0.9790	0.8684	0.5841	0.2277	0.0321	0.0004
	13	1.0000	1.0000	1.0000	0.9997	0.9935	0.9423	0.7500	0.3920	0.0867	0.0024
	14	1.0000	1.0000	1.0000	1.0000	0.9984	0.9793	0.8744	0.5836	0.1958	0.0113
	15	1.0000	1.0000	1.0000	1.0000	0.9997	0.9941	0.9490	0.7625	0.3704	0.0432
	16	1.0000	1.0000	1.0000	1.0000	1.0000	0.9987	0.9840	0.8929	0.5886	0.1330
	17	1.0000	1.0000	1.0000	1.0000	1.0000	0.9998	0.9964	0.9645	0.7939	0.3231
	18	1.0000	1.0000	1.0000	1.0000	1.0000	1.0000	0.9995	0.9924	0.9308	0.6083
	19	1.0000	1.0000	1.0000	1.0000	1.0000	1.0000	1.0000	0.9992	0.9885	0.8784
	20	1.0000	1.0000	1.0000	1.0000	1.0000	1.0000	1.0000	1.0000	1.0000	1.0000

Table III Poisson Probability Sums $\sum\limits_{x=0}^{r} p(x; \mu)$

r	\multicolumn{9}{c}{μ}								
	0.1	0.2	0.3	0.4	0.5	0.6	0.7	0.8	0.9
0	0.9048	0.8187	0.7408	0.6730	0.6065	0.5488	0.4966	0.4493	0.4066
1	0.9953	0.9825	0.9631	0.9384	0.9098	0.8781	0.8442	0.8088	0.7725
2	0.9998	0.9989	0.9964	0.9921	0.9856	0.9769	0.9659	0.9526	0.9371
3	1.0000	0.9999	0.9997	0.9992	0.9982	0.9966	0.9942	0.9909	0.9865
4		1.0000	1.0000	0.9999	0.9998	0.9996	0.9992	0.9986	0.9977
5				1.0000	1.0000	1.0000	0.9999	0.9998	0.9997
6							1.0000	1.0000	1.0000

r	\multicolumn{9}{c}{μ}								
	1.0	1.5	2.0	2.5	3.0	3.5	4.0	4.5	5.0
0	0.3679	0.2231	0.1353	0.0821	0.0498	0.0302	0.0183	0.0111	0.0067
1	0.7358	0.5578	0.4060	0.2873	0.1991	0.1359	0.0916	0.0611	0.0404
2	0.9197	0.8088	0.6767	0.5438	0.4232	0.3208	0.2381	0.1736	0.1247
3	0.9810	0.9344	0.8571	0.7576	0.6472	0.5366	0.4335	0.3423	0.2650
4	0.9963	0.9814	0.9473	0.8912	0.8153	0.7254	0.6288	0.5321	0.4405
5	0.9994	0.9955	0.9834	0.9580	0.9161	0.8576	0.7851	0.7029	0.6160
6	0.9999	0.9991	0.9955	0.9858	0.9665	0.9347	0.8893	0.8311	0.7622
7	1.0000	0.9998	0.9989	0.9958	0.9881	0.9733	0.9489	0.9134	0.8666
8		1.0000	0.9998	0.9989	0.9962	0.9901	0.9786	0.9597	0.9319
9			1.0000	0.9997	0.9989	0.9967	0.9919	0.9829	0.9682
10				0.9999	0.9997	0.9990	0.9972	0.9933	0.9863
11				1.0000	0.9999	0.9997	0.9991	0.9976	0.9945
12					1.0000	0.9999	0.9997	0.9992	0.9980
13						1.0000	0.9999	0.9997	0.9993
14							1.0000	0.9999	0.9998
15								1.0000	0.9999
16									1.0000

Table III Poisson Probability Sums $\sum_{x=0}^{r} p(x; \mu)$ (Continued)

	μ								
r	5.5	6.0	6.5	7.0	7.5	8.0	8.5	9.0	9.5
0	0.0041	0.0025	0.0015	0.0009	0.0006	0.0003	0.0002	0.0001	0.0001
1	0.0266	0.0174	0.0113	0.0073	0.0047	0.0030	0.0019	0.0012	0.0008
2	0.0884	0.0620	0.0430	0.0296	0.0203	0.0138	0.0093	0.0062	0.0042
3	0.2017	0.1512	0.1118	0.0818	0.0591	0.0424	0.0301	0.0212	0.0149
4	0.3575	0.2851	0.2237	0.1730	0.1321	0.0996	0.0744	0.0550	0.0403
5	0.5289	0.4457	0.3690	0.3007	0.2414	0.1912	0.1496	0.1157	0.0885
6	0.6860	0.6063	0.5265	0.4497	0.3782	0.3134	0.2562	0.2068	0.1649
7	0.8095	0.7440	0.6728	0.5987	0.5246	0.4530	0.3856	0.3239	0.2687
8	0.8944	0.8472	0.7916	0.7291	0.6620	0.5925	0.5231	0.4557	0.3918
9	0.9462	0.9161	0.8774	0.8305	0.7764	0.7166	0.6530	0.5874	0.5218
10	0.9747	0.9574	0.9332	0.9015	0.8622	0.8159	0.7634	0.7060	0.6453
11	0.9890	0.9799	0.9661	0.9466	0.9208	0.8881	0.8487	0.8030	0.7520
12	0.9955	0.9912	0.9840	0.9730	0.9573	0.9362	0.9091	0.8758	0.8364
13	0.9983	0.9964	0.9929	0.9872	0.9784	0.9658	0.9486	0.9261	0.8981
14	0.9994	0.9986	0.9970	0.9943	0.9897	0.9827	0.9726	0.9585	0.9400
15	0.9998	0.9995	0.9988	0.9976	0.9954	0.9918	0.9862	0.9780	0.9665
16	0.9999	0.9998	0.9996	0.9990	0.9980	0.9963	0.9934	0.9889	0.9823
17	1.0000	0.9999	0.9998	0.9996	0.9992	0.9984	0.9970	0.9947	0.9911
18		1.0000	0.9999	0.9999	0.9997	0.9994	0.9987	0.9976	0.9957
19			1.0000	1.0000	0.9999	0.9997	0.9995	0.9989	0.9980
20					1.0000	0.9999	0.9998	0.9996	0.9991
21						1.0000	0.9999	0.9998	0.9996
22							1.0000	0.9999	0.9999
23								1.0000	0.9999
24									1.0000

Table III Poisson Probability Sums $\sum_{x=0}^{r} p(x; \mu)$ (Continued)

r	μ 10.0	11.0	12.0	13.0	14.0	15.0	16.0	17.0	18.0
0	0.0000	0.0000	0.0000						
1	0.0005	0.0002	0.0001	0.0000	0.0000				
2	0.0028	0.0012	0.0005	0.0002	0.0001	0.0000	0.0000		
3	0.0103	0.0049	0.0023	0.0010	0.0005	0.0002	0.0001	0.0000	0.0000
4	0.0293	0.0151	0.0076	0.0037	0.0018	0.0009	0.0004	0.0002	0.0001
5	0.0671	0.0375	0.0203	0.0107	0.0055	0.0028	0.0014	0.0007	0.0003
6	0.1301	0.0786	0.0458	0.0259	0.0142	0.0076	0.0040	0.0021	0.0010
7	0.2202	0.1432	0.0895	0.0540	0.0316	0.0180	0.0100	0.0054	0.0029
8	0.3328	0.2320	0.1550	0.0998	0.0621	0.0374	0.0220	0.0126	0.0071
9	0.4579	0.3405	0.2424	0.1658	0.1094	0.0699	0.0433	0.0261	0.0154
10	0.5830	0.4599	0.3472	0.2517	0.1757	0.1185	0.0774	0.0491	0.0304
11	0.6968	0.5793	0.4616	0.3532	0.2600	0.1848	0.1270	0.0847	0.0549
12	0.7916	0.6887	0.5760	0.4631	0.3585	0.2676	0.1931	0.1350	0.0917
13	0.8645	0.7813	0.6815	0.5730	0.4644	0.3632	0.2745	0.2009	0.1426
14	0.9165	0.8540	0.7720	0.6751	0.5704	0.4657	0.3675	0.2808	0.2081
15	0.9513	0.9074	0.8444	0.7636	0.6694	0.5681	0.4667	0.3715	0.2867
16	0.9730	0.9441	0.8987	0.8355	0.7559	0.6641	0.5660	0.4677	0.3750
17	0.9857	0.9678	0.9370	0.8905	0.8272	0.7489	0.6593	0.5640	0.4686
18	0.9928	0.9823	0.9626	0.9302	0.8826	0.8195	0.7423	0.6550	0.5622
19	0.9965	0.9907	0.9787	0.9573	0.9235	0.8752	0.8122	0.7363	0.6509
20	0.9984	0.9953	0.9884	0.9750	0.9521	0.9170	0.8682	0.8055	0.7307
21	0.9993	0.9977	0.9939	0.9859	0.9712	0.9469	0.9108	0.8615	0.7991
22	0.9997	0.9990	0.9970	0.9924	0.9833	0.9673	0.9418	0.9047	0.8551
23	0.9999	0.9995	0.9985	0.9960	0.9907	0.9805	0.9633	0.9367	0.8989
24	1.0000	0.9998	0.9993	0.9980	0.9950	0.9888	0.9777	0.9594	0.9317
25		0.9999	0.9997	0.9990	0.9974	0.9938	0.9869	0.9748	0.9554
26		1.0000	0.9999	0.9995	0.9987	0.9967	0.9925	0.9848	0.9718
27			0.9999	0.9998	0.9994	0.9983	0.9959	0.9912	0.9827
28			1.0000	0.9999	0.9997	0.9991	0.9978	0.9950	0.9897
29				1.0000	0.9999	0.9996	0.9989	0.9973	0.9941
30					0.9999	0.9998	0.9994	0.9986	0.9967
31					1.0000	0.9999	0.9997	0.9993	0.9982
32						1.0000	0.9999	0.9996	0.9990
33							0.9999	0.9998	0.9995
34							1.0000	0.9999	0.9998
35								1.0000	0.9999
36									0.9999
37									1.0000

Table IV
Areas Under the Normal Curve

Area

z	0.00	0.01	0.02	0.03	0.04	0.05	0.06	0.07	0.08	0.09
−3.4	0.0003	0.0003	0.0003	0.0003	0.0003	0.0003	0.0003	0.0003	0.0003	0.0002
−3.3	0.0005	0.0005	0.0005	0.0004	0.0004	0.0004	0.0004	0.0004	0.0004	0.0003
−3.2	0.0007	0.0007	0.0006	0.0006	0.0006	0.0006	0.0006	0.0005	0.0005	0.0005
−3.1	0.0010	0.0009	0.0009	0.0009	0.0008	0.0008	0.0008	0.0008	0.0007	0.0007
−3.0	0.0013	0.0013	0.0013	0.0012	0.0012	0.0011	0.0011	0.0011	0.0010	0.0010
−2.9	0.0019	0.0018	0.0017	0.0017	0.0016	0.0016	0.0015	0.0015	0.0014	0.0014
−2.8	0.0026	0.0025	0.0024	0.0023	0.0023	0.0022	0.0021	0.0021	0.0020	0.0019
−2.7	0.0035	0.0034	0.0033	0.0032	0.0031	0.0030	0.0029	0.0028	0.0027	0.0026
−2.6	0.0047	0.0045	0.0044	0.0043	0.0041	0.0040	0.0039	0.0038	0.0037	0.0036
−2.5	0.0062	0.0060	0.0059	0.0057	0.0055	0.0054	0.0052	0.0051	0.0049	0.0048
−2.4	0.0082	0.0080	0.0078	0.0075	0.0073	0.0071	0.0069	0.0068	0.0066	0.0064
−2.3	0.0107	0.0104	0.0102	0.0099	0.0096	0.0094	0.0091	0.0089	0.0087	0.0084
−2.2	0.0139	0.0136	0.0132	0.0129	0.0125	0.0122	0.0119	0.0116	0.0113	0.0110
−2.1	0.0179	0.0174	0.0170	0.0166	0.0162	0.0158	0.0154	0.0150	0.0146	0.0143
−2.0	0.0228	0.0222	0.0217	0.0212	0.0207	0.0202	0.0197	0.0192	0.0188	0.0183
−1.9	0.0287	0.0281	0.0274	0.0268	0.0262	0.0256	0.0250	0.0244	0.0239	0.0233
−1.8	0.0359	0.0352	0.0344	0.0336	0.0329	0.0322	0.0314	0.0307	0.0301	0.0294
−1.7	0.0446	0.0436	0.0427	0.0418	0.0409	0.0401	0.0392	0.0384	0.0375	0.0367
−1.6	0.0548	0.0537	0.0526	0.0516	0.0505	0.0495	0.0485	0.0475	0.0465	0.0455
−1.5	0.0668	0.0655	0.0643	0.0630	0.0618	0.0606	0.0594	0.0582	0.0571	0.0559
−1.4	0.0808	0.0793	0.0778	0.0764	0.0749	0.0735	0.0722	0.0708	0.0694	0.0681
−1.3	0.0968	0.0951	0.0934	0.0918	0.0901	0.0885	0.0869	0.0853	0.0838	0.0823
−1.2	0.1151	0.1131	0.1112	0.1093	0.1075	0.1056	0.1038	0.1020	0.1003	0.0985
−1.1	0.1357	0.1335	0.1314	0.1292	0.1271	0.1251	0.1230	0.1210	0.1190	0.1170
−1.0	0.1587	0.1562	0.1539	0.1515	0.1492	0.1469	0.1446	0.1423	0.1401	0.1379
−0.9	0.1841	0.1814	0.1788	0.1762	0.1736	0.1711	0.1685	0.1660	0.1635	0.1611
−0.8	0.2119	0.2090	0.2061	0.2033	0.2005	0.1977	0.1949	0.1922	0.1894	0.1867
−0.7	0.2420	0.2389	0.2358	0.2327	0.2296	0.2266	0.2236	0.2206	0.2177	0.2148
−0.6	0.2743	0.2709	0.2676	0.2643	0.2611	0.2578	0.2546	0.2514	0.2483	0.2451
−0.5	0.3085	0.3050	0.3015	0.2981	0.2946	0.2912	0.2877	0.2843	0.2810	0.2776
−0.4	0.3446	0.3409	0.3372	0.3336	0.3300	0.3264	0.3228	0.3192	0.3156	0.3121
−0.3	0.3821	0.3783	0.3745	0.3707	0.3669	0.3632	0.3594	0.3557	0.3520	0.3483
−0.2	0.4207	0.4168	0.4129	0.4090	0.4052	0.4013	0.3974	0.3936	0.3897	0.3859
−0.1	0.4602	0.4562	0.4522	0.4483	0.4443	0.4404	0.4364	0.4325	0.4286	0.4247
−0.0	0.5000	0.4960	0.4920	0.4880	0.4840	0.4801	0.4761	0.4721	0.4681	0.4641
0.0	0.5000	0.5040	0.5080	0.5120	0.5160	0.5199	0.5239	0.5279	0.5319	0.5359
0.1	0.5398	0.5438	0.5478	0.5517	0.5557	0.5596	0.5636	0.5675	0.5714	0.5753
0.2	0.5793	0.5832	0.5871	0.5910	0.5948	0.5987	0.6026	0.6064	0.6103	0.6141
0.3	0.6179	0.6217	0.6255	0.6293	0.6331	0.6368	0.6406	0.6443	0.6480	0.6517
0.4	0.6554	0.6591	0.6628	0.6664	0.6700	0.6736	0.6772	0.6808	0.6844	0.6879
0.5	0.6915	0.6950	0.6985	0.7019	0.7054	0.7088	0.7123	0.7157	0.7190	0.7224
0.6	0.7257	0.7291	0.7324	0.7357	0.7389	0.7422	0.7454	0.7486	0.7517	0.7549
0.7	0.7580	0.7611	0.7642	0.7673	0.7704	0.7734	0.7764	0.7794	0.7823	0.7852
0.8	0.7881	0.7910	0.7939	0.7967	0.7995	0.8023	0.8051	0.8078	0.8106	0.8133
0.9	0.8159	0.8186	0.8212	0.8238	0.8264	0.8289	0.8315	0.8340	0.8365	0.8389
1.0	0.8413	0.8438	0.8461	0.8485	0.8508	0.8531	0.8554	0.8577	0.8599	0.8621
1.1	0.8643	0.8665	0.8686	0.8708	0.8729	0.8749	0.8770	0.8790	0.8810	0.8830
1.2	0.8849	0.8869	0.8888	0.8907	0.8925	0.8944	0.8962	0.8980	0.8997	0.9015
1.3	0.9032	0.9049	0.9066	0.9082	0.9099	0.9115	0.9131	0.9147	0.9162	0.9177
1.4	0.9192	0.9207	0.9222	0.9236	0.9251	0.9265	0.9278	0.9292	0.9306	0.9319
1.5	0.9332	0.9345	0.9357	0.9370	0.9382	0.9394	0.9406	0.9418	0.9429	0.9441
1.6	0.9452	0.9463	0.9474	0.9484	0.9495	0.9505	0.9515	0.9525	0.9535	0.9545
1.7	0.9554	0.9564	0.9573	0.9582	0.9591	0.9599	0.9608	0.9616	0.9625	0.9633
1.8	0.9641	0.9649	0.9656	0.9664	0.9671	0.9678	0.9686	0.9693	0.9699	0.9706
1.9	0.9713	0.9719	0.9726	0.9732	0.9738	0.9744	0.9750	0.9756	0.9761	0.9767
2.0	0.9772	0.9778	0.9783	0.9788	0.9793	0.9798	0.9803	0.9808	0.9812	0.9817
2.1	0.9821	0.9826	0.9830	0.9834	0.9838	0.9842	0.9846	0.9850	0.9854	0.9857
2.2	0.9861	0.9864	0.9868	0.9871	0.9875	0.9878	0.9881	0.9884	0.9887	0.9890
2.3	0.9893	0.9896	0.9898	0.9901	0.9904	0.9906	0.9909	0.9911	0.9913	0.9916
2.4	0.9918	0.9920	0.9922	0.9925	0.9927	0.9929	0.9931	0.9932	0.9934	0.9936
2.5	0.9938	0.9940	0.9941	0.9943	0.9945	0.9946	0.9948	0.9949	0.9951	0.9952
2.6	0.9953	0.9955	0.9956	0.9957	0.9959	0.9960	0.9961	0.9962	0.9963	0.9964
2.7	0.9965	0.9966	0.9967	0.9968	0.9969	0.9970	0.9971	0.9972	0.9973	0.9974
2.8	0.9974	0.9975	0.9976	0.9977	0.9977	0.9978	0.9979	0.9979	0.9980	0.9981
2.9	0.9981	0.9982	0.9982	0.9983	0.9984	0.9984	0.9985	0.9985	0.9986	0.9986
3.0	0.9987	0.9987	0.9987	0.9988	0.9988	0.9989	0.9989	0.9989	0.9990	0.9990
3.1	0.9990	0.9991	0.9991	0.9991	0.9992	0.9992	0.9992	0.9992	0.9993	0.9993
3.2	0.9993	0.9993	0.9994	0.9994	0.9994	0.9994	0.9994	0.9995	0.9995	0.9995
3.3	0.9995	0.9995	0.9995	0.9996	0.9996	0.9996	0.9996	0.9996	0.9996	0.9997
3.4	0.9997	0.9997	0.9997	0.9997	0.9997	0.9997	0.9997	0.9997	0.9997	0.9998

Table V*
Critical Values of the t Distribution

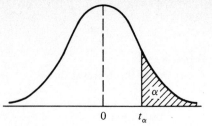

ν	α				
	0.10	0.05	0.025	0.01	0.005
1	3.078	6.314	12.706	31.821	63.657
2	1.886	2.920	4.303	6.965	9.925
3	1.638	2.353	3.182	4.541	5.841
4	1.533	2.132	2.776	3.747	4.604
5	1.476	2.015	2.571	3.365	4.032
6	1.440	1.943	2.447	3.143	3.707
7	1.415	1.895	2.365	2.998	3.499
8	1.397	1.860	2.306	2.896	3.355
9	1.383	1.833	2.262	2.821	3.250
10	1.372	1.812	2.228	2.764	3.169
11	1.363	1.796	2.201	2.718	3.106
12	1.356	1.782	2.179	2.681	3.055
13	1.350	1.771	2.160	2.650	3.012
14	1.345	1.761	2.145	2.624	2.977
15	1.341	1.753	2.131	2.602	2.947
16	1.337	1.746	2.120	2.583	2.921
17	1.333	1.740	2.110	2.567	2.898
18	1.330	1.734	2.101	2.552	2.878
19	1.328	1.729	2.093	2.539	2.861
20	1.325	1.725	2.086	2.528	2.845
21	1.323	1.721	2.080	2.518	2.831
22	1.321	1.717	2.074	2.508	2.819
23	1.319	1.714	2.069	2.500	2.807
24	1.318	1.711	2.064	2.492	2.797
25	1.316	1.708	2.060	2.485	2.787
26	1.315	1.706	2.056	2.479	2.779
27	1.314	1.703	2.052	2.473	2.771
28	1.313	1.701	2.048	2.467	2.763
29	1.311	1.699	2.045	2.462	2.756
inf.	1.282	1.645	1.960	2.326	2.576

* From Table IV of R. A. Fisher, *Statistical Methods for Research Workers*, published by Oliver & Boyd Ltd., Edinburgh, by permission of the author and publishers.

Table VI* Critical Values of the Chi-Square Distribution

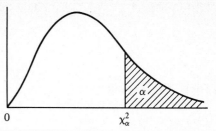

ν	α							
	0.995	0.99	0.975	0.95	0.05	0.025	0.01	0.005
1	0.0^4393	0.0^3157	0.0^3982	0.0^2393	3.841	5.024	6.635	7.879
2	0.0100	0.0201	0.0506	0.103	5.991	7.378	9.210	10.597
3	0.0717	0.115	0.216	0.352	7.815	9.348	11.345	12.838
4	0.207	0.297	0.484	0.711	9.488	11.143	13.277	14.860
5	0.412	0.554	0.831	1.145	11.070	12.832	15.086	16.750
6	0.676	0.872	1.237	1.635	12.592	14.449	16.812	18.548
7	0.989	1.239	1.690	2.167	14.067	16.013	18.475	20.278
8	1.344	1.646	2.180	2.733	15.507	17.535	20.090	21.955
9	1.735	2.088	2.700	3.325	16.919	19.023	21.666	23.589
10	2.156	2.558	3.247	3.940	18.307	20.483	23.209	25.188
11	2.603	3.053	3.816	4.575	19.675	21.920	24.725	26.757
12	3.074	3.571	4.404	5.226	21.026	23.337	26.217	28.300
13	3.565	4.107	5.009	5.892	22.362	24.736	27.688	29.819
14	4.075	4.660	5.629	6.571	23.685	26.119	29.141	31.319
15	4.601	5.229	6.262	7.261	24.996	27.488	30.578	32.801
16	5.142	5.812	6.908	7.962	26.296	28.845	32.000	34.267
17	5.697	6.408	7.564	8.672	27.587	30.191	33.409	35.718
18	6.265	7.015	8.231	9.390	28.869	31.526	34.805	37.156
19	6.844	7.633	8.907	10.117	30.144	32.852	36.191	38.582
20	7.434	8.260	9.591	10.851	31.410	34.170	37.566	39.997
21	8.034	8.897	10.283	11.591	32.671	35.479	38.932	41.401
22	8.643	9.542	10.982	12.338	33.924	36.781	40.289	42.796
23	9.260	10.196	11.689	13.091	35.172	38.076	41.638	44.181
24	9.886	10.856	12.401	13.848	36.415	39.364	42.980	45.558
25	10.520	11.524	13.120	14.611	37.652	40.646	44.314	46.928
26	11.160	12.198	13.844	15.379	38.885	41.923	45.642	48.290
27	11.808	12.879	14.573	16.151	40.113	43.194	46.963	49.645
28	12.461	13.565	15.308	16.928	41.337	44.461	48.278	50.993
29	13.121	14.256	16.047	17.708	42.557	45.722	49.588	52.336
30	13.787	14.953	16.791	18.493	43.773	46.979	50.892	53.672

* Abridged from Table 8 of *Biometrika Tables for Statisticians*, Vol. I, by permission of E. S. Pearson and the Biometrika Trustees.

Table VII* Critical
Values of the F Distribution

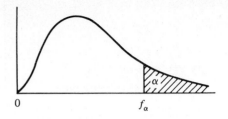

$$f_{0.05}(\nu_1, \nu_2)$$

ν_2	ν_1								
	1	2	3	4	5	6	7	8	9
1	161.4	199.5	215.7	224.6	230.2	234.0	236.8	238.9	240.5
2	18.51	19.00	19.16	19.25	19.30	19.33	19.35	19.37	19.38
3	10.13	9.55	9.28	9.12	9.01	8.94	8.89	8.85	8.81
4	7.71	6.94	6.59	6.39	6.26	6.16	6.09	6.04	6.00
5	6.61	5.79	5.41	5.19	5.05	4.95	4.88	4.82	4.77
6	5.99	5.14	4.76	4.53	4.39	4.28	4.21	4.15	4.10
7	5.59	4.74	4.35	4.12	3.97	3.87	3.79	3.73	3.68
8	5.32	4.46	4.07	3.84	3.69	3.58	3.50	3.44	3.39
9	5.12	4.26	3.86	3.63	3.48	3.37	3.29	3.23	3.18
10	4.96	4.10	3.71	3.48	3.33	3.22	3.14	3.07	3.02
11	4.84	3.98	3.59	3.36	3.20	3.09	3.01	2.95	2.90
12	4.75	3.89	3.49	3.26	3.11	3.00	2.91	2.85	2.80
13	4.67	3.81	3.41	3.18	3.03	2.92	2.83	2.77	2.71
14	4.60	3.74	3.34	3.11	2.96	2.85	2.76	2.70	2.65
15	4.54	3.68	3.29	3.06	2.90	2.79	2.71	2.64	2.59
16	4.49	3.63	3.24	3.01	2.85	2.74	2.66	2.59	2.54
17	4.45	3.59	3.20	2.96	2.81	2.70	2.61	2.55	2.49
18	4.41	3.55	3.16	2.93	2.77	2.66	2.58	2.51	2.46
19	4.38	3.52	3.13	2.90	2.74	2.63	2.54	2.48	2.42
20	4.35	3.49	3.10	2.87	2.71	2.60	2.51	2.45	2.39
21	4.32	3.47	3.07	2.84	2.68	2.57	2.49	2.42	2.37
22	4.30	3.44	3.05	2.82	2.66	2.55	2.46	2.40	2.34
23	4.28	3.42	3.03	2.80	2.64	2.53	2.44	2.37	2.32
24	4.26	3.40	3.01	2.78	2.62	2.51	2.42	2.36	2.30
25	4.24	3.39	2.99	2.76	2.60	2.49	2.40	2.34	2.28
26	4.23	3.37	2.98	2.74	2.59	2.47	2.39	2.32	2.27
27	4.21	3.35	2.96	2.73	2.57	2.46	2.37	2.31	2.25
28	4.20	3.34	2.95	2.71	2.56	2.45	2.36	2.29	2.24
29	4.18	3.33	2.93	2.70	2.55	2.43	2.35	2.28	2.22
30	4.17	3.32	2.92	2.69	2.53	2.42	2.33	2.27	2.21
40	4.08	3.23	2.84	2.61	2.45	2.34	2.25	2.18	2.12
60	4.00	3.15	2.76	2.53	2.37	2.25	2.17	2.10	2.04
120	3.92	3.07	2.68	2.45	2.29	2.17	2.09	2.02	1.96
∞	3.84	3.00	2.60	2.37	2.21	2.10	2.01	1.94	1.88

Table VII Critical Values of the F Distribution (Continued)

$$f_{0.05}(v_1, v_2)$$

					v_1					
v_2	10	12	15	20	24	30	40	60	120	∞
1	241.9	243.9	245.9	248.0	249.1	250.1	251.1	252.2	253.3	254.3
2	19.40	19.41	19.43	19.45	19.45	19.46	19.47	19.48	19.49	19.50
3	8.79	8.74	8.70	8.66	8.64	8.62	8.59	8.57	8.55	8.53
4	5.96	5.91	5.86	5.80	5.77	5.75	5.72	5.69	5.66	5.63
5	4.74	4.68	4.62	4.56	4.53	4.50	4.46	4.43	4.40	4.36
6	4.06	4.00	3.94	3.87	3.84	3.81	3.77	3.74	3.70	3.67
7	3.64	3.57	3.51	3.44	3.41	3.38	3.34	3.30	3.27	3.23
8	3.35	3.28	3.22	3.15	3.12	3.08	3.04	3.01	2.97	2.93
9	3.14	3.07	3.01	2.94	2.90	2.86	2.83	2.79	2.75	2.71
10	2.98	2.91	2.85	2.77	2.74	2.70	2.66	2.62	2.58	2.54
11	2.85	2.79	2.72	2.65	2.61	2.57	2.53	2.49	2.45	2.40
12	2.75	2.69	2.62	2.54	2.51	2.47	2.43	2.38	2.34	2.30
13	2.67	2.60	2.53	2.46	2.42	2.38	2.34	2.30	2.25	2.21
14	2.60	2.53	2.46	2.39	2.35	2.31	2.27	2.22	2.18	2.13
15	2.54	2.48	2.40	2.33	2.29	2.25	2.20	2.16	2.11	2.07
16	2.49	2.42	2.35	2.28	2.24	2.19	2.15	2.11	2.06	2.01
17	2.45	2.38	2.31	2.23	2.19	2.15	2.10	2.06	2.01	1.96
18	2.41	2.34	2.27	2.19	2.15	2.11	2.06	2.02	1.97	1.92
19	2.38	2.31	2.23	2.16	2.11	2.07	2.03	1.98	1.93	1.88
20	2.35	2.28	2.20	2.12	2.08	2.04	1.99	1.95	1.90	1.84
21	2.32	2.25	2.18	2.10	2.05	2.01	1.96	1.92	1.87	1.81
22	2.30	2.23	2.15	2.07	2.03	1.98	1.94	1.89	1.84	1.78
23	2.27	2.20	2.13	2.05	2.01	1.96	1.91	1.86	1.81	1.76
24	2.25	2.18	2.11	2.03	1.98	1.94	1.89	1.84	1.79	1.73
25	2.24	2.16	2.09	2.01	1.96	1.92	1.87	1.82	1.77	1.71
26	2.22	2.15	2.07	1.99	1.95	1.90	1.85	1.80	1.75	1.69
27	2.20	2.13	2.06	1.97	1.93	1.88	1.84	1.79	1.73	1.67
28	2.19	2.12	2.04	1.96	1.91	1.87	1.82	1.77	1.71	1.65
29	2.18	2.10	2.03	1.94	1.90	1.85	1.81	1.75	1.70	1.64
30	2.16	2.09	2.01	1.93	1.89	1.84	1.79	1.74	1.68	1.62
40	2.08	2.00	1.92	1.84	1.79	1.74	1.69	1.64	1.58	1.51
60	1.99	1.92	1.84	1.75	1.70	1.65	1.59	1.53	1.47	1.39
120	1.91	1.83	1.75	1.66	1.61	1.55	1.50	1.43	1.35	1.25
∞	1.83	1.75	1.67	1.57	1.52	1.46	1.39	1.32	1.22	1.00

Table VII Critical Values of the F Distribution (Continued)

$$f_{0.01}(v_1, v_2)$$

v_2	\multicolumn{9}{c}{v_1}
	1	2	3	4	5	6	7	8	9
1	4052	4999.5	5403	5625	5764	5859	5928	5981	6022
2	98.50	99.00	99.17	99.25	99.30	99.33	99.36	99.37	99.39
3	34.12	30.82	29.46	28.71	28.24	27.91	27.67	27.49	27.35
4	21.20	18.00	16.69	15.98	15.52	15.21	14.98	14.80	14.66
5	16.26	13.27	12.06	11.39	10.97	10.67	10.46	10.29	10.16
6	13.75	10.92	9.78	9.15	8.75	8.47	8.26	8.10	7.98
7	12.25	9.55	8.45	7.85	7.46	7.19	6.99	6.84	6.72
8	11.26	8.65	7.59	7.01	6.63	6.37	6.18	6.03	5.91
9	10.56	8.02	6.99	6.42	6.06	5.80	5.61	5.47	5.35
10	10.04	7.56	6.55	5.99	5.64	5.39	5.20	5.06	4.94
11	9.65	7.21	6.22	5.67	5.32	5.07	4.89	4.74	4.63
12	9.33	6.93	5.95	5.41	5.06	4.82	4.64	4.50	4.39
13	9.07	6.70	5.74	5.21	4.86	4.62	4.44	4.30	4.19
14	8.86	6.51	5.56	5.04	4.69	4.46	4.28	4.14	4.03
15	8.68	6.36	5.42	4.89	4.56	4.32	4.14	4.00	3.89
16	8.53	6.23	5.29	4.77	4.44	4.20	4.03	3.89	3.78
17	8.40	6.11	5.18	4.67	4.34	4.10	3.93	3.79	3.68
18	8.29	6.01	5.09	4.58	4.25	4.01	3.84	3.71	3.60
19	8.18	5.93	5.01	4.50	4.17	3.94	3.77	3.63	3.52
20	8.10	5.85	4.94	4.43	4.10	3.87	3.70	3.56	3.46
21	8.02	5.78	4.87	4.37	4.04	3.81	3.64	3.51	3.40
22	7.95	5.72	4.82	4.31	3.99	3.76	3.59	3.45	3.35
23	7.88	5.66	4.76	4.26	3.94	3.71	3.54	3.41	3.30
24	7.82	5.61	4.72	4.22	3.90	3.67	3.50	3.36	3.26
25	7.77	5.57	4.68	4.18	3.85	3.63	3.46	3.32	3.22
26	7.72	5.53	4.64	4.14	3.82	3.59	3.42	3.29	3.18
27	7.68	5.49	4.60	4.11	3.78	3.56	3.39	3.26	3.15
28	7.64	5.45	4.57	4.07	3.75	3.53	3.36	3.23	3.12
29	7.60	5.42	4.54	4.04	3.73	3.50	3.33	3.20	3.09
30	7.56	5.39	4.51	4.02	3.70	3.47	3.30	3.17	3.07
40	7.31	5.18	4.31	3.83	3.51	3.29	3.12	2.99	2.89
60	7.08	4.98	4.13	3.65	3.34	3.12	2.95	2.82	2.72
120	6.85	4.79	3.95	3.48	3.17	2.96	2.79	2.66	2.56
∞	6.63	4.61	3.78	3.32	3.02	2.80	2.64	2.51	2.41

Table VII Critical Values of the F Distribution (Continued)

$$f_{0.01}(\nu_1, \nu_2)$$

ν_2	\multicolumn{10}{c}{ν_1}									
	10	12	15	20	24	30	40	60	120	∞
1	6056	6106	6157	6209	6235	6261	6287	6313	6339	6366
2	99.40	99.42	99.43	99.45	99.46	99.47	99.47	99.48	99.49	99.50
3	27.23	27.05	26.87	26.69	26.60	26.50	26.41	26.32	26.22	26.13
4	14.55	14.37	14.20	14.02	13.93	13.84	13.75	13.65	13.56	13.46
5	10.05	9.89	9.72	9.55	9.47	9.38	9.29	9.20	9.11	9.02
6	7.87	7.72	7.56	7.40	7.31	7.23	7.14	7.06	6.97	6.88
7	6.62	6.47	6.31	6.16	6.07	5.99	5.91	5.82	5.74	5.65
8	5.81	5.67	5.52	5.36	5.28	5.20	5.12	5.03	4.95	4.86
9	5.26	5.11	4.96	4.81	4.73	4.65	4.57	4.48	4.40	4.31
10	4.85	4.71	4.56	4.41	4.33	4.25	4.17	4.08	4.00	3.91
11	4.54	4.40	4.25	4.10	4.02	3.94	3.86	3.78	3.69	3.60
12	4.30	4.16	4.01	3.86	3.78	3.70	3.62	3.54	3.45	3.36
13	4.10	3.96	3.82	3.66	3.59	3.51	3.43	3.34	3.25	3.17
14	3.94	3.80	3.66	3.51	3.43	3.35	3.27	3.18	3.09	3.00
15	3.80	3.67	3.52	3.37	3.29	3.21	3.13	3.05	2.96	2.87
16	3.69	3.55	3.41	3.26	3.18	3.10	3.02	2.93	2.84	2.75
17	3.59	3.46	3.31	3.16	3.08	3.00	2.92	2.83	2.75	2.65
18	3.51	3.37	3.23	3.08	3.00	2.92	2.84	2.75	2.66	2.57
19	3.43	3.30	3.15	3.00	2.92	2.84	2.76	2.67	2.58	2.49
20	3.37	3.23	3.09	2.94	2.86	2.78	2.69	2.61	2.52	2.42
21	3.31	3.17	3.03	2.88	2.80	2.72	2.64	2.55	2.46	2.36
22	3.26	3.12	2.98	2.83	2.75	2.67	2.58	2.50	2.40	2.31
23	3.21	3.07	2.93	2.78	2.70	2.62	2.54	2.45	2.35	2.26
24	3.17	3.03	2.89	2.74	2.66	2.58	2.49	2.40	2.31	2.21
25	3.13	2.99	2.85	2.70	2.62	2.54	2.45	2.36	2.27	2.17
26	3.09	2.96	2.81	2.66	2.58	2.50	2.42	2.33	2.23	2.13
27	3.06	2.93	2.78	2.63	2.55	2.47	2.38	2.29	2.20	2.10
28	3.03	2.90	2.75	2.60	2.52	2.44	2.35	2.26	2.17	2.06
29	3.00	2.87	2.73	2.57	2.49	2.41	2.33	2.23	2.14	2.03
30	2.98	2.84	2.70	2.55	2.47	2.39	2.30	2.21	2.11	2.01
40	2.80	2.66	2.52	2.37	2.29	2.20	2.11	2.02	1.92	1.80
60	2.63	2.50	2.35	2.20	2.12	2.03	1.94	1.84	1.73	1.60
120	2.47	2.34	2.19	2.03	1.95	1.86	1.76	1.66	1.53	1.38
∞	2.32	2.18	2.04	1.88	1.79	1.70	1.59	1.47	1.32	1.00

Table VIII* $\Pr(U \leq u | H_0$ is true) in the Wilcoxon Two-Sample Test

$n_2 = 3$

		n_1	
u	1	2	3
0	0.250	0.100	0.050
1	0.500	0.200	0.100
2	0.750	0.400	0.200
3		0.600	0.350
4			0.500
5			0.650

$n_2 = 4$

			n_1	
u	1	2	3	4
0	0.200	0.067	0.028	0.014
1	0.400	0.133	0.057	0.029
2	0.600	0.267	0.114	0.057
3		0.400	0.200	0.100
4		0.600	0.314	0.171
5			0.429	0.243
6			0.571	0.343
7				0.443
8				0.557

$n_2 = 5$

			n_1		
u	1	2	3	4	5
0	0.167	0.047	0.018	0.008	0.004
1	0.333	0.095	0.036	0.016	0.008
2	0.500	0.190	0.071	0.032	0.016
3	0.667	0.286	0.125	0.056	0.028
4		0.429	0.196	0.095	0.048
5		0.571	0.286	0.143	0.075
6			0.393	0.206	0.111
7			0.500	0.278	0.155
8			0.607	0.365	0.210
9				0.452	0.274
10				0.548	0.345
11					0.421
12					0.500
13					0.579

* Reproduced from H. B. Mann and D. R. Whitney, "On a test of whether one of two random variables is stochastically larger than the other," *Ann. Math. Statist.*, vol. 18, pp. 52–54 (1947), by permission of the authors and the publisher.

	n_1							
u	1	2	3	4	5	6	7	8
0	0.111	0.022	0.006	0.002	0.001	0.000	0.000	0.000
1	0.222	0.044	0.012	0.004	0.002	0.001	0.000	0.000
2	0.333	0.089	0.024	0.008	0.003	0.001	0.001	0.000
3	0.444	0.133	0.042	0.014	0.005	0.002	0.001	0.001
4	0.556	0.200	0.067	0.024	0.009	0.004	0.002	0.001
5		0.267	0.097	0.036	0.015	0.006	0.003	0.001
6		0.356	0.139	0.055	0.023	0.010	0.005	0.002
7		0.444	0.188	0.077	0.033	0.015	0.007	0.003
8		0.556	0.248	0.107	0.047	0.021	0.010	0.005
9			0.315	0.141	0.064	0.030	0.014	0.007
10			0.387	0.184	0.085	0.041	0.020	0.010
11			0.461	0.230	0.111	0.054	0.027	0.014
12			0.539	0.285	0.142	0.071	0.036	0.019
13				0.341	0.177	0.091	0.047	0.025
14				0.404	0.217	0.114	0.060	0.032
15				0.467	0.262	0.141	0.076	0.041
16				0.533	0.311	0.172	0.095	0.052
17					0.362	0.207	0.116	0.065
18					0.416	0.245	0.140	0.080
19					0.472	0.286	0.168	0.097
20					0.528	0.331	0.198	0.117
21						0.377	0.232	0.139
22						0.426	0.268	0.164
23						0.475	0.306	0.191
24						0.525	0.347	0.221
25							0.389	0.253
26							0.433	0.287
27							0.478	0.323
28							0.522	0.360
29								0.399
30								0.439
31								0.480
32								0.520

$$n_2 = 6$$

u	n_1					
	1	2	3	4	5	6
0	0.143	0.036	0.012	0.005	0.002	0.001
1	0.286	0.071	0.024	0.010	0.004	0.002
2	0.428	0.143	0.048	0.019	0.009	0.004
3	0.571	0.214	0.083	0.033	0.015	0.008
4		0.321	0.131	0.057	0.026	0.013
5		0.429	0.190	0.086	0.041	0.021
6		0.571	0.274	0.129	0.063	0.032
7			0.357	0.176	0.089	0.047
8			0.452	0.238	0.123	0.066
9			0.548	0.305	0.165	0.090
10				0.381	0.214	0.120
11				0.457	0.268	0.155
12				0.545	0.331	0.197
13					0.396	0.242
14					0.465	0.294
15					0.535	0.350
16						0.409
17						0.469
18						0.531

$$n_2 = 7$$

u	n_1						
	1	2	3	4	5	6	7
0	0.125	0.028	0.008	0.003	0.001	0.001	0.000
1	0.250	0.056	0.017	0.006	0.003	0.001	0.001
2	0.375	0.111	0.033	0.012	0.005	0.002	0.001
3	0.500	0.167	0.058	0.021	0.009	0.004	0.002
4	0.625	0.250	0.092	0.036	0.015	0.007	0.003
5		0.333	0.133	0.055	0.024	0.011	0.006
6		0.444	0.192	0.082	0.037	0.017	0.009
7		0.556	0.258	0.115	0.053	0.026	0.013
8			0.333	0.158	0.074	0.037	0.019
9			0.417	0.206	0.101	0.051	0.027
10			0.500	0.264	0.134	0.069	0.036
11			0.583	0.324	0.172	0.090	0.049
12				0.394	0.216	0.117	0.064
13				0.464	0.265	0.147	0.082
14				0.538	0.319	0.183	0.104
15					0.378	0.223	0.130
16					0.438	0.267	0.159
17					0.500	0.314	0.191
18					0.562	0.365	0.228
19						0.418	0.267
20						0.473	0.310
21						0.527	0.355
22							0.402
23							0.451
24							0.500
25							0.549

Table IX* Critical Values of U in the Wilcoxon Two-Sample Test
One-Tailed Test at $\alpha = 0.001$ or Two-Tailed Test at $\alpha = 0.002$

n_1	\multicolumn{12}{c}{n_2}											
	9	10	11	12	13	14	15	16	17	18	19	20
1												
2												
3									0	0	0	0
4		0	0	0	1	1	1	2	2	3	3	3
5	1	1	2	2	3	3	4	5	5	6	7	7
6	2	3	4	4	5	6	7	8	9	10	11	12
7	3	5	6	7	8	9	10	11	13	14	15	16
8	5	6	8	9	11	12	14	15	17	18	20	21
9	7	8	10	12	14	15	17	19	21	23	25	26
10	8	10	12	14	17	19	21	23	25	27	29	32
11	10	12	15	17	20	22	24	27	29	32	34	37
12	12	14	17	20	23	25	28	31	34	37	40	42
13	14	17	20	23	26	29	32	35	38	42	45	48
14	15	19	22	25	29	32	36	39	43	46	50	54
15	17	21	24	28	32	36	40	43	47	51	55	59
16	19	23	27	31	35	39	43	48	52	56	60	65
17	21	25	29	34	38	43	47	52	57	61	66	70
18	23	27	32	37	42	46	51	56	61	66	71	76
19	25	29	34	40	45	50	55	60	66	71	77	82
20	26	32	37	42	48	54	59	65	70	76	82	88

One-Tailed Test at $\alpha = 0.01$ or Two-Tailed Test at $\alpha = 0.02$

n_1	\multicolumn{12}{c}{n_2}											
	9	10	11	12	13	14	15	16	17	18	19	20
1												
2					0	0	0	0	0	0	1	1
3	1	1	1	2	2	2	3	3	4	4	4	5
4	3	3	4	5	5	6	7	7	8	9	9	10
5	5	6	7	8	9	10	11	12	13	14	15	16
6	7	8	9	11	12	13	15	16	18	19	20	22
7	9	11	12	14	16	17	19	21	23	24	26	28
8	11	13	15	17	20	22	24	26	28	30	32	34
9	14	16	18	21	23	26	28	31	33	36	38	40
10	16	19	22	24	27	30	33	36	38	41	44	47
11	18	22	25	28	31	34	37	41	44	47	50	53
12	21	24	28	31	35	38	42	46	49	53	56	60
13	23	27	31	35	39	43	47	51	55	59	63	67
14	26	30	34	38	43	47	51	56	60	65	69	73
15	28	33	37	42	47	51	56	61	66	70	75	80
16	31	36	41	46	51	56	61	66	71	76	82	87
17	33	38	44	49	55	60	66	71	77	82	88	93
18	36	41	47	53	59	65	70	76	82	88	94	100
19	38	44	50	56	63	69	75	82	88	94	101	107
20	40	47	53	60	67	73	80	87	93	100	107	114

* Adapted and abridged from Tables 1, 3, 5, and 7 of D. Auble, "Extended tables for the Mann-Whitney statistic," *Bulletin of the Institute of Educational Research at Indiana University*, vol. 1, no. 2(1953), by permission of the director.

One-Tailed Test at $\alpha = 0.025$ or Two-Tailed Test at $\alpha = 0.05$

n_1	9	10	11	12	13	14	15	16	17	18	19	20
1												
2	0	0	0	1	1	1	1	1	2	2	2	2
3	2	3	3	4	4	5	5	6	6	7	7	8
4	4	5	6	7	8	9	10	11	11	12	13	13
5	7	8	9	11	12	13	14	15	17	18	19	20
6	10	11	13	14	16	17	19	21	22	24	25	27
7	12	14	16	18	20	22	24	26	28	30	32	34
8	15	17	19	22	24	26	29	31	34	36	38	41
9	17	20	23	26	28	31	34	37	39	42	45	48
10	20	23	26	29	33	36	39	42	45	48	52	55
11	23	26	30	33	37	40	44	47	51	55	58	62
12	26	29	33	37	41	45	49	53	57	61	65	69
13	28	33	37	41	45	50	54	59	63	67	72	76
14	31	36	40	45	50	55	59	64	67	74	78	83
15	34	39	44	49	54	59	64	70	75	80	85	90
16	37	42	47	53	59	64	70	75	81	86	92	98
17	39	45	51	57	63	67	75	81	87	93	99	105
18	42	48	55	61	67	74	80	86	93	99	106	112
19	45	52	58	65	72	78	85	92	99	106	113	119
20	48	55	62	69	76	83	90	98	105	112	119	127

One-Tailed Test at $\alpha = 0.05$ or Two-Tailed Test at $\alpha = 0.10$

n_1	9	10	11	12	13	14	15	16	17	18	19	20
1											0	0
2	1	1	1	2	2	2	3	3	3	4	4	4
3	3	4	5	5	6	7	7	8	9	9	10	11
4	6	7	8	9	10	11	12	14	15	16	17	18
5	9	11	12	13	15	16	18	19	20	22	23	25
6	12	14	16	17	19	21	23	25	26	28	30	32
7	15	17	19	21	24	26	28	30	33	35	37	39
8	18	20	23	26	28	31	33	36	39	41	44	47
9	21	24	27	30	33	36	39	42	45	48	51	54
10	24	27	31	34	37	41	44	48	51	55	58	62
11	27	31	34	38	42	46	50	54	57	61	65	69
12	30	34	38	42	47	51	55	60	64	68	72	77
13	33	37	42	47	51	56	61	65	70	75	80	84
14	36	41	46	51	56	61	66	71	77	82	87	92
15	39	44	50	55	61	66	72	77	83	88	94	100
16	42	48	54	60	65	71	77	83	89	95	101	107
17	45	51	57	64	70	77	83	89	96	102	109	115
18	48	55	61	68	75	82	88	95	102	109	116	123
19	51	58	65	72	80	87	94	101	109	116	123	130
20	54	62	69	77	84	92	100	107	115	123	130	138

Table X* Critical Values of W in the Wilcoxon Test for Paired Observations

n	One-sided $\alpha = 0.01$ Two-sided $\alpha = 0.02$	One-sided $\alpha = 0.025$ Two-sided $\alpha = 0.05$	One-sided $\alpha = 0.05$ Two-sided $\alpha = 0.10$
5			1
6		1	2
7	0	2	4
8	2	4	6
9	3	6	8
10	5	8	11
11	7	11	14
12	10	14	17
13	13	17	21
14	16	21	26
15	20	25	30
16	24	30	36
17	28	35	41
18	33	40	47
19	38	46	54
20	43	52	60
21	49	59	68
22	56	66	75
23	62	73	83
24	69	81	92
25	77	90	101
26	85	98	110
27	93	107	120
28	102	117	130
29	111	127	141
30	120	137	152

* Reproduced from F. Wilcoxon and R. A. Wilcox, *Some Rapid Approximate Statistical Procedures*, American Cyanamid Company, Pearl River, N.Y., 1964, by permission of the American Cyanamid Company.

Table XI* Critical Values for Cochran's Test

k \ n	2	3	4	5	6	7	8	9	10	11	17	37	145	∞
2	0.9999	0.9950	0.9794	0.9586	0.9373	0.9172	0.8988	0.8823	0.8674	0.8539	0.7949	0.7067	0.6062	0.5000
3	0.9933	0.9423	0.8831	0.8335	0.7933	0.7606	0.7335	0.7107	0.6912	0.6743	0.6059	0.5153	0.4230	0.3333
4	0.9676	0.8643	0.7814	0.7212	0.6761	0.6410	0.6129	0.5897	0.5702	0.5536	0.4884	0.4057	0.3251	0.2500
5	0.9279	0.7885	0.6957	0.6329	0.5875	0.5531	0.5259	0.5037	0.4854	0.4697	0.4094	0.3351	0.2644	0.2000
6	0.8828	0.7218	0.6258	0.5635	0.5195	0.4866	0.4608	0.4401	0.4229	0.4084	0.3529	0.2858	0.2229	0.1667
7	0.8376	0.6644	0.5685	0.5080	0.4659	0.4347	0.4105	0.3911	0.3751	0.3616	0.3105	0.2494	0.1929	0.1429
8	0.7945	0.6152	0.5209	0.4627	0.4226	0.3932	0.3704	0.3522	0.3373	0.3248	0.2779	0.2214	0.1700	0.1250
9	0.7544	0.5727	0.4810	0.4251	0.3870	0.3592	0.3378	0.3207	0.3067	0.2950	0.2514	0.1992	0.1521	0.1111
10	0.7175	0.5358	0.4469	0.3934	0.3572	0.3308	0.3106	0.2945	0.2813	0.2704	0.2297	0.1811	0.1376	0.1000
12	0.6528	0.4751	0.3919	0.3428	0.3099	0.2861	0.2680	0.2535	0.2419	0.2320	0.1961	0.1535	0.1157	0.0833
15	0.5747	0.4069	0.3317	0.2882	0.2593	0.2386	0.2228	0.2104	0.2002	0.1918	0.1612	0.1251	0.0934	0.0667
20	0.4799	0.3297	0.2654	0.2288	0.2048	0.1877	0.1748	0.1646	0.1567	0.1501	0.1248	0.0960	0.0709	0.0500
24	0.4247	0.2871	0.2295	0.1970	0.1759	0.1608	0.1495	0.1406	0.1338	0.1283	0.1060	0.0810	0.0595	0.0417
30	0.3632	0.2412	0.1913	0.1635	0.1454	0.1327	0.1232	0.1157	0.1100	0.1054	0.0867	0.0658	0.0480	0.0333
40	0.2940	0.1915	0.1508	0.1281	0.1135	0·1033	0.0957	0.0898	0.0853	0.0816	0.0668	0.0503	0.0363	0.0250
60	0.2151	0.1371	0.1069	0.0902	0.0796	0.0722	0.0668	0.0625	0.0594	0.0567	0.0461	0.0344	0.0245	0.0167
120	0.1225	0.0759	0.0585	0.0489	0.0429	0.0387	0.0357	0.0334	0.0316	0.0302	0.0242	0.0178	0.0125	0.0083
∞	0	0	0	0	0	0	0	0	0	0	0	0	0	0

* Reproduced with permission from C. Eisenhart, M. W. Hastay, W. A. Wallis, *Techniques of Statistical Analysis*, Chapter 15, McGraw-Hill Book Company, Inc., New York, 1947.

Table XI Critical Values for Cochran's Test (Continued)

k \ n	2	3	4	5	6	7	8	9	10	11	17	37	145	∞
2	0.9985	0.9750	0.9392	0.9057	0.8772	0.8534	0.8332	0.8159	0.8010	0.7880	0.7341	0.6602	0.5813	0.5000
3	0.9669	0.8709	0.7977	0.7457	0.7071	0.6771	0.6530	0.6333	0.6167	0.6025	0.5466	0.4748	0.4031	0.3333
4	0.9065	0.7679	0.6841	0.6287	0.5895	0.5598	0.5365	0.5175	0.5017	0.4884	0.4366	0.3720	0.3093	0.2500
5	0.8412	0.6838	0.5981	0.5441	0.5065	0.4783	0.4564	0.4387	0.4241	0.4118	0.3645	0.3066	0.2513	0.2000
6	0.7808	0.6161	0.5321	0.4803	0.4447	0.4184	0.3980	0.3817	0.3682	0.3568	0.3135	0.2612	0.2119	0.1667
7	0.7271	0.5612	0.4800	0.4307	0.3974	0.3726	0.3535	0.3384	0.3259	0.3154	0.2756	0.2278	0.1833	0.1429
8	0.6798	0.5157	0.4377	0.3910	0.3595	0.3362	0.3185	0.3043	0.2926	0.2829	0.2462	0.2022	0.1616	0.1250
9	0.6385	0.4775	0.4027	0.3584	0.3286	0.3067	0.2901	0.2768	0.2659	0.2568	0.2226	0.1820	0.1446	0.1111
10	6.6020	0.4450	0.3733	0.3311	0.3029	0.2823	0.2666	0.2541	0.2439	0.2353	0.2032	0.1655	0.1308	0.1000
12	0.5410	0.3924	0.3264	0.2880	0.2624	0.2439	0.2299	0.2187	0.2098	0.2020	0.1737	0.1403	0.1100	0.0833
15	0.4709	0.3346	0.2758	0.2419	0.2195	0.2034	0.1911	0.1815	0.1736	0.1671	0.1429	0.1144	0.0889	0.0667
20	0.3894	0.2705	0.2205	0.1921	0.1735	0.1602	0.1501	0.1422	0.1357	0.1303	0.1108	0.0879	0.0675	0.0500
24	0.3434	0.2354	0.1907	0.1656	0.1493	0.1374	0.1286	0.1216	0.1160	0.1113	0.0942	0.0743	0.0567	0.0417
30	0.2929	0.1980	0.1593	0.1377	0.1237	0.1137	0.1061	0.1002	0.0958	0.0921	0.0771	0.0604	0.0457	0.0333
40	0.2370	0.1576	0.1259	0.1082	0.0968	0.0887	0.0827	0.0780	0.0745	0.0713	0.0595	0.0462	0.0347	0.0250
60	0.1737	0.1131	0.0895	0.0765	0.0682	0.0623	0.0583	0.0552	0.0520	0.0497	0.0411	0.0316	0.0234	0.0167
120	0.0998	0.0632	0.0495	0.0419	0.0371	0.0337	0.0312	0.0292	0.0279	0.0266	0.0218	0.0165	0.0120	0.0083
∞	0	0	0	0	0	0	0	0	0	0	0	0	0	0

Table XII* Least Significant Studentized Ranges r_p
$\alpha = 0.05$

ν	2	3	4	5	6	7	8	9	10
					p				
1	17.97	17.97	17.97	17.97	17.97	17.97	17.97	17.97	17.97
2	6.085	6.085	6.085	6.085	6.085	6.085	6.085	6.085	6.085
3	4.501	4.516	4.516	4.516	4.516	4.516	4.516	4.516	4.516
4	3.927	4.013	4.033	4.033	4.033	4.033	4.033	4.033	4.033
5	3.635	3.749	3.797	3.814	3.814	3.814	3.814	3.814	3.814
6	3.461	3.587	3.649	3.680	3.694	3.697	3.697	3.697	3.697
7	3.344	3.477	3.548	3.588	3.611	3.622	3.626	3.626	3.626
8	3.261	3.399	3.475	3.521	3.549	3.566	3.575	3.579	3.579
9	3.199	3.339	3.420	3.470	3.502	3.523	3.536	3.544	3.547
10	3.151	3.293	3.376	3.430	3.465	3.489	3.505	3.516	3.522
11	3.113	3.256	3.342	3.397	3.435	3.462	3.480	3.493	3.501
12	3.082	3.225	3.313	3.370	3.410	3.439	3.459	3.474	3.484
13	3.055	3.200	3.289	3.348	3.389	3.419	3.442	3.458	3.470
14	3.033	3.178	3.268	3.329	3.372	3.403	3.426	3.444	3.457
15	3.014	3.160	3.250	3.312	3.356	3.389	3.413	3.432	3.446
16	2.998	3.144	3.235	3.298	3.343	3.376	3.402	3.422	3.437
17	2.984	3.130	3.222	3.285	3.331	3.366	3.392	3.412	3.429
18	2.971	3.118	3.210	3.274	3.321	3.356	3.383	3.405	3.421
19	2.960	3.107	3.199	3.264	3.311	3.347	3.375	3.397	3.415
20	2.950	3.097	3.190	3.255	3.303	3.339	3.368	3.391	3.409
24	2.919	3.066	3.160	3.226	3.276	3.315	3.345	3.370	3.390
30	2.888	3.035	3.131	3.199	3.250	3.290	3.322	3.349	3.371
40	2.858	3.006	3.102	3.171	3.224	3.266	3.300	3.328	3.352
60	2.829	2.976	3.073	3.143	3.198	3.241	3.277	3.307	3.333
120	2.800	2.947	3.045	3.116	3.172	3.217	3.254	3.287	3.314
∞	2.772	2.918	3.017	3.089	3.146	3.193	3.232	3.265	3.294

* Abridged from H. L. Harter, "Critical values for Duncan's new multiple range test," *Biometrics*, vol. 16, no. 4(1960), by permission of the author and the editor.

ν	p								
	2	3	4	5	6	7	8	9	10
1	90.03	90.03	90.03	90.03	90.03	90.03	90.03	90.03	90.03
2	14.04	14.04	14.04	14.04	14.04	14.04	14.04	14.04	14.04
3	8.261	8.321	8.321	8.321	8.321	8.321	8.321	8.321	8.321
4	6.512	6.677	6.740	6.756	6.756	6.756	6.756	6.756	6.756
5	5.702	5.893	5.989	6.040	6.065	6.074	6.074	6.074	6.074
6	5.243	5.439	5.549	5.614	5.655	5.680	5.694	5.701	5.703
7	4.949	5.145	5.260	5.334	5.383	5.416	5.439	5.454	5.464
8	4.746	4.939	5.057	5.135	5.189	5.227	5.256	5.276	5.291
9	4.596	4.787	4.906	4.986	5.043	5.086	5.118	5.142	5.160
10	4.482	4.671	4.790	4.871	4.931	4.975	5.010	5.037	5.058
11	4.392	4.579	4.697	4.780	4.841	4.887	4.924	4.952	4.975
12	4.320	4.504	4.622	4.706	4.767	4.815	4.852	4.883	4.907
13	4.260	4.442	4.560	4.644	4.706	4.755	4.793	4.824	4.850
14	4.210	4.391	4.508	4.591	4.654	4.704	4.743	4.775	4.802
15	4.168	4.347	4.463	4.547	4.610	4.660	4.700	4.733	4.760
16	4.131	4.309	4.425	4.509	4.572	4.622	4.663	4.696	4.724
17	4.099	4.275	4.391	4.475	4.539	4.589	4.630	4.664	4.693
18	4.071	4.246	4.362	4.445	4.509	4.560	4.601	4.635	4.664
19	4.046	4.220	4.335	4.419	4.483	4.534	4.575	4.610	4.639
20	4.024	4.197	4.312	4.395	4.459	4.510	4.552	4.587	4.617
24	3.956	4.126	4.239	4.322	4.386	4.437	4.480	4.516	4.546
30	3.889	4.056	4.168	4.250	4.314	4.366	4.409	4.445	4.477
40	3.825	3.988	4.098	4.180	4.244	4.296	4.339	4.376	4.408
60	3.762	3.922	4.031	4.111	4.174	4.226	4.270	4.307	4.340
120	3.702	3.858	3.965	4.044	4.107	4.158	4.202	4.239	4.272
∞	3.643	3.796	3.900	3.978	4.040	4.091	4.135	4.172	4.205

Table XIII* Sample Size for the t Test of the Mean

Value of $\Delta = \dfrac{\mu - \mu_0}{\sigma}$	Level of t-test																				
Single-sided test → Double-sided test →	$\alpha = 0.005$ / $\alpha = 0.01$					$\alpha = 0.01$ / $\alpha = 0.02$					$\alpha = 0.025$ / $\alpha = 0.05$					$\alpha = 0.05$ / $\alpha = 0.1$					
$\beta =$	0.01	0.05	0.1	0.2	0.5	0.01	0.05	0.1	0.2	0.5	0.01	0.05	0.1	0.2	0.5	0.01	0.05	0.1	0.2	0.5	
0.05																					0.05
0.10																					0.10
0.15																				122	0.15
0.20										139					99					70	0.20
0.25					110					90				128	64			139	101	45	0.25
0.30				134	78				115	63			119	90	45		122	97	71	32	0.30
0.35			125	99	58			109	85	47		109	88	67	34		90	72	52	24	0.35
0.40		115	97	77	45		101	85	66	37	117	84	68	51	26	101	70	55	40	19	0.40
0.45		92	77	62	37	110	81	68	53	30	93	67	54	41	21	80	55	44	33	15	0.45
0.50	100	75	63	51	30	90	66	55	43	25	76	54	44	34	18	65	45	36	27	13	0.50
0.55	83	63	53	42	26	75	55	46	36	21	63	45	37	28	15	54	38	30	22	11	0.55
0.60	71	53	45	36	22	63	47	39	31	18	53	38	32	24	13	46	32	26	19	9	0.60
0.65	61	46	39	31	20	55	41	34	27	16	46	33	27	21	12	39	28	22	17	8	0.65
0.70	53	40	34	28	17	47	35	30	24	14	40	29	24	19	10	34	24	19	15	8	0.70
0.75	47	36	30	25	16	42	31	27	21	13	35	26	21	16	9	30	21	17	13	7	0.75
0.80	41	32	27	22	14	37	28	24	19	12	31	22	19	15	9	27	19	15	12	6	0.80
0.85	37	29	24	20	13	33	25	21	17	11	28	21	17	13	8	24	17	14	11	6	0.85
0.90	34	26	22	18	12	29	23	19	16	10	25	19	16	12	7	21	15	13	10	5	0.90
0.95	31	24	20	17	11	27	21	18	14	9	23	17	14	11	7	19	14	11	9	5	0.95
1.00	28	22	19	16	10	25	19	16	13	9	21	16	13	10	6	18	13	11	8	5	1.00
1.1	24	19	16	14	9	21	16	14	12	8	18	13	11	9	6	15	11	9	7		1.1
1.2	21	16	14	12	8	18	14	12	10	7	15	12	10	8	5	13	10	8	6		1.2
1.3	18	15	13	11	8	16	13	11	9	6	14	10	9	7		11	8	7	6		1.3
1.4	16	13	12	10	7	14	11	10	9	6	12	9	8	7		10	8	7	5		1.4
1.5	15	12	11	9	7	13	10	9	8	6	11	8	7	6		9	7	6			1.5
1.6	13	11	10	8	6	12	10	9	7	5	10	8	7	6		8	6	6			1.6
1.7	12	10	9	8	6	11	9	8	7		9	7	6	5		8	6	5			1.7
1.8	12	10	9	8	6	10	8	7	7		8	7	6			7	6				1.8
1.9	11	9	8	7	6	10	8	7	6		8	6	6			7	5				1.9
2.0	10	8	8	7	5	9	7	7	6		7	6	5			6					2.0
2.1	10	8	7	7		8	7	6	6		7	6				6					2.1
2.2	9	8	7	6		8	7	6	5		7	6				6					2.2
2.3	9	7	7	6		8	6	6			6	5				5					2.3
2.4	8	7	7	6		7	6	6			6										2.4
2.5	8	7	6	6		7	6	6			6										2.5
3.0	7	6	6	5		6	5	5			5										3.0
3.5	6	5	5			5															3.5
4.0	6																				4.0

* Reproduced with permission from O. L. Davies, *Design and Analysis of Industrial Experiments*, Oliver & Boyd Ltd., Edinburgh, 1956.

Table XIV* Sample Size for the t Test of the Difference Between Two Means

	Level of t-test																				
Single-sided test	$\alpha = 0.005$					$\alpha = 0.01$					$\alpha = 0.025$					$\alpha = 0.05$					
Double-sided test	$\alpha = 0.01$					$\alpha = 0.02$					$\alpha = 0.05$					$\alpha = 0.1$					
$\beta =$	0.01	0.05	0.1	0.2	0.5	0.01	0.05	0.1	0.2	0.5	0.01	0.05	0.1	0.2	0.5	0.01	0.05	0.1	0.2	0.5	$\beta =$
0.05																					0.05
0.10																					0.10
0.15																					0.15
0.20																				137	0.20
0.25															124					88	0.25
0.30										123					87					61	0.30
0.35					110					90					64				102	45	0.35
0.40					85					70				100	50			108	78	35	0.40
0.45				118	68				101	55			105	79	39		108	86	62	28	0.45
0.50				96	55			106	82	45		106	86	64	32		88	70	51	23	0.50
0.55			101	79	46		106	88	68	38		87	71	53	27	112	73	58	42	19	0.55
0.60		101	85	67	39		90	74	58	32	104	74	60	45	23	89	61	49	36	16	0.60
0.65		87	73	57	34	104	77	64	49	27	88	63	51	39	20	76	52	42	30	14	0.65
0.70	100	75	63	50	29	90	66	55	43	24	76	55	44	34	17	66	45	36	26	12	0.70
0.75	88	66	55	44	26	79	58	48	38	21	67	48	39	29	15	57	40	32	23	11	0.75
0.80	77	58	49	39	23	70	51	43	33	19	59	42	34	26	14	50	35	28	21	10	0.80
0.85	69	51	43	35	21	62	46	38	30	17	52	37	31	23	12	45	31	25	18	9	0.85
0.90	62	46	39	31	19	55	41	34	27	15	47	34	27	21	11	40	28	22	16	8	0.90
0.95	55	42	35	28	17	50	37	31	24	14	42	30	25	19	10	36	25	20	15	7	0.95
1.00	50	38	32	26	15	45	33	28	22	13	38	27	23	17	9	33	23	18	14	7	1.00
1.1	42	32	27	22	13	38	28	23	19	11	32	23	19	14	8	27	19	15	12	6	1.1
1.2	36	27	23	18	11	32	24	20	16	9	27	20	16	12	7	23	16	13	10	5	1.2
1.3	31	23	20	16	10	28	21	17	14	8	23	17	14	11	6	20	14	11	9	5	1.3
1.4	27	20	17	14	9	24	18	15	12	8	20	15	12	10	6	17	12	10	8	4	1.4
1.5	24	18	15	13	8	21	16	14	11	7	18	13	11	9	5	15	11	9	7	4	1.5
1.6	21	16	14	11	7	19	14	12	10	6	16	12	10	8	5	14	10	8	6	4	1.6
1.7	19	15	13	10	7	17	13	11	9	6	14	11	9	7	4	12	9	7	6	3	1.7
1.8	17	13	11	10	6	15	12	10	8	5	13	10	8	6	4	11	8	7	5		1.8
1.9	16	12	11	9	6	14	11	9	8	5	12	9	7	6	4	10	7	6	5		1.9
2.0	14	11	10	8	6	13	10	9	7	5	11	8	7	6	4	9	7	6	4		2.0
2.1	13	10	9	8	5	12	9	8	7	5	10	8	6	5	3	8	6	5	4		2.1
2.2	12	10	8	7	5	11	9	7	6	4	9	7	6	5		8	6	5	4		2.2
2.3	11	9	8	7	5	10	8	7	6	4	9	7	6	5		7	5	5	4		2.3
2.4	11	9	8	6	5	10	8	7	6	4	8	6	5	4		7	5	4	4		2.4
2.5	10	8	7	6	4	9	7	6	5	4	8	6	5	4		6	5	4	3		2.5
3.0	8	6	6	5	4	7	6	5	4	3	6	5	4	4		5	4	3			3.0
3.5	6	5	5	4	3	6	5	4	4		5	4	4	3		4	3				3.5
4.0	6	5	4	4		5	4	4	3		4	4	3			4					4.0

Value of $\Delta = \dfrac{\mu_1 - \mu_2}{\sigma}$

* Reproduced with permission from O. L. Davies, *Design and Analysis of Industrial Experiments*, Oliver & Boyd Ltd., Edinburgh, 1956.

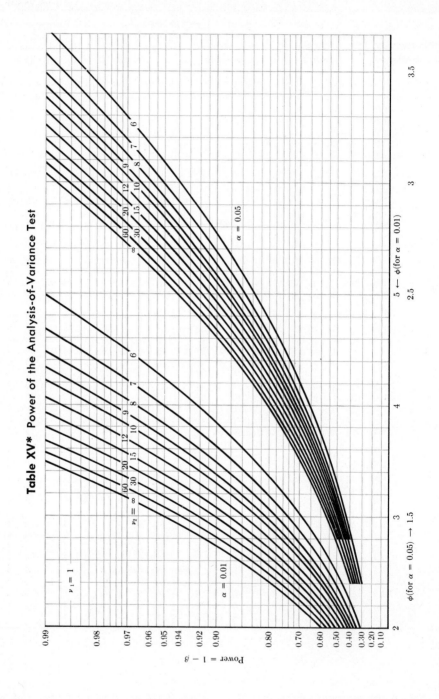

Table XV* Power of the Analysis-of-Variance Test

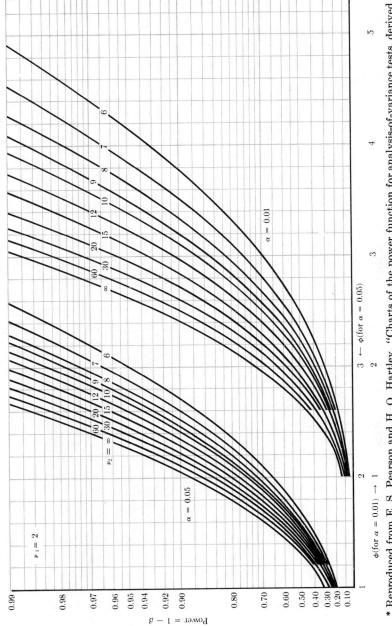

Power $= 1 - \beta$

$\nu_1 = 2$

$\alpha = 0.05$

$\alpha = 0.01$

ϕ(for $\alpha = 0.01$) $\rightarrow$ 1

ϕ(for $\alpha = 0.05$)

$\nu_2 =$

* Reproduced from E. S. Pearson and H. O. Hartley, "Charts of the power function for analysis-of-variance tests, derived from the non-central F distribution," *Biometrika*, vol. 38, p. 112(1951), by permission of the editor.

Table XV Power of the Analysis-of-Variance Test (Continued)

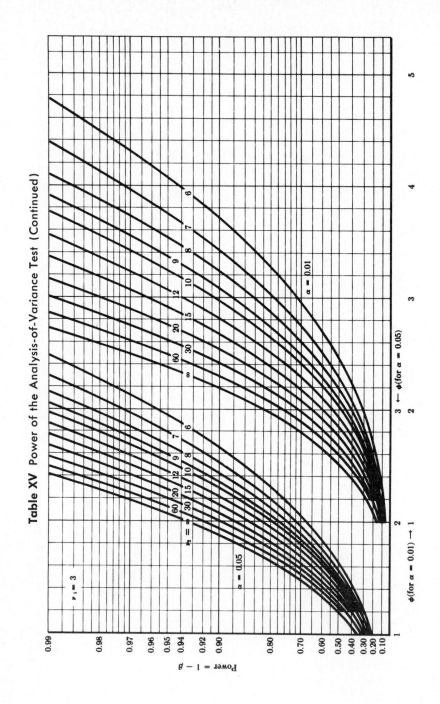

480

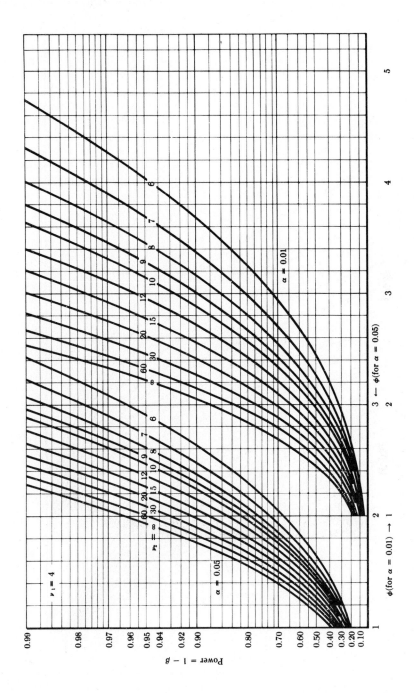

$\nu_1 = 4$

$\alpha = 0.05$

$\alpha = 0.01$

ϕ(for $\alpha = 0.01$) $\rightarrow$ 1

ϕ(for $\alpha = 0.05$)

Power = $1 - \beta$

Table XV Power of the Analysis-of-Variance Test (Continued)

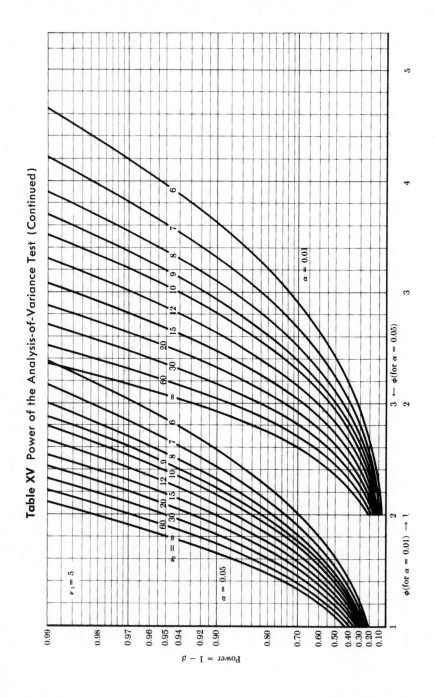

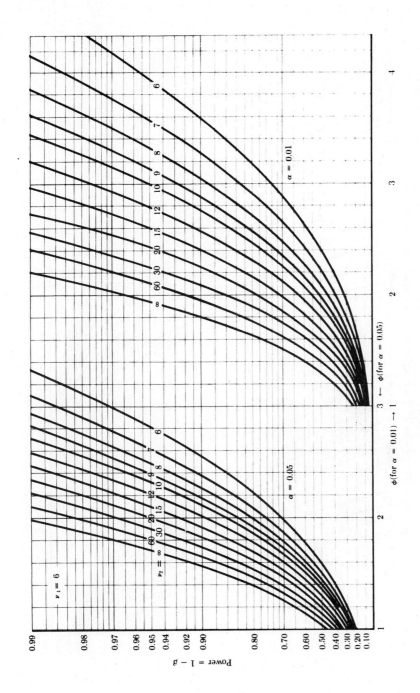

Table XV Power of the Analysis-of-Variance Test (Continued)

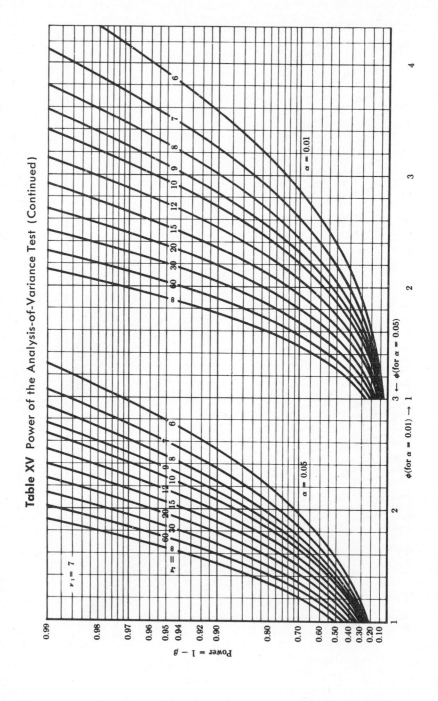

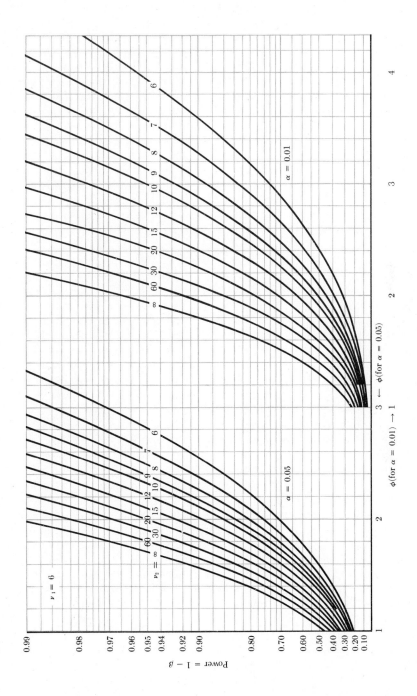

ANSWERS TO EXERCISES

CHAPTER 1

1. (a) $\{7, 14, 21, 28, 35, 42, 49\}$.
   (b) $\{-3, 2\}$.
   (c) $\{$H1, H2, H3, H4, H5, H6, T1, T2, T3, T4, T5, T6$\}$.
   (d) $\{$North America, South America, Europe, Asia, Africa, Australia, Antarctica$\}$.
   (e) $\varnothing$.

2. (a) T. (b) F. (c) F. (d) T. (e) T. (f) T.

3. $A = C$.

4. 16; $\{p, q, r\}$, $\{p, r, s\}$, $\{p, q, s\}$, $\{q, r, s\}$.

5. $\{$air, land$\}$, $\{$air, sea$\}$, $\{$land, sea$\}$, $\{$air$\}$, $\{$land$\}$, $\{$sea$\}$, $\varnothing$.

8. (a) $\{0, 2, 3, 4, 5, 6, 8\}$. (b) $\varnothing$. (c) $\{0, 1, 6, 7, 8, 9\}$. (d) $\{1, 3, 5, 6, 7, 9\}$.
   (e) $\{0, 1, 6, 7, 8, 9\}$. (f) $\{2, 4\}$.

9. (a) $\{$nitrogen, potassium, uranium, oxygen$\}$.
   (b) $\{$copper, sodium, zinc, oxygen$\}$.
   (c) $\{$copper, sodium, nitrogen, potassium, uranium, zinc$\}$.
   (d) $\{$copper, uranium, zinc$\}$. (e) $\varnothing$. (f) $\{$oxygen$\}$.

10. $P \cup Q = \{z \mid z < 9\}$; $P \cap Q = \{z \mid 1 < z < 5\}$.

12. (a)

			Red			
Green	*1*	*2*	*3*	*4*	*5*	*6*
1	(1,1)	(1,2)	(1,3)	(1,4)	(1,5)	(1,6)
2	(2,1)	(2,2)	(2,3)	(2,4)	(2,5)	(2,6)
3	(3,1)	(3,2)	(3,3)	(3,4)	(3,5)	(3,6)
4	(4,1)	(4,2)	(4,3)	(4,4)	(4,5)	(4,6)
5	(5,1)	(5,2)	(5,3)	(5,4)	(5,5)	(5,6)
6	(6,1)	(6,2)	(6,3)	(6,4)	(6,5)	(6,6)

    (b) $A = \{(1,1), (1,2), (1,3), (2,1), (3,1), (2,2)\}$.

    (c) $B = \{(1,6), (2,6), (3,6), (4,6), (5,6), (6,6),$
          $(6,1), (6,2), (6,3), (6,4), (6,5)\}$.

    (d) $C = \{(2,1), (2,2), (2,3), (2,4), (2,5), (2,6)\}$.

13. (a) $S = \{HH, HT, T1, T2, T3, T4, T5, T6\}$.

    (b) $A = \{T1, T2, T3\}$.

    (c) $B = \varnothing$.

14. (a) $S = \{YYY, YYN, YNY, NYY, YNN, NYN, NNY, NNN\}$.

    (b) $E = \{YYY, YYN, YNY, NYY\}$.

    (c) One possible event: "The second woman interviewed uses brand X."

15. (a) 120. (b) 72.

16. 24.

17. 256.

18. 50,400.

19. 15,120.

20. (a) 100. (b) 48. (c) 48.

21. 120.

22. 144.

23. 120.

24. 210.

25. 280.

26. 6930.

27. (a) 56. (b) 30. (c) 10.

28. 8,211,173,256.

29. 180.

30. 70.

31. $Pr(C) = 1/2; Pr(A') = 3/4$.

32. 1/6.

33. $S = \{\$10, \$25, \$100\}$; 9/10.

34. 1/9; 1/6.

35. (a) 33/54,145. (b) 33/66,640.

36. (a) 3/8. (b) 9/28.

37. 46/221.

38. 1/4.

39. (a) 0.9. (b) 0.6. (c) 0.5.

40. 19/64; 45/64.

41. (a) 0.0001. (b) 0.9999.

42. 5/8.

43. 38/63.

44. (a) 2/11. (b) 5/11.

45. 2/9.

46. 0.04.

47. (a) 1/4. (b) 5/27. (c) 1/20.

48. (a) 0.35. (b) 0.875. (c) 0.55.

49. 3/10.

50. 15/23.

51. 0.174.

CHAPTER 2

1. Discrete; continuous; continuous; discrete; discrete; continuous.

2.

x	0	1	2	3
$Pr(X = x)$	8/27	4/9	2/9	1/27

3. $f(x) = \dfrac{\binom{5}{x}\binom{5}{4-x}}{\binom{10}{4}}$, $x = 0, 1, 2, 3, 4$.

4. $f(x) = 1/6$, $x = 1, 2, \ldots, 6$.

5.

x	0	1	2
$Pr(X = x)$	1/5	3/5	1/5

6.

x	0	1	2	3
$Pr(X = x)$	1/27	2/9	4/9	8/27

7.
$$F(x) = \begin{cases} 0 & \text{for } x < 0 \\ 1/5 & \text{for } 0 \le x < 1 \\ 4/5 & \text{for } 1 \le x < 2 \\ 1 & \text{for } x \ge 2. \end{cases}$$
(a) 3/5. (b) 4/5.

9.
$$F(x) = \begin{cases} 0 & \text{for } x < 0 \\ 1/27 & \text{for } 0 \le x < 1 \\ 7/27 & \text{for } 1 \le x < 2 \\ 19/27 & \text{for } 2 \le x < 3 \\ 1 & \text{for } x \ge 3. \end{cases}$$
(a) 2/3. (b) 8/27.

11. (b) 1/4. (c) 0.3.

12. (a) 16/27. (b) 1/3.

13. $F(x) = (x - 1)/2; 1/4.$

14. $F(x) = (x + 4)(x - 2)/27; 1/3.$

15. (d) 55; 79.

16. (d) 4.5.

17. (a) $f(1,0) = 3/70, f(2,0) = 9/70, f(3,0) = 3/70$
$f(0,1) = 2/70, f(1,1) = 18/70, f(2,1) = 18/70, f(3,1) = 2/70$
$f(0,2) = 3/70, f(1,2) = 9/70, f(2,2) = 3/70.$
(b) 1/2.

18. (a) 135/1024. (b) 1/2.

19. (a) 1/50. (b) 13/75. (c) 14/25. (d) 8/15.

20. (a)

y	0	1	2
$f(y \mid 2)$	3/10	3/5	1/10

(b) 3/10.

21.

x	1	2	3
$g(x)$	1/3	19/36	5/36

y	1	2	3
$h(y)$	1/4	14/45	79/180

x	1	2	3
$f(x \mid 1)$	0	2/3	1/3
$f(x \mid 2)$	9/14	5/14	0
$f(x \mid 3)$	24/79	45/79	10/79

y	1	2	3
$f(y \mid 1)$	0	3/5	2/5
$f(y \mid 2)$	6/19	4/19	9/19
$f(y \mid 3)$	3/5	0	2/5

22.

x	2	4
$g(x)$	0.40	0.60

y	1	3	5
$h(y)$	0.25	0.50	0.25

X and Y are independent.

23. Independent.

24. Dependent.

25. (a) $g(y,z) = 2yz^2/9, 0 < y < 1, 0 < z < 3.$
(b) $h(y) = 2y, 0 < y < 1.$
(c) 7/162.

26. (a) Dependent.
(b) 1/3.

27. 1.

28. $200.

29. $1.08.

30. 3/4.

31. 1/3.

32. $(\ln 4)/\pi$.

33. 76/3.

34. $420.

35. 1/6.

36. 35.2.

37. 8/3.

38. (a) 209. (b) 65/4.

39. (a) 7. (b) 0. (c) 12.25.

40. (a) -2.60. (b) 9.60.

41. 2/5.

42. 3.041.

43. $\mu = 1.125; \sigma^2 = 0.502$.

44. 1/18.

45. 0.9839.

46. $-3/14$.

47. -0.1244.

48. 1/36.

49. $-1/144$.

51. (a) 175/12. (b) 175/6.

52. 68.

53. 52.

54. (a) At least 3/4. (b) At least 8/9.

55. (a) At most 4/9. (b) At least 5/9.
(c) At least 21/25. (d) 10.

CHAPTER 3

1. $f(x) = 1/10, x = 1, 2, \ldots, 10; Pr(X < 4) = 3/10$.

2. $f(x) = 1/25, x = 1, 2, \ldots, 25$.

3. $\mu = 5.5; \sigma^2 = 8.25$.

4. 0.4219.

5. Uniform and binomial.

6. 0.3134.

7. 0.1240.

8. 0.1035.

9. 0.0006.

10. 63/64.

11. $f(x) = \binom{5}{x}\left(\frac{1}{4}\right)^x\left(\frac{3}{4}\right)^{5-x}$, $x = 0, 1, 2, \ldots, 5; \mu \pm 2\sigma = 1.25 \pm 1.936$.

12. Four-engine plane.

13. Two-engine plane when $q = 1/2$; either plane when $q = 1/3$.

14. $3.75; \mu \pm 2\sigma = 3.75 \pm 3.35$.

15. 15/128.

16. 0.0095.

17. 21/256.

18. 53/65.

19. 0.9517.

20. (a) 77/115 (b) 3/25.

21. (a) 0.6815. (b) 0.1106.

22. 5/14.

23. 0.0025.

24. $f(x) = \dfrac{\binom{4}{x}\binom{2}{3-x}}{\binom{6}{3}}$, $x = 1, 2, 3$;

$Pr(2 \leq X \leq 3) = 4/5$.

25. (a) 1/6. (b) 203/210.

26. 1.2.

27. $3.25; 0.52 - 5.98$.

28. 0.2131.

29. 0.9453.

30. 0.0129.

31. 4/33.

32. 17/63.

33. 0.1008.

34. (a) 0.1429. (b) 0.1353.

35. (a) 0.1512. (b) 0.4015.

36. (a) 0.3840. (b) 0.0067.

37. 0.6288.

38. 0.2657.

39. 0 − 8.

40. 0.0515.

41. 0.0651.

42. 63/64.

43. 0.1172.

44. (a) 0.0630. (b) 0.9730.

CHAPTER 4

1. (a) 0.0913. (b) 0.9849. (c) 0.3362. (d) 39.244. (e) 46.756.

2. (a) 0.9192. (b) 0.9821. (c) 0.6106. (d) 208.42. (e) 188.5; 211.5.

3. (a) 0.1151. (b) 16.375. (c) 0.5403. (d) 20.55.

4. (a) 0.0548. (b) 0.4514. (c) 23. (d) 6.66 ounces.

5. (a) 0.0062. (b) 0.6826. (c) 3.986.

6. (a) 64. (b) 86.

7. (a) 17. (b) 524. (c) 72. (d) 26.

8. 62.

9. (a) 57.11%. (b) $4.23.

10. (a) 0.0401. (b) 0.0244.

11. (a) 19.36%. (b) 39.70%.

12. 26.

13. (a) 0.0023. (b) 0.2146. (c) 0.0520.

14. 6.238 years.

15. 0.0018.

16. (a) 0.7925. (b) 0.0352. (c) 0.0101.

17. (a) 0.8643. (b) 0.2978. (c) 0.0796.

18. 0.9515.

19. (a) 0.0846. (b) 0.1630.

20. 0.1179.

21. 0.1357.

22. 0.4356.

23. $2.8e^{-1.8} - 3.4e^{-2.4} = 0.1545$.

24. $4e^{-3} = 0.1992$.

26. $\sum_{x=4}^{6} \binom{6}{x} (1 - e^{-3/4})^x (e^{-3/4})^{6-x} = 0.3968$.

27. 0.0350.

29. $e^{-4} = 0.0183$.

CHAPTER 5

1. $g(y) = 1/3$, $y = 1, 3, 5$.

2. $g(y) = \binom{3}{\sqrt{y}} \left(\frac{2}{5}\right)^{\sqrt{y}} \left(\frac{3}{5}\right)^{3 - \sqrt{y}}$, $y = 0, 1, 4, 9$.

3. $g(y_1, y_2) = \left(\dfrac{y_1 + y_2}{2}, \dfrac{y_1 - y_2}{2}, 2 - y_1\right) \left(\dfrac{1}{4}\right)^{(y_1 + y_2)/2} \left(\dfrac{1}{3}\right)^{(y_1 - y_2)/2} \left(\dfrac{5}{12}\right)^{2 - y_1}$;

$y_1 = 0, 1, 2; y_2 = -2, -1, 0, 1, 2; y_2 \leq y_1; y_1 + y_2 = 0, 2, 4$.

4.

y	1	2	3	4	6
$h(y)$	1/18	2/9	1/6	2/9	1/3

6. $g(y) = 1/6y^{1/3}, 0 < y < 8.$
7. Gamma distribution with $\alpha = 3/2$ and $\beta = m/2b.$
9. $h(w) = 6 + 6w - 12w^{1/2}, 0 < w < 1.$
10. $g(y) = 1/2 \sqrt{y}, 0 < y < 1.$
11. $g(y) = 2/9 \sqrt{y}, 0 < y < 1,$
    $= (\sqrt{y} + 1)/9 \sqrt{y}, 1 < y < 4.$
13. $\mu = 1/p, \ \sigma^2 = q/p^2.$
15. 0.9306.
17. $\bar{x} = 2; \tilde{x} = 1.5; m = 1.$
18. $\bar{x} = 5.875; \tilde{x} = 6.5; m = 4$ and 7.
19. $\bar{x} = 25.2; \tilde{x} = 17; m = 11.$
20. Range $= 5; s = 1.57.$
21. $s = 0.55.$
24. (a) 3.367. (b) 13.468. (c) 3.367.
25. 0.3159.
26. 100.
27. Adjust the machine.
28. (a) $\mu_{\bar{X}} = 68.5; \sigma_{\bar{X}} = 0.54.$ (b) 161. (c) 0.
31. 0.7064.
32. (a) 0.0772. (b) 0.2814.
33. (a) 34.805. (b) 16.047. (c) 13.277.
34. (a) 0.05. (b) 0.94.
35. Not valid.
37. (a) 2.110. (b) $-2.764.$ (c) 1.714.
38. No; $\mu > 20.$
41. 0.05.
39. Yes.
42. 0.99.
40. (a) 2.71. (b) 3.51. (c) 2.92.
    (d) 0.47. (e) 0.34.

CHAPTER 6

2. $765 < \mu < 795.$
3. $7.24 < \mu < 7.56.$
4. (a) $67.61 < \mu < 69.39.$
   (b) $e < 0.89.$
5. (a) $13,882 < \mu < 15,118.$
   (b) $e < 618.$
6. 68.
7. 11.
8. 28.
9. $0.978 < \mu < 1.033.$
10. $30.69 < \mu < 34.91.$
11. $15.63 < \mu < 21.57.$
12. $47.722 < \mu < 49.278.$
13. $2.9 < \mu_1 - \mu_2 < 7.1.$
14. $6.56 < \mu_1 - \mu_2 < 11.24.$
15. $0.00109 < \mu_1 - \mu_2 < 0.00131;$
    yes.
16. $0.3 < \mu_1 - \mu_2 < 9.7.$
17. $1.5 < \mu_1 - \mu_2 < 12.5.$
18. $-4036 < \mu_1 - \mu_2 < 1836.$
19. $-11.9 < \mu_{II} - \mu_I < 36.5.$
20. $-0.7 < \mu_D < 6.3.$
21. $-1795 < \mu_D < 445.$
22. $3.2 < \mu_D < 13.1.$
23. (a) $0.498 < p < 0.642.$
    (b) $e < 0.072.$
24. (a) $0.1442 < p < 0.1998.$
    (b) $e < 0.0278.$
25. $0.017 < p < 0.143.$
26. (a) 0.85. (b) $0.739 < p < 0.961.$
    (c) No.
27. 2586.
28. 160.
29. 9604.

30. 16,577.

31. $-0.0136 < p_F - p_M < 0.0636$.

32. $0.016 < p_A - p_B < 0.164$.

33. $0.0011 < p_1 - p_2 < 0.0869$.

34. $0.0284 < p_1 - p_2 < 0.1416$.

35. $0.00022 < \sigma^2 < 0.00357$.

36. $3.430 < \sigma < 6.587$.

37. $1.410 < \sigma < 6.385$.

38. $1.258 < \sigma^2 < 5.410$.

39. $0.600 < \sigma_1/\sigma_2 < 2.819$; yes.

40. $0.236 < \sigma_1^2/\sigma_2^2 < 1.877$.

41. $0.016 < \sigma_1^2/\sigma_2^2 < 0.454$.

42. $p^* = 0.0923$ for $x = 0$
      $= 0.1066$ for $x = 1$
      $= 0.1183$ for $x = 2$.

43. $p^* = 0.0982$ for $x = 0$
      $= 0.1075$ for $x = 1$
      $= 0.1157$ for $x = 2$.

44. $8.077 < \mu < 8.692$.

45. $f(\mu \mid x_1, x_2, \ldots, x_{25}) = \dfrac{1}{\sqrt{2\pi}\,13.706} e^{-1/2[(\mu-780)/20]^2}$, $770 < \mu < 830$.

47. $R(\hat{P}; p) = pq/n$.

48.
$$R(\Theta; \theta) = \begin{cases} 0 & \text{for } \theta = 0 \\ 2/3 & \text{for } \theta = 1 \\ 2/3 & \text{for } \theta = 2 \\ 0 & \text{for } \theta = 3. \end{cases}$$

49.
$$R(\Theta_2; \theta) = \begin{cases} 0 & \text{for } \theta = 0 \\ 1/3 & \text{for } \theta = 1 \\ 1 & \text{for } \theta = 2 \\ 0 & \text{for } \theta = 3. \end{cases}$$

50. $\hat{\Theta}_1$.

51. $\hat{\Theta}_2$.

CHAPTER 7

1. $\alpha = 0.0853$; $\beta = 0.8287$; $\beta = 0.7817$; not a good test procedure.

2. $\alpha = 0.0548$; $\beta = 0.3504$; $\beta = 0.6177$; $\beta = 0.8281$.

3. $\alpha = 0.0146$; $\beta = 0.0126$.

4. $\alpha = 0.0386$; $\beta = 0.2779$; $\beta = 0.5$.

5. $z = -1.643$; accept H_0.

6. $z = -1.25$; accept H_0.

7. $z = 3.143$; $\mu > 68.5$.

8. $z = 10.417$; $\mu > 12,000$.

9. $t = 0.771$; accept H_0.

10. $t = 2.776$; $\mu > 30$.

11. $t = 1.296$; valid claim.

12. $t = 1.781$; accept H_0.

13. $z = 4.222$; $\mu_1 > \mu_2$.

14. $z = -2.603$; $\mu_A - \mu_B < 12$.

15. $z = 2.448$; $\mu_1 - \mu_2 > \$500$.

16. $t = 6.575$; $\mu_1 > \mu_2$.

17. $t = 1.501$; yes.

18. $t = -0.850$; accept H_0.

19. $t' = 0.215$; accept H_0.

20. $t = 1.821$; accept H_0.

21. $t = -2.109$; accept H_0.

22. $t = -0.910$; accept H_0.

23. $\chi^2 = 18.120$; accept H_0.

24. $\chi^2 = 10.735$; accept H_0.

25. $\chi^2 = 17.530$; reject H_0.

26. $\chi^2 = 17.188$; accept H_0.

27. $f = 1.325$; accept H_0.

28. $f = 0.748$; accept H_0.

29. $f = 0.086$; reject H_0.

30. 22.

31. 80.

32. 12.

33. 10.

34. 68.

35. $\beta = 0.045$.

36. Accept H_0.

37. $u = 43.5$; $\mu_I = \mu_{II}$.

38. $\mu_1 = \mu_2$.

39. Accept H_0.

40. $w = 8.5$; accept H_0.

41. $w = 4.5$; accept H_0.
42. $w = 9.5$; accept H_0.
43. $z = -1.443$; $p = 0.6$.
44. Valid claim.
45. $z = 1.443$; accept H_0.
46. $z = 1.339$; valid estimate.
47. $z = 2.395$; yes.
48. $z = 1.878$; yes.
49. $z = 1.109$; no.
50. $z = 2.090$; yes.

51. $\chi^2 = 4.467$; yes.
52. $\chi^2 = 6.76$; no.
53. $\chi^2 = 1.667$; accept H_0.
54. $\chi^2 = 2.326$; accept H_0.
55. $\chi^2 = 10.000$; reject H_0.
56. $\chi^2 = 2.571$; accept H_0.
57. $\chi^2 = 9.613$; not normal.
58. $\chi^2 = 14.464$; not independent.
59. $\chi^2 = 9.048$; independent.
60. $\chi^2 = 6.239$; accept H_0.

CHAPTER 8

1. (a) $\hat{y} = 12.2784 + 0.6703x$.
   (b) 20.32.
2. (a) $\hat{y} = 6.4136 + 1.8091x$.
   (b) 9.5795.
3. (a) $\hat{y} = 42.5830 - 0.6859x$.
   (b) 25.7784.
4. (a) $\hat{y} = (27.6)(0.804)^x$.
   (b) 11.5 milligrams.
5. (a) $\hat{\gamma} = 2.660$; $\hat{C} = 2.63 \times 10^6$.
   (b) 22.9 pounds per square inch.
8. $t = 3.07$; reject H_0.
9. $4.323 < \alpha < 8.505$; $0.445 < \beta < 3.173$.
10. $21.959 < \alpha < 63.207$; $-1.478 < \beta < 0.106$.
11. $24.444 < \mu_{Y|24.5} < 27.113$; $21.888 < y_0 < 29.669$.
13. $7.808 < y_0 < 10.808$.
14. (a) $b = \sum_{i=1}^{n} x_i y_i / \sum_{i=1}^{n} x_i^2$. (b) $\hat{y} = 2.003x$.
15. $\hat{y} = 0.349 + 1.929x$; $t = 1.38$; accept H_0.
16. $E(B) = \beta + \gamma \sum_{i=1}^{n} (x_{1i} - \bar{x}_1)x_{2i} / \sum_{i=1}^{n} (x_{1i} - \bar{x}_1)^2$.
17. $f = 9.00$; reject H_0.
18. (a) $\hat{y} = 5.8253 + 0.567x$. (b) 34.205 grams. (c) $t = -0.163$; accept H_0.
    (d) $f = 1.58$; regression is linear.
19. (a) $\hat{y} = 3.1266 + 1.8429x$. (b) $f = 2.60$; accept H_0.
    (c) $1.2722 < \beta < 2.4136$.
20. $r = -0.526$.
21. $r = 0.240$.
22. $r = 0.679$; $z = 2.62$; reject H_0.

CHAPTER 9

1. $\hat{y} = 4.4882 - 0.0394x_1 + 0.6378x_2$.
2. $\hat{y} = 55.2266 - 0.0378x_1 + 1.6816x_2$.
3. $\hat{y} = 0.5801 + 2.712x_1 + 2.05x_2$.
4. $\hat{y} = 19.98518 + 0.30363x_1 + 0.59635x_2 - 0.49706x_3 - 0.70377x_4$.

5. 34.3699.

6. 0.00106.

7. $\hat{\sigma}^2_{B_1} = 0.000071$; $\hat{\sigma}^2_{B_2} = 0.063523$; $\hat{\sigma}_{B_1 B_2} = -0.001134$.

8. (a) $\hat{\sigma}^2_{B_2} = 0.00002$. (b) $\hat{\sigma}_{B_1 B_4} = -0.000003$.

9. $26.2353 < y_0 < 57.1515$; $34.8580 < \mu_{Y|2500, 48.0} < 48.5288$.

10. $16.7749 < y_0 < 16.9291$; $16.8260 < \mu_{Y|8, 2, 6.0, 10.3, 5.8} < 16.8780$.

11. $t = -4.48$; reject H_0.

12. $t = 3.38$; reject H_0.

13. $R^2 = 0.9999$.

14. $f = 12{,}598$; regression is significant.

15. $f = 12{,}589$; regression is significant.

16. $f = 11{,}040$; reject H_0.

17. $f = 20.07$; reject H_0.

18. (a) $\hat{y} = 9.9 + 0.575x_1 + 0.550x_2 + 1.150x_3$.
    (b) $f = 7.69$ for β_1; not significant.
    $f = 7.04$ for β_2; not significant.
    $f = 30.77$ for β_3; significant at the 0.01 level.

CHAPTER 10

1. (a) $\hat{y} = 8.697 - 2.341x + 0.288x^2$. (b) 5.2.

2. $\hat{y} = 141.6112 - 0.2819x + 0.0003x^2$.

3. $f = 8.16$; reject H_0.

4. (a) $\hat{y} = 30.033 + 0.9667(x - 20)/5 - 0.5095\{[(x - 20)/5]^2 - 2\}$.
    (b) $f(\text{linear}) = 13.74$; significant at the 0.01 level.
    $f(\text{quadratic}) = 5.34$; significant at the 0.05 level.
    (c) $f = 0.02$; model is adequate. Yes.

5. $29.930 < \mu_{Y|19.5} < 31.970$.

6. (a) $\hat{y} = 18.7583 - 0.808\, p_1(x_1) + 0.5 p_1(x_2) - 0.558\, p_2(x_2)$
    $- 0.05\, p_1(x_1) p_2(x_2)$.
    (b) $f = 27.63$ for γ_1; significant at the 0.01 level.
    $f = 7.05$ for γ_2; significant at the 0.05 level.
    $f = 26.36$ for γ_{22}; significant at the 0.01 level.
    $f = 0.07$ for γ_{12}; not significant.

CHAPTER 11

3. $f = 0.307$; no significant difference.

4. $f = 0.464$; no significant difference.

5. $b = 0.189$; variances are equal.

6. (a) $b = 3.80$; variances are equal.
    (b) $f = 13.33$; reject H_0.
    (c) (1) $f = 14.27$; reject H_0. (2) $f = 23.23$; reject H_0.
    (3) $f = 2.48$; accept H_0.

7. (a) $f = 7.10$; reject H_0. (b) Blend 4 differs significantly from all others.

8. (a) $f(\text{blocks}) = 8.30$; significant at the 0.05 level.

   $f(\text{fertilizer}) = 6.11$; significant at the 0.05 level.

   (b) (1) $f = 17.37$; significant at the 0.01 level.

   (2) $f = 0.96$; not significant.

10. $f(\text{blocks} = 4.86$; significant. $f(\text{treatments}) = 3.33$; no significant difference.

11. $\hat{\sigma}_\alpha^2 = 28.91$; $s^2 = 8.32$.   **14.** 9.

12. $\hat{\sigma}_\alpha^2 = 1.08$; $\hat{\sigma}_\beta^2 = 2.25$.   **15.** No; 16.

CHAPTER 12

1. (a) $f = 1.63$; accept H_0.

   (b) $f(\text{temperatures}) = 8.13$; reject H_0.

   $f(\text{ovens}) = 5.18$; reject H_0.

2. (a) $f = 5.29$; reject H_0. (b) $f = 1.97$; accept H_0.

   (c) $f = 33.94$; reject H_0. (d) $f = 0.498$; accept H_0.

3. (a) AB: $f = 3.83$; reject H_0.   (b) A: $f = 0.54$; accept H_0.

   AC: $f = 3.79$; reject H_0. B: $f = 6.85$; reject H_0.

   BC: $f = 1.31$; accept H_0. C: $f = 2.15$; accept H_0.

   ABC: $f = 1.63$; accept H_0.

4. (a) $f = 2.66$; accept H_0.

   (b) A: $f = 3.37$; reject H_0.

   B: $f = 5.67$; reject H_0.

   C: $f = 4.85$; reject H_0.

   The pooled error includes BC.

5. (a) $f = 1.49$; accept H_0.

   (b) $f(\text{operators}) = 12.45$; reject H_0.

   $f(\text{filters}) = 8.39$; reject H_0.

   (c) $\hat{\sigma}_\alpha^2 = 0.17$ (filters);

   $\hat{\sigma}_\beta^2 = 0.3514$ (operators);

   $s^2 = 0.1867$.

6. (a) $\hat{\sigma}_\beta^2, \hat{\sigma}_\gamma^2, \hat{\sigma}_{\alpha\gamma}^2$ are significant. (b) $\hat{\sigma}_\gamma^2, \hat{\sigma}_{\alpha\gamma}^2$ are significant.

7. Yes.   **8.** 0.59.

CHAPTER 13

1. SSA $= 2.6667$, SSB $= 170.6667$, SSC $= 104.1667$, SS(AB) $= 1.5000$,

   SS(AC) $= 42,6667$, SS(BC) $= 0.0000$, SS(ABC) $= 1.5000$.

2. Significant effects:

   A: $f = 1940.64$

   B: $f = 4411.62$

   C: $f = 1098.38$

   D: $f = 458.50$

   CD: $f = 5.38$.

3. Significant effects:

A: $f = 57.85$ D: $f = 44.72$ AD: $f = 4.85$ CD: $f = 6.52$.
B: $f = 7.52$ AB: $f = 6.94$ BC: $f = 10.96$
C: $f = 127.87$ AC: $f = 7.08$ BD: $f = 4.85$

Insignificant effects:

ABC: $f = 1.26$ BCD: $f = 1.20$
ABD: $f = 1.14$ $ABCD$: $f = 0.87$.
ACD: $f = 1.20$

4. A, B, C, AC, BC, and ABC each with 1 degree of freedom can be tested using an error mean square with 12 degrees of freedom.

5. (a) A: $f = 1.55$; not significant.
   B: $f = 1.27$; not significant.
   C: $f = 3.49$; not significant.
   D: $f = 0.75$; not significant.

   (b) ABC.

6.

Block 1	Block 2	Block 3	Block 4
(1)	c	d	a
ab	abc	ac	b
acd	ad	bc	cd
bcd	bd	abd	abcd

CD is also confounded.

7. (a)

Block 1	Block 2	Block 3	Block 4
(1)	a	b	ab
c	ac	bc	abc
ae	e	abe	be
bd	abd	d	ad
ace	ce	abce	bce
bcd	abcd	cd	acd
abde	bde	ade	de
abcde	bcde	acde	cde

   (b) BD.

   (c) SSA $= 21.9453$; SSC $= 2.4753$; SS(E) $= 1.0878$;
   SSB $= 40.2753$; SSD $= 7.7028$.

8.

Block		Block		Block	
1	2	1	2	1	2
abc	ab	abc	ab	(1)	a
a	ac	a	ac	c	b
b	bc	b	bc	ab	ac
c	(1)	c	(1)	abc	bc

Replicate 1   Replicate 2   Replicate 3
ABC confounded ABC confounded AB confounded

Analysis of Variance

Source of variation	Degrees of freedom
Blocks	5
A	1
B	1
C	1
AB	1'
AC	1
BC	1
ABC	1'
Error	11
Total	23

Relative information on
$ABC = 1/3$.
Relative information on
$AB = 2/3$.

9. (a) ABC; $ABCD$.

(b)

$A: f = 3.35$	$AC: f = 0.54$	$ABC: f = 1.94$
$B: f = 0.84$	$AD: f = 1.20$	$ABD: f = 0.54$
$C: f = 7.53$	$BC: f = 0.84$	$ACD: f = 0.03$
$D: f = 13.38$	$BD: f = 0.54$	$BCD: f = 0.30$
$AB: f = 0.84$	$CD: f = 0.30$	$ABCD: f = 0.89$

10.

Block	
1	2
(1)	a
ab	b

Replicate 1

Block	
1	2
a	(1)
ab	b

Replicate 2

Block	
1	2
(1)	ab
a	b

Replicate 3

11. $A \equiv CDE$ $AB \equiv BCDE$ $BD \equiv ABCE$
$B \equiv ABCDE$ $AC \equiv DE$ $BE \equiv ABCD$
$C \equiv ADE$ $AD \equiv CE$ $ABC \equiv BDE$
$D \equiv ACE$ $AE \equiv CD$ $ABD \equiv BCE$
$E \equiv ACD$ $BC \equiv ABDE$ $ABE \equiv BCD$.

12. $A \equiv BCD \equiv ABDEF \equiv CEF$
$B \equiv ACD \equiv DEF \equiv ABCEF$
$C \equiv ABD \equiv BCDEF \equiv AEF$
$D \equiv ABC \equiv BEF \equiv ACDEF$
$E \equiv ABCDE \equiv BDF \equiv ACF$
$F \equiv ABCDF \equiv BDE \equiv ACE$.

13. Interactions sacrificed: $ABCD$, $CDEFG$, BDF, $ABEFG$, ACF, $BCEG$, $ADEG$.

14.

$A: f = 0.48$	$E: f = 5.39$
$B: f = 1.35$	$F: f = 1.09$
$C: f = 3.03$	$G: f = 4.36$.
$D: f = 1.94$	

INDEX